Comparative Biochemistry of
Photoreactive Systems

SYMPOSIA ON
Comparative Biology

OF THE KAISER FOUNDATION RESEARCH INSTITUTE

VOLUME 1

Comparative Biochemistry of Photoreactive Systems
    Edited by MARY BELLE ALLEN

1960

VOLUME 2

The Lower Metazoa — Comparative Biology and Phylogeny
    Edited by BENJAMIN G. CHITWOOD AND M. M. J. LAVOIPIERRE

In preparation

Comparative Biochemistry of Photoreactive Systems

EDITED BY
MARY BELLE ALLEN

Director
Laboratory of Comparative Biology
The Kaiser Foundation Research Institute
Richmond, California

1960

ACADEMIC PRESS • New York and London

Copyright ©, 1960, by Academic Press Inc.

ALL RIGHTS RESERVED

NO PART OF THIS BOOK MAY BE REPRODUCED IN ANY FORM,
BY PHOTOSTAT, MICROFILM, OR ANY OTHER MEANS,
WITHOUT WRITTEN PERMISSION FROM THE PUBLISHERS.

ACADEMIC PRESS INC.
111 FIFTH AVENUE
NEW YORK 3, N. Y.

United Kingdom Edition
Published by
ACADEMIC PRESS INC. (LONDON) LTD.
17 OLD QUEEN STREET, LONDON S.W. 1

Library of Congress Catalog Card Number 60-14262

PRINTED IN THE UNITED STATES OF AMERICA

CONTRIBUTORS

M. B. ALLEN, *Laboratory of Comparative Biology, Kaiser Foundation Research Institute, Richmond, California*

J. M. ANDERSON, *Lawrence Radiation Laboratory, University of California, Berkeley, California*

SELINA BENDIX, *Laboratory of Comparative Biology, Kaiser Foundation Research Institute, Richmond, California*

U. BLASS, *Lawrence Radiation Laboratory, University of California, Berkeley, California*[1]

L. R. BLINKS, *Hopkins Marine Station of Stanford University, Pacific Grove, California*

LAWRENCE BOGORAD, *Department of Botany, University of Chicago, Chicago, Illinois*

J. S. BROWN, *Department of Plant Biology, Carnegie Institution of Washington, Stanford, California*

M. CALVIN, *Department of Chemistry and Lawrence Radiation Laboratory, University of California, Berkeley, California*

C. O. CHICHESTER, *Department of Food Science and Technology, University of California, Davis, California*

ELLSWORTH C. DOUGHERTY, *Laboratory of Comparative Biology, Kaiser Foundation Research Institute, Richmond, California*

DENIS L. FOX, *Division of Marine Biochemistry, Scripps Institution of Oceanography, University of California, La Jolla, California*

C. STACY FRENCH, *Department of Plant Biology, Carnegie Institution of Washington, Stanford, California*

T. W. GOODWIN, *Department of Agricultural Biochemistry, Institute of Rural Science, University College of Wales, Aberystwyth, Wales*

HELEN M. HABERMANN, *Department of Plant Biology, Carnegie Institution of Washington, Stanford, California*[2]

---

[1] *Present address:* Sandoz A. G., Basel, Switzerland.
[2] *Present address:* Department of Biological Sciences, Goucher College, Towson, Baltimore, Maryland.

FRANCIS T. HAXO, *Scripps Institution of Oceanography, University of California, La Jolla, California*

STERLING B. HENDRICKS, *Mineral Nutrition Laboratory, Soil and Water Conservation Research Division, Agricultural Research Service, U. S. Department of Agriculture, Beltsville, Maryland*

A. S. HOLT, *Division of Applied Biology, National Research Council, Ottawa, Canada*

MARTIN D. KAMEN, *Graduate Department of Biochemistry, Brandeis University, Waltham, Massachusetts*

G. MACKINNEY, *Department of Food Science and Technology, University of California, Berkeley, California*

GUY C. MCLEOD, *Department of Plant Biology, Carnegie Institution of Washington, Stanford, California*

N. MILLOTT, *Department of Zoology, Bedford College, University of London, London, England*

H. V. MORLEY, *Division of Applied Biology, National Research Council, Ottawa, Canada*[3]

JACK MYERS, *Department of Plant Biology, Carnegie Institution of Washington, Stanford, California*[4]

WHEELER J. NORTH, *Division of Marine Biochemistry and Institute of Marine Resources, Scripps Institution of Oceanography, University of California, La Jolla, California*

COLM Ó HEOCHA, *Chemistry Department, University College, Galway, Ireland*

MICHAEL H. PROCTOR, *Department of Bacteriology, University of Wisconsin, Madison, Wisconsin*[5]

STANLEY SCHER, *Hopkins Marine Station of Stanford University, Pacific Grove, California*[6]

---

[3] *Present address:* Department of Chemistry, McMaster University, Hamilton, Ontario, Canada.

[4] *Present address:* Departments of Botany and Zoology, University of Texas, Austin, Texas.

[5] *Present address:* Department of Biochemistry, Cambridge University, Cambridge, England.

[6] *Present address:* Laboratory of Comparative Biology, Kaiser Foundation Research Institute, Richmond, California.

JAMES H. C. SMITH, *Department of Plant Biology, Carnegie Institution of Washington, Stanford, California*

R. Y. STANIER, *Department of Bacteriology, University of California, Berkeley, California*

J. R. VALLENTYNE, *Department of Zoology, Cornell University, Ithaca, New York*

WOLF VISHNIAC, *Department of Biology, Brookhaven National Laboratory, Upton, New York*[7]

G. WEBER, *Department of Biochemistry, University of Sheffield, Sheffield, England*

FREDERICK T. WOLF, *Department of Biology, Vanderbilt University, Nashville, Tennessee, and Biology Division, Oak Ridge National Laboratory, Oak Ridge, Tennessee*

JEROME J. WOLKEN, *Biophysical Research Laboratory, Eye and Ear Hospital, and the University of Pittsburgh Medical School, Pittsburgh, Pennsylvania*

---

[7] On leave until September 1, 1960 from Department of Microbiology, Yale University.

PREFACE

During the last few years there have been marked advances in our knowledge of photoreactive pigments. Gaps in the known pattern of pigment distribution have been filled and the structure of several pigment molecules has been elucidated. The most striking progress, however, has been made in the biochemistry of photoreactive systems. Whereas only a few years ago the investigator of such systems could only make guesses about their functioning by studying the outside of an impenetrable box that contained the photoreactive mechanism, and thus could speak only of hypothetical reactions and catalysts, today it is possible in some cases to disassemble the box and study the working of its component parts. The hypothetical catalysts and substances are being replaced by known intermediates and enzyme reactions. Moreover, application of physical techniques is yielding much new information on the photochemical reactions involved in photobiological processes.

In any such rapidly developing field the literature tends to be in bits and pieces which only the specialist actively working in the field can put together. It was therefore the aim of the symposium on which this book is based to bring together in one place and at one time a discussion of the various advances that have been made during the past few years. Authors have been encouraged to present an integrated treatment of their work and its relation to other developments in the field, rather than a summary of their latest research results, although a few brief papers on especially significant new findings have been included.

The papers in this volume were presented at the First Annual Symposium on Comparative Biology of the Kaiser Foundation Research Institute. The symposium was supported in part by the National Science Foundation, whose help was greatly appreciated. Future symposia will cover other areas of basic biological research in which lively development is occurring.

It is a pleasure to thank Dr. Clifford H. Keene, Director of the Kaiser Foundation Research Institute and Program Director for The Kaiser Foundation and Kaiser Foundation Hospitals for his interest in and support of this symposium. Thanks are also due to all those who made this meeting possible, including the section chairmen and, last but not least, the secretarial staff that ably handled the manuscript and discussions.

MARY BELLE ALLEN

July, 1960

CONTENTS

CONTRIBUTORS v

PREFACE ix

1. Algal Carotenoids
   T. W. GOODWIN 1

2. Pigments of Plant Origin in Animal Phyla
   DENIS L. FOX 11

3. Native and Extractable Forms of Chlorophyll in Various Algal Groups
   M. B. ALLEN, C. S. FRENCH, AND J. S. BROWN 33

4. Pteridine Pigments in Microorganisms and Higher Plants
   FREDERICK T. WOLF 53

5. On the Existence of Two Chlorophylls in Green Bacteria
   R. Y. STANIER 69

6. A New Leaf Pigment
   HELEN M. HABERMANN 73

7. Fossil Pigments
   J. R. VALLENTYNE 83

8. Pigments in Phototaxis
   SELINA BENDIX 107

9. Is Pigmentation a Clue to Protistan Phylogeny?
   ELLSWORTH C. DOUGHERTY AND MARY BELLE ALLEN 129

10. Photoreceptors: Comparative Studies
    JEROME J. WOLKEN 145

11. Recent Studies of Chlorophyll Chemistry
    A. S. HOLT AND H. V. MORLEY 169

12. Chemical Studies of Phycoerythrins and Phycocyanins
    COLM Ó HEOCHA 181

13. Biosynthesis of Carotenoids
    G. MACKINNEY AND C. O. CHICHESTER 205

14. Biosynthesis and Possible Relations among the Carotenoids and between Chlorophyll *a* and *b*
    J. M. ANDERSON, U. BLASS, AND M. CALVIN 215

15. The Biosynthesis of Protochlorophyll
    LAWRENCE BOGORAD 227

16. Protochlorophyll Transformations
    JAMES H. C. SMITH 257

17. The Photosensitivity of Sea Urchins
    N. MILLOTT 279

18. Sensitivity to Light in the Sea Anemone *Metridium senile* (L.): Duration of the Sensitized State
    WHEELER J. NORTH 295

19. The Photoreactions Controlling Photoperiodism and Related Responses
    STERLING B. HENDRICKS 303

20. Hematin Compounds in Photosynthesis
    MARTIN D. KAMEN 323

21. The Wavelength Dependence of Photosynthesis and the Role of Accessory Pigments
    FRANCIS T. HAXO 339

22. Automatic Recording of Photosynthesis Action Spectra Used to Measure the Emerson Enhancement Effect
    C. STACY FRENCH, JACK MYERS, AND GUY C. MCLEOD 361

23. Chromatic Transients in the Photosynthesis of Green, Brown, and Red Algae
    L. R. BLINKS 367

24. Chemical Participation of Chlorophyll in Photosynthesis
    WOLF VISHNIAC 377

25. Studies with Photosynthetic Bacteria: Anaerobic Oxidation of Aromatic Compounds
    STANLEY SCHER AND MICHAEL H. PROCTOR 387

26. Fluorescence Parameters and Photosynthesis
    G. WEBER 395

AUTHOR INDEX 413

SUBJECT INDEX 423

1

Algal Carotenoids

T. W. GOODWIN

*Department of Agricultural Biochemistry, Institute of Rural Science,
University College of Wales, Aberystwyth, Wales*

Plastid Pigments

It is now well established that all photosynthetic organisms so far examined contain carotenoids in addition to chlorophylls and, following the work of Stanier and his colleagues (Griffiths *et al.*, 1955; Stanier and Cohen-Bazire, 1957), it would appear unlikely that any photosynthetic organism exists without having carotenoids associated with chlorophylls. The two pigment groups are spatially closely associated in the cell, for they are attached either to the same protein or very similar proteins in the grana or the lamellae of the chloroplasts in higher plants and algae, and in the chromatophores of the photosynthetic bacteria (these correspond to the grana of higher plants)—see Goodwin (1959). The qualitative distribution of the major carotenoids in photosynthetic organisms is summarized in Table I.

All higher plants synthesize the same four major carotenoids: β-carotene, lutein, violaxanthin, and neoxanthin, and with minor exceptions, four classes of algae, the Chlorophyceae, Xanthophyceae (Heterokontae), Rhodophyceae, and Euglenineae (Euglenophyceae) follow the same pattern.* The Phaeophyceae and the Bacillariophyceae are characterized by the presence of large amounts of fucoxanthin, which replaces lutein as the major xanthophyll. The structure of this carotenoid is still unknown, but it is unique in its alkali-lability and in possessing a C=C=C grouping (Torto and Weedon, 1955). A further characteristic of brown algae and diatoms is that the conjugation of fucoxanthin with its apoprotein causes a marked bathochromic shift (up to 40 mμ.) in its absorption spectrum. This is the main cause of the characteristic color of brown algae and diatoms. Apart from the alleged existence of traces of a β-carotene-protein complex in which the bathochromic shift is about 60 mμ (Nishimura and Takamatsa, 1957), this is the only case in photo-

---

* Two reports which came to the notice of the author after completion of the manuscript (Strain, 1958; Krinsky, 1960) suggest that this pattern is not always followed in the Heterokontae and Euglenophyceae.

TABLE I

Carotenoid Distribution in Various Algal Classes[a],[b] (+, present; —, absent; ?, possibly present in traces)

Pigments	Chloro-phyceae	Xantho-phyceae	Bacillario-phyceae	Chryso-phyceae	Phaeo-phyceae	Rhodo-phyceae	Dino-phyceae	Eugle-nineae	Cyano-phyceae	Crypto-phyceae
Carotenes										
α-Carotene	+[c]	—	—	—	—	+	—	—	—	—
β-Carotene	+[d]	+	+	+	+	+	+	+	+	+
γ-Carotene	+[e]	—	—	—	—	—	—	—	—	—
ε-Carotene	?	—	+	—	—	—	—	—	+	—
Flavacene	—	—	—	—	—	—	—	—	+	—
Xanthophylls										
Echinenone	—	—	—	—	—	—	—	—	+	—
Lutein	+	+	?	+	+?	+	—	+	?	—
Zeaxanthin	+	—	—	—	+	—	—	?	—	+[g]
Violaxanthin	+	+	+	—	+?	+?	—	—	—	—
Flavoxanthin	—	+	—	—	—	—	—	—	—	—
Neoxanthin	+	+	—	+	+	—	—	+	—	—
Fucoxanthin	—	—	+	—	+?	—	—	—	—	—
Diatoxanthin	—	—	+	—	—	—	—	—	—	—
Diadinoxanthin	—	—	+	—	—	—	+	—	—	—
Dinoxanthin	—	—	—	—	—	—	+	—	—	—
Peridinin	—	—	—	—	—	—	+	—	—	—
Myxoxanthophyll	—	—	—	—	—	—	—	—	+	—
Siphonaxanthin	+[c]	—	—	—	—	—	—	—	—	—
Astaxanthin	+[d]	—	—	—	—	—	—	+[f]	—	—
Oscillaxanthin	—	—	—	—	—	—	—	—	+[f]	—

[a] Occasional variations from this general picture are discussed in the text.
[b] No information exists on the carotenoids of the Chloromonadineae.
[c] The main pigments of the Siphonales.
[d] The main pigments (hematochrome) of encysted flagellates.
[e] Present in reproductive regions of certain genera.
[f] Not present in every species.
[g] The pigment present may be diatoxanthin.

synthetic tissues of such a carotenoprotein, although many occur in the lower animals, especially in marine invertebrates.

The carotenoid distribution in the blue-green algae (Cyanophyceae) is unique; the main pigments present being β-carotene, echinenone (4-oxo-β-carotene) and myxoxanthophyll (unknown structure) (Goodwin, 1957). The last two, which replace the normal plastid carotenoids, are not found in any other photosynthetic plants. Echinenone is also found in certain marine invertebrates, and recently it has been reported that one of the main pigments of a strain of *Mycobacterium phlei* is probably myxoxanthophyll (Schlegel, 1958). Because of the specificity of distribution of echinenone in the plant world, its quantitative determination in lake deposits may be a means of determining the growth of blue-green algae in the past (Vallentyne, 1954).

For some time it had been thought that the photosynthetic pigments of the blue-green algae were located in small particles corresponding to bacterial chromatophores. These are now considered to be artifacts, and the chloroplasts of blue-green algae are said to resemble the grana-free, lamellated chloroplasts of *Chlorella* (Elbers et al., 1957), except for the lack of a definite bounding membrane.

A further characteristic of blue-green algae is the presence of various phycobilin pigments; members of this group of pigments are also present in the red algae (Rhodophyceae). However, the similarity between these two classes does not extend to the carotenoids. The carotenoids of the red algae are essentially the same as those in the Chlorophyceae; there is, however, an interesting exception for *Phycodrys sinuosa* is said to lack β-carotene (Larsen and Haug, 1956).

A study of carotenoid distribution in pure culture of members of the Chrysophyceae has only recently been undertaken; this was made possible by the isolation of pure strains of a number of Chrysophyceae and the development of media suitable for their large scale culture. Dr. M. B. Allen kindly sent us cultures of *Ochromonas danica* and *Prymnesium parvum;* and Mrs. S. Phagpolngarm has shown that they have the carotenoid mixture characteristic of the brown algae and diatoms. Fucoxanthin represents about 75% of the total carotenoids, the remainder being mainly β-carotene together with small amounts of an unknown pigment more strongly adsorbed on a chromatographic column than fucoxanthin. This work confirms the very early work of Heilbron (1942), who found β-carotene and fucoxanthin in a mixed culture of *Apistonema carteri, Thallochrysis litoralis,* and *Gloeschrysis maritima.*

The class Cryptophyceae has also been neglected for much the same reason as the Chrysophyceae, but recently Haxo and Fork (1959) have been able to examine the carotenoids in a pure culture of *Cryptomonas ovata*. They found that α-carotene was the main carotene and that zeaxanthin, or possibly the closely related diatoxanthin, was the main xanthophyll. This is a carotenoid distribution as distinctive as that found in the blue-green algae.

Perhaps the most interesting of the recently examined algae is *Cyanidium caldarium*, which taxonomically can be considered to be a green alga, although it produced phycocyanin and only chlorophyll *a* (Allen, 1954, 1959; Hirose, 1950), a close relationship with the blue-green algae thus being suggested. Again owing to the kindness of Dr. M. B. Allen, we have recently been able to examine the carotenoids of *C. caldarium;* these are characteristic of neither green nor blue-green algae, but consist of β-carotene (55%), zeaxanthin (32%) and an unidentified xanthophyll (13%). Thus it shares with *Cryptomonas ovata* the ability to synthesize comparatively large amounts of zeaxanthin. These observations are not incompatible with the view that *Cyanidium caldarium* might be a transitional form between the cryptomonads and the green algae, because it is possible that the cryptomonads represent a primitive flagellate group from which green algae have developed (M. B. Allen, personal communication). The high relative amount of β-carotene in *C. caldarium* is reminiscent of the blue-green algae (Goodwin, 1957), although *Chlamydomonas reinhardii* also contains a high proportion of β-carotene (Sager and Zalokar, 1958).

Extraplastidic Pigments

Extraplastidic carotenoids frequently occur in (*a*) the gametes of multicellular algae and (*b*) the cytoplasm of many unicellular flagellates. Typical examples of specific carotenoid accumulation in reproductive regions are: (i) both male and female gametes of *Ulva* spp. accumulate large amounts of γ-carotene (Haxo and Clendenning, 1953); and (ii) the male gametes of *Fucus* spp. and *Ascophyllum nodosum* are bright orange owing to the presence of β-carotene, while the olive green color of the ova is produced by a mixture of fucoxanthin and chlorophylls (Heilbron, 1942). Many flagellates turn bright red when they go into a resting state and encyst; this is due to the accumulation in the cytoplasm of large amounts of either β-carotene (*Trentepohlia aurea* and *Dunaliella salina*) or astaxanthin (3,3′-dioxo-4,4′-dihydroxy-β-carotene) (*Haematococcus pluvialis*)—see Goodwin and Jamikorn (1954a).

It has often been assumed that the bright red color of the eye-spot (stigma) of various flagellates, including the Euglenineae, is astaxanthin. There is no experimental evidence for this and so far astaxanthin has not been detected in flagellates*—even in *E. gracilis* grown in the presence of streptomycin, which inhibits the synthesis of the plastid pigments but has little if any effect on the synthesis of the eyespot pigment (Goodwin and Jamikorn, 1954b; Wolken, 1956). It is of course possible that trace amounts present in the eyespot escaped detection.

The Siphonales (Chlorophyceae) appear to be an interesting deviation from normalcy; according to Strain, α-carotene is the main hydrocarbon pigment and an esterified pigment, siphonaxanthin (structure unknown), is the main xanthophyll (Strain, 1951). It would be instructive to know if these pigments are plastidic or not, because xanthophyll esters are not known in the plastids of any other algae or in the chloroplasts of higher plants.

Pigment Mutants

The first fundamental work on algal mutants in which carotenoid biosynthesis was deranged was carried out by Claes (1954, 1956, 1957, 1958, 1959). She isolated four x-ray mutants of *Chlorella vulgaris* in which chlorophyll synthesis was almost completely blocked and the synthesis of plastid carotenoids was interrupted to varying extents (Table II).

TABLE II
Polyenes Synthesized by Mutant Strains of *Chlorella vulgaris*

Strain	Polyenes
Wild	β-Carotene, lutein, violaxanthin, neoxanthin
5/871	Phytoene only
5/515	Phytoene, phytofluene, ζ-carotene; no xanthophylls
9a	Phytoene, phytofluene, ζ-carotene and an unidentified carotene; normal xanthophylls
5/520	*In the dark:* phytoene, phytofluene, ζ-carotene, protetrahydrolycopene, prolycopene
	In the light: usual plastid carotenoids

This interruption is accompanied by the accumulation of the partly saturated acyclic polyenes. Three of the four mutants are killed by simultaneous exposure to light and oxygen, while the fourth (5/520) produces different carotenoids according to whether it is cultured in the presence or absence of light.

Through the courtesy of Dr. M. B. Allen we have been able to examine a number of ultraviolet-induced mutants of *Chlorella pyrenoidosa* in which carotenoid synthesis has been interrupted. The results obtained by Mrs. S. Phagpolngarm are outlined in Table III. It is clear that G 34,

---

* Except as described in the preceding paragraph.

which accumulates only phytoene, closely resembles the *C. vulgaris* mutant 5/871; both are also light-sensitive in the presence of oxygen. The amount of phytoene produced by G 34 is approximately the same as the total amount of carotenoid synthesized by the wild type *C. pyrenoidosa*. G 44 accumulates mainly phytofluene and ζ-carotene, but

TABLE III
Polyenes Synthesized by Mutant Strains of *Chlorella pyrenoidosa*

Strain	Polyenes
Wild	β-Carotene, lutein, violaxanthin, neoxanthin
G 34	Phytoene only
G 41	Usual plastid carotenoids
G 44	Phytofluene, ζ-carotene; a mixture of 9 unidentified xanthophylls

unlike Claes's mutants 5/515 and 9a it synthesizes no phytoene. Furthermore, xanthophyll production was not completely inhibited (compare mutant 5/515), but it was reduced and qualitatively altered (compare mutant 9a); about 20% of the total pigments were xanthophylls and they consisted of approximately equal amounts of nine unidentified pigments. Mutant G-41 is mentioned here because, although it has its chlorophyll synthesis impaired, its carotenoid metabolism is only very slightly disturbed, if at all.

Sager and Zalokar (1958) have described an interesting mutant of *Chlamydomonas reinhardi* in which carotenoid synthesis was some 200–500 times less than that of the wild strain. Only carotenes were synthesized, with β-carotene predominating. Significantly, no accumulation of phytoene or phytofluene was observed.

Chlorotic Substrains

When the normal strain of *Euglena gracilis* v. *bacillaris* is grown in the presence of streptomycin it produces almost colorless cells in which all the chlorophylls have disappeared and the amount of the carotenoids is reduced to a level equal to or below that of the normal strain grown in the dark (Goodwin and Jamikorn, 1954b). These changes are due to destruction of the chloroplasts, which makes them indistinguishable from mitochondria (Provasoli *et al.*, 1948). This change is permanent and can also be brought about by treatment with pyribenzamine or by high temperatures. A number of such permanent chlorotic substrains isolated by Gross and Jahn (1958) have recently been examined and the results are summarized in Table IV. From this table it will be seen: (*a*) that one substrain (PBZ-GA) produces no detectable carotenoids; (*b*) that in all other strains β-carotene is accompanied by phytofluene and "a ζ-carotene-like fraction" (neither of these polyenes is present in the wild strain); (c) that xanthophylls were present only in traces in SM-P, PBZ-G1, and PBZ-G2, but in greater amounts in HB-G and PBZ-G3 [in these last two substrains zeaxanthin was also present in addition to very small amounts of lutein, and echinenone was identified in HB-G; so we have at

TABLE IV
Carotenoid Distribution in Permanently Bleached Sub-strains of *Euglena gracilis*

Sub-strain	Xanthophylls	Hydrocarbon polyenes
Wild type	Lutein, violaxanthin, neoxanthin	β-Carotene
SM-L1		
SMP	Lutein and violaxanthin in	
PBZ-G1	minute traces	Phytofluene
PBZ-G2		β-Carotene
		ζ-Carotene
PBZ-G3[a]	Zeaxanthin plus traces of lutein	
HB-G[a]	Zeaxanthin, echinenone, and lutein	

[a] In PBZ-G3 and HB-G the xanthophylls predominated over the hydrocarbon polyenes.

last a microorganism (albeit an artificially produced one) other than a blue-green alga which produces echinenone]; and (d) that the concentration of carotenoids in all the substrains is much less than in the wild strain.

Formation of Carotenoids in Algae

The biochemical details of carotenoid synthesis in microorganisms will be dealt with later in this symposium; it is sufficient to say here that the general pathway in algae is probably the same as that in other carotenogenic organisms, because [2-C^{14}] mevalonate, a specific intermediate in terpenoid biosynthesis, is incorporated into β-carotene in both *Chlorella vulgaris* and *Euglena gracilis* (Goodwin, unpublished observations). Here, however, the author will briefly discuss some interesting observations concerning the role of light in carotenoid formation in algae.

Normal Strains

Light has no effect on carotenoid synthesis in *Chlorella vulgaris* which, when grown heterotrophically in the dark, produces as high a concentration of pigments as do cells grown autotrophically in the light (Goodwin, 1954); in other words, in this organism light has no effect on chloroplast formation. In *Euglena gracilis,* on the other hand, growth in darkness results in complete failure to synthesize chlorophyll; carotenoids are synthesized, but only to the extent of one-fifth of the synthesis in the light. Furthermore only lutein is present, β-carotene, neoxanthin, and violaxanthin being absent or present in only minute traces. In these respects "etiolated" *E. gracilis* resembles etiolated maize seedlings. However, on illumination of *E. gracilis*, there is a lag period of some 14 hours before the chloroplast pigments, the chlorophylls and the carotenoids, begin to be synthesized; in maize seedlings the synthesis begins immediately. We are only just starting our investigations into this problem, and at the moment of writing we have done little more than make the observation that the lag period, which exists with chlorophyll pigments (Brawerman and Chargaff, 1959), also exists for the carotenoids and that the formation of these two pigments runs parallel. The difference between maize seedlings and *E. gracilis* is that leaves grown in the dark contain colorless structural units (leucoplasts), which are considered the precursors of chloroplasts (see Wolken, 1959). On illumination, the synthesis of pigments is immediately triggered off and the true, functional chloroplasts immediately begin to appear. In *E. gracilis,* on the other hand, no structures corresponding to the leucoplasts have yet been observed in cells grown in the dark. The lag period in pigment production is probably caused by the time required for the re-organization of cellular material into structural units corresponding to leucoplasts; these can be formed in the absence of growth and without the net synthesis of protein and ribonucleic acid (Brawerman and Chargaff, 1959).

Mutants

Using the fascinating *Chlorella* mutant 5/520 (see Table I) Claes (1958, 1959) has made some important observations on the effect of light on carotenogenesis.

(A) If dark-grown cells are incubated anaerobically in the light, α- and β-carotenes are formed while some of the more saturated polyenes (ζ-carotene, phytofluene) which have accumulated in the dark, disappear. This is strongly suggestive that α- and β-carotenes are formed from phytofluene via ζ-carotene; this is discussed in detail in another paper in this volume.*

(B) If dark-grown cells are incubated aerobically in the light, both carotenes and xanthophylls are formed, indicating the need for oxygen in the formation of xanthophylls. Analyzing these observations in more detail, Claes (1959) has recently found that if dark-grown cells are first illuminated anaerobically and subsequently incubated aerobically in the

---

* See Paper No. 13 by G. Mackinney and C. O. Chichester.

dark, xanthophylls are rapidly synthesized. This indicates that xanthophyll precursors accumulate during the anaerobic/light phase and that these are oxidized to xanthophylls by a dark reaction. As the amount of hydrocarbon carotenes disappearing during this latter phase just equals the amount of xanthophylls appearing, α- and β-carotenes are strong candidates for the position of immediate precursors of the xanthophylls. That is, the oxygen function is inserted after the unsaturated polyene chain is completed; this agrees with the observations on the photosynthetic bacteria *Rhodospirillum rubrum* and *Rhodopseudomonas spheroides*.

(C) Short term (15 hours) illumination with blue light of dark-grown cells of mutant 5/520 in the presence of oxygen causes the isomerization of prolycopene and protetrahydrolycopene to the all-*trans* compounds; presumably the blue light absorbed by the carotenoids supplied the energy for the isomerization. In the presence of N_2, but not in the presence of O_2, red light has the same effect as blue light. This reaction is apparently sensitized by the trace amounts of chlorophylls present in the dark-grown cells, because *in vitro* studies have shown that in light petroleum solutions containing prolycopene (or protetrahydrolycopene) and chlorophyll *a*, red light stimulates isomerization to the all-*trans* forms in an atmosphere of nitrogen, but that the presence of oxygen inhibits this change (Claes and Nakayama, 1959).

REFERENCES

Allen, M. B. (1954). *Rappt. et Commun. 8ᵉ Congr. Intern. Botan.* **7** p. 41.
Allen, M. B. (1959). *Arch. Mikrobiol.* **32**, 270-277.
Brawerman, G., and Chargaff, E. (1959). *Biochim. et Biophys. Acta* **31**, 164, 172, 178.
Claes, H. (1954). *Z. Naturforsch.* **9b**, 461.
Claes, H. (1956). *Z. Naturforsch.* **11b**, 260.
Claes, H. (1957). *Z. Naturforsch.* **12b**, 401.
Claes, H. (1958). *Z. Naturforsch.* **13b**, 222.
Claes, H. (1959). *Z. Naturforsch.* **14b**, 4.
Claes, H., and Nakayama, T. O. M. (1959). *Nature* **183**, 1053.
Elbers, P. F., Minnaert, K., and Thomas, J. B. (1957). *Acta Botan. Neerl.* **6**, 345.
Goodwin, T. W. (1954). *Experientia.* **10**, 213.
Goodwin, T. W. (1957). *J. Gen. Microbiol.* **17**, 467.
Goodwin, T. W. (1959). *Advances in Enzymol.* **21**, 296.
Goodwin, T. W., and Gross, J. A. (1958). *J. Protozool.* **5**, 292.
Goodwin, T. W., and Jamikorn, M. (1954a). *Biochem. J.* **57**, 376.
Goodwin, T. W., and Jamikorn, M. (1954b). *J. Protozool.* **1**, 216.
Griffiths, M., Sistrom, W. R., Cohen-Bazire, G., and Stanier, R. Y. (1955). *Nature* **176**, 1211.
Gross, J. A., and Jahn, T. L. (1958). *J. Protozool.* **5**, 126.
Haxo, F. T., and Clendenning, K. A. (1953). *Biol. Bull.* **105**, 103.

Haxo, F. T., and Fork, D. C. (1959). *Nature* **184**, 1061.
Heilbron, I. M. (1942). *J. Chem. Soc.* p. 79.
Hirose, H. (1950). *Botan. Mag.* (*Tokyo*) **63**, 107-111.
Krinsky, N. I. (1960). *Federation Proc.* **19**, 329.
Larsen, B., and Haug, A. (1956). *Acta Chem. Scand.* **10**, 470.
Nishimura, M., and Takamatsu, K. (1957). *Nature* **180**, 699.
Provasoli, L., Hutner, S. H., and Schatz, A. (1948). *Proc. Soc. Exptl. Biol. Med.* **69**, 279.
Sager, R., and Zalokar, M. (1958). *Nature* **182**, 98.
Schlegel, H. G. (1958). *Arch. Mikrobiol.* **31**, 231.
Stanier, R. Y., and Cohen-Bazire, G. (1957). *In* "Microbial Ecology" (R. E. O. Williams and C. C. Spicer, eds.), p. 56. Cambridge Univ. Press, London and New York.
Strain, H. H. (1951). *In* "A Manual of Phycology" (G. M. Smith, ed.), pp. 243-262. Chronica Botanica, Waltham, Massachusetts.
Strain, H. H. (1958). "Chloroplast Pigments and Chromatographic Analysis," 180 pp. Pennsylvania State Univ. Press, University Park, Pennsylvania.
Torto, F. G., and Weedon, B. C. L. (1955). *Chem. & Ind.* (*London*) 1219.
Vallentyne, J. R. (1954). *Science* **119**, 605.
Wolken, J. J. (1956). *J. Protozool.* **3**, 211.
Wolken, J. J. (1959). *Ann. Rev. Plant Physiol.* **10**, 71.

Discussion

MACKINNEY: In the *Chlorella pyrenoidosa* mutants is there at the same time a moderate to severe upset of chlorophyll synthesis?

ALLEN: All the mutants that Professor Goodwin studied lacked chlorophyll b; there were also changes in the amount of chlorophyll present, especially in mutant G-34.

PRINGSHEIM: I should like to mention that there are a number of "colorless" flagellates that contain yellow pigments whose nature is not known yet. In *Polytoma* there are spores which are deep orange color and in *Polytomella* there are cysts which again are orange. It might be useful to investigate these. There are also colorless flagellates that retain their eyespots. These should be useful for investigation of the pigments of eyespots.

2

Pigments of Plant Origin in Animal Phyla*

DENIS L. FOX

Division of Marine Biochemistry, Scripps Institution of Oceanography, University of California, La Jolla, California

The observed contrasts in size, morphology, and complexity among living species are fundamentally of biochemical origin. But it is equally true that certain basic physiological properties of all cells rest upon common chemical foundations. Responsible for interspecific differences are the constituent proteins, which vary widely through permutations in the relative proportions and the linear bonding sequences of two or three dozen amino acids.

The "green universe of plants" and its contained "red kingdom of animals" utilize much the same basic fuels in respiration, and store similar chemical types of reserves. And, whereas essentially all organisms, so far as we know, seem capable of synthesizing the tetrapyrrole nucleus, the green plants have elaborated therefrom chlorophyll compounds and have thus become autotrophic, while the vertebrates and some invertebrate animals build hemoglobin as an oxygen-conveyer. Other porphyrins, such as the cytochromes, catalase, and other peroxidases are biocatalysts common to both plants and animals. Moreover, the porphyrin skeleton, once synthesized, possesses considerable chemical stability. Identifiable derivatives of hems and especially of chlorophylls may persist as biochemical fossils of great antiquity. A paper concerning these, as well as others such as carotenoids, will be presented in more detail by another contributor to this symposium.†

Colored molecules offer much of interest in the realm of comparative biochemistry, and invite plant and animal biologists to common meeting grounds. In the plant world, for example, there are countless instances of the heritable capacity to synthesize carotenoid biochromes in certain tissues, e.g., as in poppies with white or yellow corollae, red versus pale yellow tomatoes or grapefruit, or orange contrasted with colorless fungal mycelia, as in *Neurospora* mutants. Again, the elaboration of carotenoids may depend, in kind or in degree, upon environmental factors (Good-

---

* Contribution, Scripps Institution of Oceanography, University of California, La Jolla, New Series.
† See Paper No. 7 by J. R. Vallentyne.

win, 1955). Successive blooms of the California poppy may be observed to exhibit paler yellow-orange to yellow corollae with the advance of warm summer days, in contrast to the rich orange hues of the first spring blossoms; and all are familiar with the accumulation of carotenoids in many ripening fruits. The "red-snow" alga *Hematococcus pluvialis* has been demonstrated to elaborate large quantities of β-carotene when supplied with acetate (Lwoff and Lwoff, 1930). The halophilic algal flagellate *Dunaliella salina,* cosmopolitan in natural or industrial salterns the world over, is brick-red in saturated NaCl solutions, contains much β-carotene, little α-carotene and but traces of lutein and chlorophyll; in less concentrated media (12 to 15% NaCl), the cells are green with much chlorophyll, and contain far less β-carotene and minor quantities of α-carotene and lutein (Fox and Sargent, 1938).

Animals, deriving their carotenoids directly or indirectly from plant sources, exert different kinds of fractionation upon these molecules, whether through selective control of their assimilation within the lumen or walls of the alimentary tract, or in other tissues. Differences may reflect kinds, concentrations or activity of oxidase (perhaps lipoxidase) systems within the animal.

While there are indeed but few if any biochromes synthesized *de novo* only by animals,* there are several classes, in addition to the tetrapyrroles cited above, that are generated by both animals and plants, e.g., the indigoids, arising from the metabolism of tryptophan, the melanins deriving from that of tyrosine, and the pterins which are pyrimidopyrazine compounds. Beyond these and a few other examples, however, there is a wide array of pigments synthesized only by plants. Of these, a few are assimilated by animals in their nutrition. Passing references will be given to the flavins and quinones, after which we shall return to the prominence of the carotenoids.

Flavins

Riboflavin, variously called lacto-, cyto- or ovoflavin, is benzisoalloxazine-6,7-dimethyl-9, D-riboflavin, synthesized by most higher plants, notably in growing leaves, and by a number of microorganisms. Its physical and chemical properties have been well established and have been recorded rather extensively elsewhere (see Fox, 1953, pp. 282 *et seq.*) and its occurrence in very minute amounts in nearly all animal

---

* Even the few non-hem metalloproteins of animals, e.g., hemocyanin, hemerythrin, etc., can hardly be regarded as more than borderline exceptions, since animals depend upon ultimate plant sources for certain essential amino acids involved in protein synthesis.

tissues is well known; likewise the fact that, as one of the B vitamins, it is an indispensable factor in cellular respiration. Moreover, riboflavin and the flavoproteins may occupy a significant position in photoreception, leading to photokinesis in certain plants (Goodwin, 1959), and perhaps are involved in the biochemistry of vision. Riboflavin is known to exercise some deterring influence over the incidence of cataract (Hogan, 1939).

Flavoproteins in the skin of certain fishes (e.g., eels) are localized in areas bearing carotenoids and melanophores, but are absent from the white, guanine-laden parts. While these observations have suggested a possible intermediary role of flavins in melanin formation, the demonstrated photoinactivation of tyrosinase and other biocatalysts by riboflavin indicates that it may possibly control rather than accelerate photically-induced melanogenesis (Fox, 1953, p. 287).

Quinones

Naphthoquinones

The naphthoquinones, encountered in many plants, are of various yellow, orange, red, or purple colors. Their appearance in animals is very limited; numerous species of echinoid (sea urchins, sand dollars, and the like) display them as red, purple, or in some instances green polyhydroxy derivatives known as echinochromes, in calcareous spines, in shell, and in thin ectodermal and endodermal tissues. The same pigments are to be found in the elaeocytes of the body fluid, in the lining of the alimentary tract, in nervous tissues, and in some instances within the egg-jelly (see Fox, 1953, Chapter VIII).

Incidentally the sea otter *Enhydra lutris* acquires pink and purple calcareous echinochrome deposits in its skull, teeth, and other skeletal parts from the eating of large numbers of sea urchins in their natural surroundings (see Fox, 1953, p. 195).

Certain members of a second echinoderm class, the crinoids (featherstars or sea-lilies), assimilate echinochromes. The pigments have been identified both in fossil species and in the living tissues of the freeswimming crinoid *Antedon bifida* (Dimelow, 1958).

Nothing is known with certainty of the physiological position of the naphthoquinones. Their chemical similarity to vitamin K has been ascertained, and a possible role has been suggested for them as respiratory stimulants for the eggs, and as activator of the sperm of certain echinoids. The careful experiments of Millott and his colleagues (Millott, 1957; Millott and Yoshida, 1957) lead one to suspect not only that echinochromes may act synergistically with melanins in screening out injurious light rays, but also that they may share with carotenoids the role of photoeffectors in the shading reaction.

Anthraquinones

The red polyhydroxyanthraquinone carboxylic acids (as well as glucosides and methyl or ethyl esters thereof) including carminic, kermesic, and laccaic acids, are doubtless derived by the respective scale-insects cochineal, kermes, and lac (lakh) from their plant hosts. The pigment, whether in the so-called fat-body of the cochineal or in the waxy excrescence of the (female) lac insect, represents, in all likelihood, a relatively inert residue or a secretory material; it has no recognized physiological function in the animal (Fox, 1953, pp. 205-208).

Recently another red anthraquinone has been discovered, occurring in red crinoids of the genus *Comatula*. Its investigators, Sutherland and Wells (1959), have called it rhodocomatulin, but have not to date ventured a guess as to its source or its potential biological function. The very fact of seemingly extravagant biochromy in some of these animals, through pigments deriving ultimately from plants, presents an enigma. Why do a few animal species assimilate and store the pigments while others, consuming the same food, do not? And do those which store the colored labile molecules utilize them anywhere in their over-all economy, e.g., in redox systems or as photoeffectors?

Carotenoids

This ubiquitous class of plant-synthesized polyene compounds emphasizes the questions raised above. For not only are the carotenoids of virtually universal occurrence in animals from the protozoan phylum to the highest vertebrates, but they are concentrated chiefly in three types of tissue: exposed integumentary surfaces, gonads (including immature spermaries and maturing eggs), and other glandular parts. Moreover the carotenoids themselves, and certain animal-produced derivatives of some of them, i.e., the A vitamins, have long been known to possess important physiological attributes. We have been aware, since the work of McCollum and Davis, and that of Osborne and Mendel, both in 1913, of the indispensability of vitamin A as a growth-promoting compound in vertebrates. A few years later its antixerotic or anti-keratosis properties were revealed by the studies of a whole series of workers in many laboratories. It was Moore who, in 1929, established the biochemical derivation of the vitamin from certain yellow carotenoids; nearly two decades later it was first demonstrated, in England by Glover and associates (1947) and in this country by Deuel and his colleagues (Mattson *et al.*, 1947; Wiese *et al.*, 1947) that the site of conversion of carotene into vitamin A is in the small intestine (of rats). Moreover, dietary carotene is converted into vitamin A by the cod *Gadus callarias* and by the lobster *Homarus americanus* (Neilands, 1947); and Kon and his associates in Reading have demonstrated the storage of relatively high concentrations of vitamin A, nearly exclusively in the eye-stalks, by numerous marine microcrustaceans of the euphausiid family, including "krill," the forms consumed in vast numbers by whales, which subsequently store vitamin-A-rich oil in their livers (Fisher *et al.*, 1955). It was demonstrated many years ago that mammals, even those storing but traces of carotenoids (e.g., swine), require them or vitamin A itself for successful reproduction.

The stimulating effects of the purified carotenoids, astaxanthin and, to a lesser extent, β-carotene, found in the egg-oil of the rainbow trout *Salmo irideus* upon the sperm of the same species, have been reported by Hartman *et al.* (1947), who found lutein and lactoflavin, both also present in the egg-oil, to be without any such effect.

Vision and Photokinesis

Wald and his associates, among others, studying visual processes during the past quarter-century, have illuminated the unique position of the A vitamins in the visual pigmentary apparatus. The remarkable role of this carotenoid-fission product, derived by and functional in animals as a photolabile purple or red protein-conjugant, would appear to be an evolutionary biochemical advance from a basic and perhaps spectrally more restricted property of full-chain-length carotenoids. For we know that various molecules of this type in some way induce photokinesis, i.e., orientation or locomotion either toward or away from light-stimuli (Goodwin, 1959). This holds in both green and achlorophyllous plants; moreover, many animals, including eyeless invertebrate species, appear to have "borrowed" the same molecules, utilizing for similar purposes the carotenoids originally synthesized only in the plant world. The photosensitivity spectrum of some of these organisms has been observed to match closely the range of maximal light absorption by the carotenoids in their tissues. An interesting example is that of the sea-star *Marthasterias glacialis*,* in which Millott and Vevers (1955) have studied the carotenoid pigments of the optic cushion, a structure resembling a kind of "saddle" across the base of the small azygos tentacular finger at the tip of each arm or ray. The optic cushion bears some 150 deeply pig-

---

* Rockstein (1956) and Rockstein and Rubenstein (1957) also report the finding of photosensitive pigments in the terminal eyespots and skin of an asteroid, *Asterias forbesi*.

mented, so-called optic cups, and yields orange and reddish-orange carotenoids, the latter being prominent in the cups themselves. The pigments are labile *in vitro* if air is present, and are readily decomposed by light. The prominent pigments in the optic cushion seemed to be β-carotene and esterified astaxanthin, although Vevers and Millott (1957) found 8 chromatographic carotenoid fractions in the general skin of this star. One fraction exhibited the properties of β-carotene, another behaved like a neutral keto-carotenoid, while a hypophasic component resembled lutein; astaxanthin also was present, both free and esterified. Another important discovery made by these workers was that no traces of vitamin A itself were detectable in the animal. There is therefore the strong implication that asteroids may utilize full-length (40 C-atom), colored carotenoids as photoreceptors in their eye-spots or optic cushions. Moreover, in view of the exclusively carnivorous habit of most asteroid species, their carotenoid supplies are acquired from the ultimate plant source in an indirect way.

Disposition of Carotenoids

Ingested carotenoids are subject to several alternative but not necessarily mutually exclusive fates in the animal body, as follows:

1. Discharge with feces, in chemically unaltered condition;
2. Assimilation and storage, chemically unchanged;
3. Assimilation and conversion into other carotenoids, e.g., A vitamins, astaxanthin, "canary xanthophyll," echinenone, or other special animal carotenoids;
4. Assimilated portions oxidatively consumed or otherwise converted to colorless substances;
5. Destruction in gut, whether by symbiotic microorganisms or by host's enzymes.

The many animals which have been included in a broad survey to date are classifiable into several groups and subgroups in accordance with their predominating metabolic disposition of ingested carotenoids (cf. Zechmeister, 1937; Fox, 1953), as follows.

I. Assimilation and storage *in statu quo*:
   1. Nonselective, e.g., man, frog (*Rana*), *Octopus*, echinoids, some insects
   2. Selective
      (a) Xanthophyll selectors: most fishes, some invertebrates, many birds
      (b) Carotene selectors: horse, cow, some sponges, some worms (e.g., *Thoracophelia*)

II. Production and storage of colored, oxidized derivatives, e.g., astaxanthin, from conventional dietary carotenoids: crustaceans, asteroid and ophiuroid echinoderms, nudibranch mollusks, some insects (locust), some coelenterates, certain birds (flamingo)

III. Excluders, by nonassimilation or through complete oxidation of carotenoids: carnivorous birds and mammals, pig, goat, guinea-pig, rabbit

Few if any generalities may be obvious in this condensed classification, but some definite trends exist. In the first place, carotene selectors tend to be herbivores or omnivores, and the same applies generally to the nonselective assimilators, although *Octopus bimaculatus,* a strict carnivore, is an exception (Fox and Crane, 1942). Moreover, the majority of animals in these two groups tend to store their carotenoids only internally and in small to moderate quantities. Certain sponges and bryozoans of brilliant colors provide exceptions to this generalization by depositing their carotenoids in the integument (although, considering the relative simplicity of their structure, they do not have much choice if they are to store the pigment at all). The vast majority of fishes examined carry solely or chiefly xanthophylls in their skin, liver, ovaries, or (rarely) flesh; carotenes, when present, are greatly in the minority, and in no published instance has the skin of any fish species yielded carotenes without xanthophylls. It will be remembered, also, that most fishes are carnivores, while many are omnivorous (Fox, 1957).

Mammals as a class are relatively poor in carotenoids; many store none of the colored members at all, although all demand a supply of vitamin or provitamin A. Some store carotenoids in plasma, liver, adipose tissue, adrenals, and corpus luteum. Of these, horses and cattle are outstanding carotene selectors (Palmer, 1922), which expel with the feces the xanthophylls not earlier degraded in the gut (Rogozinski, 1937). Other herbivorous mammals, such as sheep, goats, and swine, store no carotenoids or but the smallest traces, and, unlike cattle, secrete white instead of yellow butter-fat in their milk (Palmer, 1922). Rabbits (*Lepus cuniculus*) characteristically assimilate no carotenoids, and store white depot fat, save for a recessive strain whose adipose tissues are orange-yellow with xanthophyll (Pease, 1928; Willimott, 1928).

In contrast to numerous invertebrate carnivores, their mammalian dietary counterparts, e.g., the feline and canine families, store very little carotenoid in their bodies and, as in other mammals (save for some strains of cattle on a heavy diet of green food), none in the skin. Even omnivorous mammals such as rats are poor in carotenoids, while man remains an exception, depositing, notably in his body fat, portions of any carotenoid or combinations thereof from his recent diet. Indeed the extravagant consumption of carotenoid-rich foods, such as oranges or pumpkins, has been observed to confer upon the skin of children a distinctly yellow color, known variously as carotenemia, xanthemia, or false jaundice. The condition is not pathogenic, and is directly controllable by altering the diet (Fox, 1953, p. 174).

The blue whale, feeding upon such planktonic crustaceans as *Calanus* and *Euphausia*, stores much astaxanthin in its oil; indeed, according to Schmidt-Nielsen *et al.* (1932), an occasional reddish specimen of this whale may be seen, wherein the red carotenoid has been distributed into the tissues generally, rather than remaining only in the usual depots. The whale's supplies of vitamin A are supposedly obtained from the relatively rich amounts found in the eye-stalks of the crustaceans swallowed as food, although there is a persistent idea that the whale may, like certain other animals (see below) be capable of converting astaxanthin into vitamin A.

Secretion or Excrescence to the Exterior

Examples wherein carotenoids are secreted to the outside are to be found in some animal phyla. Many crustaceans mobilize astaxanthin from their thin, pigment-rich hypoderm into the chitinous carapace, and something similar may be observed in certain insects, notably the lady-beetle *Coccinella septempunctata*, which secretes lycopene as well as α- and β-carotenes into its elytra (or hard wing-covers). The honey-bee, *Apis mellifera*, deposits its pollen-derived carotenoids in its various products; β-carotene has been recovered from amber-colored honey, and carotenoids have been recognized also in the wax, while several fractions, including β-carotene and a lutein ester have been recovered from propolis or bee-glue (Schuette and Bott, 1928; Zechmeister, 1937).

Examples of carotenoids in the yolky parts of eggs are so numerous as to require here no more than passing reference, save for an interesting example in an insect, *Colias philodice*, whose caterpillar is either blue or green, depending upon a recessive gene which confers failure to assimilate and store yellow carotenoids from the common diet of green clover; the adults of this genotype lay colorless eggs which hatch into larvae colored blue (due to the presence of a tetrapyrrole from the degradation of chlorophyll), while the dominant green type lays yellow eggs, giving rise to larvae colored green from the presence of both the blue tetrapyrrole and the yellow carotenoid. Moreover, the latter coloration serves an incidentally useful role, for the carotenoid-assimilating, green-colored larvae enjoy a measurable degree of chromatic protection from predation by English sparrows, denied to their blue siblings against the natural background of green clover (Gerould, 1921).

Some insects secrete carotenoids into the silk of their cocoons. A well-known example is the larval stage of the silk-moth *Bombyx mori* which depends upon two genes for the spinning of yellow rather than white silk: C for assimilation of mulberry-leaf carotenoids to give yellow blood, and Y for the secretion of such xanthophylls as taraxanthin, violaxanthin, and lutein into the silk (Uda, 1919; Oku, 1929, 1930, 1932, 1933; Manunta, 1933, 1935, 1937). Another instance is that of a hemipteran or sucking-bug *Apanteles flaviconchae* which, preying upon the *Colias* caterpillars mentioned above, ultimately secretes cocoon-silk of a golden color if feeding upon the blood of the green type, and a white silk if nourished by that of the blue, carotenoid-free form (Gerould, 1921).

The cephalopod *Octopus bimaculatus*, assimilating either or both carotenes and xanthophylls, including astaxanthin, in its large "liver" or hepatopancreas, secretes into the closely adjacent ink-sac free and esterified xanthophylls but no carotenes (Fox and Crane, 1942).

Among the chordates, some ascidians or sea-squirts deposit rich stores of carotenoid, largely xanthophylls including astaxanthin, in the tunicin material of the outer cloak or test (Lederer, 1934). Certain brilliantly colored fishes, notably the marine dorado *Beryx decadactylus*, have mucus of a bright red color, due to the presence of astaxanthin, in the mouth and gills (Lederer, 1935). Among the reptiles the male spiny-tailed iguanid *Ctenosaura hemilopha* exudes a red, orange, or yellow waxy substance from its multiple femoral pits in the ventral skin of the thigh. The preponderant carotenoid is an esterified alcohol closely resembling taraxanthin (Fox, 1953, p. 158).

An account of the carotenoids secreted by bird species into their feathers would constitute an extensive chapter by itself, and must be deferred here, save for some special considerations below.

In mammals, apart from the presence of carotenoids in the milk of cattle and some other ruminants, as well as in the milk and particularly the colostrum of humans, there are but few instances of carotenoid secretion. An instance, however, is the ear-wax of cattle, the chief polyene of which appears to be β-carotene (Palmer and Eckles, 1914).

From a brief survey of the comparative secretion of carotenoids in animals, one might be tempted to draw some analogies with the known role of vitamin A in maintaining the status of mucous surfaces. Current difficulties are, however, that some fishes, crustaceans, anemones, and other animals may elaborate mucus containing carotenoids while others do not; moreover, we require far more study of the total constitution and physiology of mucus and other secreted materials before our thinking may advance much further.

Many animals store in various tissues what appears to be merely the excess of ingested carotenoids in chemically unaltered condition, e.g., among fishes, amphibians, reptiles, birds and mammals; although there remains the perplexing question of selective fractionation in many instances, the metabolic economy of carotenoids in these creatures still is obscure save for the importance of the A vitamins.

Biochemical Modification of Dietary Carotenoids

There remain the two other classes, i.e. (1) animals lacking or very poor in carotenoids, selecting but little and/or destroying essentially all of the pigment ingested save for that converted to vitamin A, and (2) the richly colored kinds, capable of converting assimilated carotenoids into partially oxidized derivatives often red in color.

We know little indeed about the biochemical fate of carotenoids consumed by animals of the first, relatively destructive category. And, while we have learned also but little of the pathways of carotenoid metabolism in the oxidative modifiers in the latter class, some account of a few empirical findings may stimulate further interest.

In a search for a possible astaxanthin precursor in the large marine crayfish *Panulirus interruptus,* marked and segregated specimens were maintained on a basic diet containing: in one group, no colored carotenoids; in a second, only carotenes; and in the last, only a taraxanthin-like xanthophyll ester. During the experimental period of many months in laboratory aquaria, several animals moulted the carapace. Cast carapaces and whole animals from the three groups were analyzed for total astaxanthin content, the values for which decreased with the passing of time; thus no evidence was forthcoming as to a potential carotenoid source of the red crustacean pigment (Fox and Kritzler, unpublished experiments, 1946-1947). But soon afterward Goodwin (1949, 1952) reported that 40% of the β-carotene, the only polyene found in newly laid eggs of the locusts *Locusta migratoria migratorioides* and *Schistocerca gregaria,* was replaced by astaxanthin by the time of hatching. He concluded that an oxidase must be present to effect this conversion, and that such conditions should ensure the equipment of the newly hatched hoppers with a means of photoreception, since no vitamin A was detected in these locusts and since astaxanthin has been cited as a likely photokinetic carotenoid in animals which do not synthesize the vitamin itself (Wald, 1943).

Indeed vitamin A, derived from certain precursors by all vertebrates, is lacking in many invertebrate species, including not only certain generally photosensitive forms, e.g., among the asteroids and echinoids, but even in kinds which possess functional eyes, notably most insects and many crustaceans. The relative obscurity of the visual processes in insects is emphasized by Fisher and Kon (1959), who remind us that no requirement for vitamin A has been detected by various workers in several species, e.g., the fruit fly *Drosophila melanogaster*, the flour beetle *Tribolium confusum*, the meal worm *Tenebrio molitor*, the clothes moth *Tineola biseliella* or the cockroach *Blatella germanica*. Likewise Goldsmith (1958) cites the failure of the grasshopper *Melanoplus* and the dragonfly *Sympetrum* to yield vitamin A in extracts of their heads; but his researches on the honeybee *Apis mellifera* revealed the presence of retinene$_1$ in the eyes (actually the whole heads) but not from the headless bodies similarly extracted. Addition of $SbCl_3$ to the retinene yielded a bright blue compound with an absorption maximum at 664 mµ, characteristic of retinene$_1$, while reduction of the terminal aldehyde group with potassium borohydride gave vitamin A which, treated with $SbCl_3$, manifested its blue product, absorbing maximally at 618 mµ. Similarly, Wolken (1959) has detected A vitamins in the heads of houseflies.

The extensive survey of Kon and his associates at Reading would suggest that in the arthropod phylum we encounter a kind of border area with respect to the synthesis of vitamin A and the compounds involved in vision; for, in view of the comparatively high concentrations of the vitamin in the eyes of numerous euphausiid species and its plentiful distribution between eyes and hepatopancreas in several groups of decapods, it is surprising that the compound should be completely lacking from most copepods examined, from several branchiopods (e.g., species of the "water-flea" *Daphnia* and the "fairy shrimp" or "brine shrimp" *Artemia salina*), and from all save 2 out of some 54 species of amphipods examined. The lack of vitamin A from adult cirripeds (barnacles) is less surprising since these sessile animals lie head-down within a shell; however they react to abrupt changes in incident light intensity by withdrawing the feeding appendages and closing the shell's aperture. Occurrence of the vitamin was found to be quite variable among the isopods and mysids examined. Moreover, in the decapod order, its presence varied not only between species but in some instances even among individuals within a species (Fisher and Kon, 1959; and Dr. Kon, personal communication, 1959). While it is certain that vitamin A serves a visual function in some Crustacea, it would appear likely that those species which do not store it consistently or which carry it in the body but not in the eyes must, like those seeing animals which completely lack the vitamin, employ instead an alternative chromophoric molecule such as astaxanthin in the visual cycle. Crustaceans commonly store astaxanthin, and notably in the eyes, whether or not accompanied therein by vitamin A. Wald (1943) has pointed out the degree of correlation between the absorption maximum of astaxanthin (ca. 490 mµ in castor oil) and the spectral range of sensitivity for photokinetic responses in several different species. In this connection, however, it will be borne in mind that the wavelength range of light perceived is likely to find more correlation with the absorption spectrum of the carotenoid-protein complex than with that of merely the pigmented moiety dissolved in an oil.

Some challenging riddles remain in the comparative biochemistry of photokinesis, and notably of vision. One of these poses the question: while all vertebrates and many invertebrates utilize vitamin A-protein conjugants in their retinal processes, how do invertebrates, closely related to this type but lacking vitamin A, effect ocular photoreception? If a colored 40-carbon carotenoid such as astaxanthin is present instead, are there photolabile, reversible carotenoid-protein conjugation systems involved, corresponding in function with the biochemical visual apparatus which is operative in the conventional A-vitamin-protein systems of many animals? We know that natural astaxanthin-protein complexes may assume any of several different colors, and that these conjugated chromoproteins may be carefully and reversibly dissociated by certain procedures. It would be interesting if it could be shown that such a colored complex, occurring in the eyes of dark-adapted, vitamin A-lacking crustacean species, might undergo photically induced reversible dissociation with changes of color (cf. Wald, 1943, p. 213).

Fisher and Goldie (1959) report that the euphausiid crustacean *Meganyctiphanes norvegica*, a prominent component of the "krill" eaten by whales, relies for its food in Loch Fyne during spring, almost exclusively upon copepods, which supply no vitamin A, but astaxanthin as the chief if not the only carotenoid. It was during that season, however, that the vitamin A concentrations in the euphausiids' eyes increased most rapidly. This species stores about 85% of the total supply of the vitamin in the inner ends of the ommatidia where the rhabdomes are situated (Fisher and Kon, 1959). It would be interesting to learn whether there might be differences in visual acuity and the color-spectrum thereof between arthropods storing only astaxanthin in the eyes and those which carry vitamin A there as well.

It has been found by Grangaud and Massonet (1955) that the mosquito-fish *Gambusia holbrooki*, rendered completely lacking in vitamin A through a restricted diet, exhibits characteristic signs of the deficiency, including emaciation and fraying of the caudal and other fins and an arciform flexure of the spine, but that the supplementation of the vitamin A-free diet with chromatographically separated astaxanthin ester at an advanced period of the regime will preclude the appearance of any aspect of the syndrome, and will later bring about restoration of vitamin A, notably in the eyes of the fish.

Thus there are the vertebrates in addition to certain insects and crustaceans which utilize vitamin A in their visual processes, and which obtain the vitamin either directly by eating other animals supplied with it, or through the splitting of any of several alternative precursors, e.g., α-, β-, or γ-carotenes, cryptoxanthin, echinenone, or 5,6-5′,6′-diepioxy-β-carotene (Fisher et al., 1954). There are arthropods, some of which seem to utilize only astaxanthin in their photoreceptive chromophores, while others may use vitamin A as well or instead. Finally, certain fishes are able to convert astaxanthin into vitamin A.

Moreover, in order to form the vitamin A or the astaxanthin from their precursors, different biochemical processes are required on the part of the respective organisms. The genesis of vitamin A from carotene, bearing some resemblance to a simple hydrolysis through the introduction of one hydrogen atom and one hydroxyl radical linked to the same carbon atom

$$(\tfrac{1}{2}\ \beta\text{-carotene or } C_{20}H_{28} \ldots \xrightarrow{\substack{H \\ OH}} \text{Vitamin A or } C_{20}H_{30}O)$$

actually involves the conversion of a hydrocarbon radical with 6 double bonds to a primary alcohol with but 5. The formation of vitamin A_2 would involve an oxidation, however, since the net effect is not only to add an oxygen atom but to remove two hydrogen atoms from the cyclohexenyl ring, thereby conferring another site of unsaturation. Concerning the elaboration of astaxanthin, an example cited above was the locust which, while developing within the egg, oxidizes β-carotene, $C_{40}H_{56}$, to its symmetrical dihydroxy-diketo derivative $C_{40}H_{52}O_4$. If a copepod, having achieved this oxidation, is consumed by the euphausiid *Meganyctiphanes norvegica*, this predator then reduces the astaxanthin to vitamin A by introducing hydrogen, thus saturating double bonds, and by removal of oxygen. The fish *Gambusia* seemingly effects the same reduction. Studies with radioisotopic carbon might tell us whether certain crustaceans, storing both vitamin A and astaxanthin in their eyes, first oxidize plant carotenoids to astaxanthin and then reduce this, with splitting, to give the vitamin. One would like to know also whether some of the lower animals might be able to utilize tetrahydroxyxanthophylls, e.g., taraxanthin, or more likely dinoxanthin or a spectrally similar xanthophyll, which occurs widely in fishes (Fox, 1953, 1957), as precursors to astaxanthin or other red, oxidized carotenoids.

Instances are known wherein fishes oxidize carotene to a yellow xanthophyll, in contradistinction to *Gambusia*'s reduction of astaxanthin to vitamin A. Early researches of Sumner and Fox (1933, 1935) indicated that the Pacific killifish *Fundulus parvipinnis,* and in all likelihood the long-jawed goby *Gillichthys mirabilis* as well, are capable of increasing their body stores of xanthophyll during growth on diets containing no xanthophylls at all but β-carotene as the only recognizable carotenoid.

We know too that somewhat similar oxidations occur in birds, of which two or three examples should suffice here. Canaries, *Serinus canaria canaria,* investigated *inter alia* by Brockmann and Völker (1934), were fed on rations free from colored carotenoids until the new feathers emerged white; the birds were then given diets supplemented by known individual carotenoids. The hydrocarbons β-carotene and lycopene were neither altered nor assimilated; violaxanthin was chemically changed in the gut but not subsequently deposited in any organs, egg-yolk, or feathers; zeaxanthin was assimilated, partly unaltered and partly modified chemically, in egg-yolk and in feathers, conferring upon the latter a golden yellow color tinged with red-orange; and lutein, assimilated and stored unchanged in liver, egg-yolk and fat, emerged in the plumage as a new, bright yellow pigment called by the authors "canary-xanthophyll," with absorption maxima reminiscent of taraxanthin (472, 443 mμ in petroleum ether) rather than its precursor, whose spectrum exhibits peaks at 477.5 and 447.5 mμ in the same solvent. This retrogression of absorption maxima toward the shorter wavelengths suggests that an oxidation must have occurred.

According to Kritzler (1943), wild bishop birds, *Euplectes franciscanus* and *E. nigroventris,* deposited in their new feathers a unique red carotenoid with absorption maxima at 495 and 450 mμ (in carbon disulfide) as a result of a diet supplemented with mixed tomato carotenoids. Failure of other common carotenoids substituted in the diet to elicit this response led to the conclusion that lycopene in the tomato was the precursor of the new pigment. A third species, *E. taha,* manifested in its new plumage no chromatic response to the tomato diet.

Recent studies by the writer (Fox, 1955; and unpublished investigations) and by Völker (1958) have shown that several species of flamingo, notably the so-called American or West Indian and most brightly colored form, *Phoenicopterus ruber,* convert the yellow carotenoids of green plants into unique red derivatives which give characteristic pink or vermilion colors to the bird's feathers, naked leg-skin, and depot fat. The mesenterial fat, plasma, and adrenals are often of an orange or apricot color, while the liver, gonad, and other glands and the egg-yolk yield much carotenoid material as well. However, most of the ingested carotenoid is destroyed in the gut, partly doubtless by symbiotic bacteria; the voided feces yield no carotenoids. Captive specimens apparently must receive a diet rich in the pigment if an excess is to be deposited in skin and plumage.

Investigations by the writer, using a flock of flamingos at the San Diego Zoo, and observations by Conway (1958, 1959; personal communications, 1958-1960) at the New York Zoo, have confirmed the ability of *Ph. ruber* to convert various carotenoids into the characteristic oxidized products; Conway administers a diet enriched with oil extracted from carrots and therefore containing β-carotene as the chief pigment, while at San Diego the flamingos receive as a part of their diet shrimps, ground crayfish shell, and red salmon flesh as the principal sources of carotenoid, in this instance astaxanthin. This carotenoid reappears, at least in part, among the unique neutral red carotenoids of feathers and leg-skin, while the egg-yolk, blood plasma, fat, and other internal tissues have yielded no astaxanthin among the rich supplies of other carotenoids. Some deeply colored feathers, grown by American flamingos on the carrot-oil supplemented diet in New York and sent out to La Jolla by Mr. Conway have been found to yield, among other carotenoid fractions: (*a*) a substantial component which remains persistently epiphasic in petroleum ether over 90 to 95% methanol, in contrast to the mere traces of this fraction recoverable from feathers of the San Diego flock (this is not carotene, but a unique derivative with absorption maximum at 448 to 450 mµ and an inflexion at 468 mµ in petroleum ether); (*b*) more surprisingly, hydrolysis of the feathers yielded also a minor quantity of astacene, reminiscent of the other flock receiving astaxanthin in their regime; moreover, pyridine extracts of the feathers yielded minor quantities of astaxanthin itself. This notwithstanding the fact that the oily mash fed to the New York flock yields no traces of astaxanthin, and it is extremely unlikely that the New York birds might have adventitious sources of this carotenoid. Both colonies stored in their feathers a major carotenoid fraction chromatographically and spectrophotometrically indistinguishable from the red carotenoid canthaxanthin, recoverable from the pink mushroom *Cantharellus cinnabarinus* (Haxo, 1950) and shown by Petracek and Zechmeister (1956) to be the neutral 4,4'-diketo-β-carotene.

The initially edematous legs of the newly hatched, downy-white-plumed flamingo chick are bright red or pink from hemoglobin beneath the thin epidermal layer. This skin turns black, however, on about the eighth day as the chick prepares to leave the shelter of the nest and the hovering parents. This rapid melanization, perhaps induced by multiple short exposures to solar rays, may, in the bird's native habitat, serve as an aid in protecting the thin skin and the delicate underlying tissues from injury due to unavoidable exposure to the blazing tropical sunlight. As time passes, a thick cuticle grows over the leg-skin; this skin later gradually loses its melanin, replacing it by red carotenoids. Whether these carotenoids, through their moderate absorption of light in the critical, erythemogenic fraction (ca. 295 to 320 mμ) may afford some continuing protection, notably over the flexible hock and over the toes, remains to be determined.

A final example may suffice for the purposes of this brief comparative survey. The oceanic siphonophore *Velella lata* exhibits striking blue colors in both mantle and tentacles, reflecting the presence of astaxanthin conjugated with protein (Fox and Haxo, 1958). Clear blue aqueous systems of the fresh mantle pigment manifest a smooth, rounded absorption profile with a broad maximum in the yellow region around 585 to 588 mμ, while like preparations from the blue tentacles exhibit a curve of very similar shape, but with maximal absorption in the orange at about 610 mμ, thus suggesting differences in the kind or ratio of protein conjugated with the same chromophore.

A typical coelenterate in its carnivorous habits, *Velella* may derive its supplies of astaxanthin from its known consumption of planktonic crustacean species and their floating eggs. The digestive tissues and the hanging medusa buds harbor an abundance of minute yellow-brown algal cells or zooxanthellae. These algae contain no astaxanthin, but a full complement of photosynthetic pigments, e.g., chlorophyll *a*, common carotenes, and several algal xanthophylls. The fresh mantle tissue, examined through dorsal and ventral surfaces and including the thin intermediary alga-laden endoderm, exhibit a rounded absorption maximum at 590 mμ, contributed by the astaxanthin-protein complex, and a secondary, minor peak at 675 mμ from the algal chlorophyll.

The members of this plant-animal association would appear to afford each other some features of mutual advantage. The algal cells within the tissues of the siphonophore receive harborage against consumption by filtering or particulate feeders, and perhaps also against their sinking into darkness; moreover, they reside in a site favorable for their assimilation of soluble metabolic wastes and digestive products of the host. And the animal's blue translucent canopy, while shading the plant cells against excessive sunlight at the ocean surface, demonstrably admits ample light for photosynthetic needs. The zooxanthellae, on their part, evolve *in situ* photosynthetic oxygen in excess of the respiratory requirements of the association. Free excess gas, perhaps relatively high in oxygen content, is therefore normally present in the chambers beneath the upper mantle surface. The siphonophore colony, consequently remaining afloat, occupies a region of adequate food supply, e.g., buoyant eggs, microscopic animals feeding upon particulate organic detritus or leptopel, often in the region of extensive surface "slicks." This floating facility, implemented by the entrapped gases, may not only reduce the likelihood of detection by nectonic predators but, augmented by the erect sail, should provide for the colony's transportation by prevailing winds toward or in company with surface concentrations of food materials.

※　※　※

Continued investigations should reveal increasing evidence of the vital importance of colored molecules, synthesized originally by plants and borrowed by animals, whether directly or through serial food-chains, often chemically modified in the process, and sometimes not without incidental usefulness to associated representatives of the original synthesizers.

Various animals may modify plant carotenoids for use in generalized or regionally focused photoreception. There remains the possibility that, in invertebrates and perhaps in some fishes, 40-carbon-atom xanthophylls may occupy roles somewhat parallel to those fulfilled by vitamin A in terrestrial vertebrates, e.g., in embryonic development and growth, in the maintenance of moist surfaces, including the elaboration and secretion of mucus and ink, and perhaps also in the deposition of chitinous skeletal structures.

Returning to the systems of invertebrates with eyes, it should be of interest to determine whether arthropod species lacking vitamin A may employ instead photolabile protein complexes of alternative polar carotenoids for perceiving light in certain spectral regions. Astaxanthin, the wide occurrence in arthropod eyes and the chemical properties of which would suggest it as a candidate for such a role, is a polar molecule, $3,3'$-dihydroxy-$4,4'$-diketo-β-carotene, capable through enolization of acting as a weak acid and thus combining with metallic ions to form salts; alternatively, behaving as a secondary keto alcohol, it is esterifiable with fatty acids. Moreover, enolic astaxanthin occurs conjugated with proteins to give various blue, green, purple, grey, red, or brown complexes; and these are reversibly dissociable by gently warming and promptly recooling in aqueous media, or alternatively by rendering their electrolytic solutions slightly acidic, followed by early neutralization. Under such conditions a blue astaxanthin-protein association will progress gradually and reversibly through blue-purple and red-purple to red, e.g., on warming the chromoprotein from lobster shells (Wald et al., 1948) or from *Velella lata* (Fox and Haxo, 1958). Denaturation of the protein moiety by heating to 100°C or by the addition of mineral acids, alcohol, acetone, or pyridine sets free the red-orange carotenoid. How chemically extensive or physiologically active photic energy might prove to be when allowed to impinge upon (dark-adapted) optical systems involving astaxanthin-protein compounds would now appear to be a demanding issue.

In the meanwhile, any direct correlation between the biochemical kind of visual equipment an animal possesses and the brightness or shadiness of its environment would appear, from surveys by Kon and his colleagues, to be nearly ruled out. But the suggested biochemical investigations might be undertaken with a view toward discovering any possible correlations between (*a*) action-spectra of acuity in visual response, (*b*) morphology of the visual apparatus, and (*c*) polarity and chemical reactivity of chromogens involved in photokinesis.

Answers derived from any such studies would, in all probability, leave us still puzzled about some insects which reportedly lack vitamin A and can be raised to normal adults of a second generation on diets lacking the vitamin or any other carotenoid (cited by Fox, 1953, p. 136).

REFERENCES

Brockmann, H., and Völker, O. (1934). *Z. physiol. Chem.* **224**, 195-215.
Conway, W. G. (1958). *Animal Kingdom* **16**, 159-173.
Conway, W. G. (1959). The Avicult. Mag., Jul.-Oct., pp. 108-111.
Dimelow, E. J. (1958). *Nature* **182**, 812.
Fisher, L. R., and Goldie, E. H. (1959). *J. Marine Biol. Assoc. United Kingdom* **38**, 291-312.
Fisher, L. R., and Kon, S. K. (1959). *Biol. Revs. Cambridge Phil. Soc.* **34**, 1-36.
Fisher, L. R., Kon, S. K., and Thompson, S. Y. (1954). *J. Marine Biol. Assoc. United Kingdom* **33**, 589-612.
Fisher, L. R., Kon, S. K., and Thompson, S. Y. (1955). *J. Marine Biol. Assoc. United Kingdom* **34**, 81-100.
Fox, D. L. (1953). "Animal Biochromes." Cambridge Univ. Press, London and New York.
Fox, D. L. (1955). *Nature* **175**, 942-943.
Fox, D. L. (1957). *In* "The Physiology of Fishes" (M. E. Brown, ed.), Vol. II, pp. 367-385. Academic Press, New York.
Fox, D. L., and Crane, S. C. (1942). *Biol. Bull.* **82**, 284-291.
Fox, D. L., and Haxo, F. T. (1958). *15th Intern. Congr. Zool. London Sect. III*, Paper 2 pp. 280-282.

Fox, D. L., and Sargent, M. C. (1938). *Chem. & Ind.* (*London*) **57**, 1111.
Gerould, J. H. (1921). *J. Exptl. Zool.* **34**, 385-415.
Glover, J., Goodwin, T. W., and Morton, R. A. (1947). *Biochem. J.* **41**, xlv.
Goldsmith, T. H. (1958). *Proc. Natl. Acad. Sci.* (*U.S.*) **44**, 123-126.
Goodwin, T. W. (1949). *Biochem. J.* **45**, 472-479.
Goodwin, T. W. (1952). *Biol. Revs. Cambridge Phil. Soc.* **27**, 439-460.
Goodwin, T. W. (1955). *Ann. Rev. Biochem.* **24**, 497-522.
Goodwin, T. W. (1959). *Advances in Enzymol.* **21**, 295-368.
Grangaud, R., and Massonet, R. (1955). *Arch. sci. physiol.* **9**, 245-256.
Hartman, M., Medem, F. G., Kuhn, R., and Bielig, H. J. (1947). *Z. Naturforsch.* **2b**, 330-349.
Haxo, F. (1950). *Botan. Gaz.* **112**, 228-232.
Hogan, A. G. (1939). *In* "The Vitamins" (M. Fishbein, ed.), pp. 280-281. Am. Med. Assoc., Chicago, Illinois.
Kritzler, H. (1943). *Physiol. Zoöl.* **16**, 241-255.
Lederer, E. (1934). *Compt. rend. soc. biol.* **117**, 1086-1088.
Lederer, E. (1935). "Les Carotenoides des Animaux." Hermann, Paris.
Lwoff, M., and Lwoff, A. (1930). *Compt. rend. soc. biol.* **105**, 454-456.
Manunta, C. (1933). *Boll. soc. ital. biol. sper.* **8**, 1278-1282.
Manunta, C. (1935). *Atti soc. nat. e mat. Modena* **66**, 104-113.
Manunta, C. (1937). *Boll. soc. ital. biol. sper.* **12**, 31-34, 626-628, 698-699.
Mattson, F. H., Mehl, J. W., and Deuel, H. J., Jr. (1947). *Arch. Biochem.* **15**, 65-73.
Millott, N. (1957). *Proc. Roy. Soc.* **129**, 263-272.
Millott, N., and Vevers, H. G. (1955). *J. Marine Biol. Assoc. United Kingdom* **34**, 279-287.
Millott, N., and Yoshida, M. (1957). *J. Exptl. Biol.* **34**, 394-401.
Neilands, J. B. (1947). *Arch. Biochem.* **13**, 415-419.
Oku, S. (1929). *Bull. Agr. Chem. Soc. Japan* **5**, 81-83.
Oku, S. (1930). *Bull. Agr. Chem. Soc. Japan* **6**, 104-105.
Oku, S. (1932). *Bull. Agr. Chem. Soc. Japan* **8**, 7-8; 89.
Oku, S. (1933). *Bull. Agr. Chem. Soc. Japan* **9**, 91-92.
Palmer, L. S. (1922). "Carotinoids and Related Pigments," pp. 125-137. Am. Chem. Soc. Monograph Ser. No. 9. Chemical Catalog Co., New York.
Palmer, L. S., and Eckles, C. H. (1914). *J. Biol. Chem.* **17**, 211-221.
Pease, M. (1928). *Z. Induktive Abstammungs- u. Vererbslehre Suppl.* **2**, 1153-1156; *Biol. Abstr.* **3**, 1121.
Petracek, F. J., and Zechmeister, L. (1956). *Arch. Biochem. Biophys.* **61**, 137-139.
Rockstein, M. (1956). *Nature* **177**, 341-342.
Rockstein, M., and Rubenstein, M. (1957). *Biol. Bull.* **113**, 353-354.
Rogozinski, F. (1937). *Bull. Inst. Acad. Cracovie*, Classe Sci., Math. Nat. B. **2**, 183-193; *Chem. Abstr.* **32**, 1760.
Schmidt-Nielsen, S., Sørensen, N. A., and Trumpy, B. (1932). *Kgl. Norske Videnskab. Selskabs Forh.* **5** (30), 118-121.
Schuette, H. A., and Bott, P. A. (1928). *J. Am. Chem. Soc.* **50**, 1998.
Sumner, F. B., and Fox, D. L. (1933). *J. Exptl. Zool.* **66**, 263-301.
Sumner, F. B., and Fox, D. L. (1935). *Proc. Natl. Acad. Sci. U. S.* **21**, 330-340.
Sutherland, M. D., and Wells, J. W. (1959). *Chem. & Ind.* (*London*) **78**, 291-292.
Uda, H. (1919). *Genetics* **4**, 395-416.

Vevers, H. G., and Millott, N. (1957). *Proc. Zool. Soc. London* **129**, 75-80.
Völker, O. (1958). *J. Ornithol.* **99**, 209-217.
Wald, G. (1943). *Vitamins and Hormones* **1**, 195-227.
Wald, G., Nathanson, N., Jencks, W. P., and Tarr, E. (1948). *Biol. Bull.* **95**, 249-250.
Wiese, C. E., Mehl, J. W., and Deuel, H. J., Jr. (1947). *Arch. Biochem.* **15**, 75-79.
Willimott, S. G. (1928). *Biochem. J.* **22**, 157-159.
Wolken, J. J. (1959). Personal communication.
Zechmeister, L. (1937). *Ergeb. Physiol. biol. Chem. u. exptl. Pharmakol.* **39**, 117-191.

Discussion

FRENCH: I am interested in your breeding experiment with the fish. I wonder if any comparable experiments have been done with lobsters; can you raise a white lobster by taking away carotenoids from his diet when he is changing shells?

Fox: We tried very hard to do that. Henry Fisk and I worked more than a year to find out what the precursor of astaxanthin was. We failed. The lobster got paler and paler, we got less and less astaxanthin in the carapaces as they were shed, but we never got to the end point. However, sometimes when they are kept for a long time in an aquarium on white food, the carapace is a kind of tan color instead of being red. It never gets white.

FRENCH: And how about the Echinida. Is it possible to raise those?

Fox: No, I have never tried that. You know they are not ideal laboratory specimens; they drop their spines. One of the first things that happens if you give them a dietary insufficiency is they begin to drop their spines and go to pieces.

FRENCH: Are these pigments essential to the animal or do they just pile them up because they have no use for them?

Fox: Both, I expect. I am on the spot here; I do not know. We do know, however, or we have every reason to suspect with all likelihood of being right, that they use some of these chromoproteins for vision.

FRENCH: Aside from vision and Vitamin A, is there any function of carotenoids known?

Fox: I do not know. I was thinking about mucus secretion and secretion of shells, but I do not know. They simply seem to occur together.

MILLOTT: The question might be answered another way. You were thinking of metabolic function. Of course, in natural history, there is an entirely different function. Carotenoids are extremely important as sexual colors.

MACKINNEY: If you had enough time and patience and raised the male and female exclusively on Vitamin A so they did not develop this sexual dichromatism, would you get normal mating, would you suspect?

Fox: Yes, I would suspect that you would. In the forms without eyes, it does not make any difference, of course. In G. K. Noble's experiments with fishes, he found out that with the female on one side of the glass and the male on the other side of the glass, she did not deposit eggs just because he was brightly colored, she deposited eggs if he were brightly colored and went through the proper ceremonies, went through a lot of movements. If there was a dull colored male who went through the right motions and a brightly colored male who was idle, she responded to the former, not the latter, although she did have color vision.

MILLOTT: I would like to refer to the rather peculiar and particular condition of starfish. We could not find Vitamin A in starfish and *Marthasterias*. We then referred to Fisher and Conrad Reddings who are experts in this matter, and Fisher has recently confirmed that he has been unable to find Vitamin A. In the meantime there was a note in *Nature* by a man called Rockstein stating that he had examined the eyespots of an American asteroid and had found there a pigment which behaved like rhodopsin; it was photosensitive, and he claimed to find tests for vitamin A. Now there is a complete impasse—there is Fisher and Vevers and myself who cannot find Vitamin A, while Rockstein has found in the eyespots a pigment of the rhodopsin type which seems to be based on Vitamin A.

Now in the case of the eyespots which Professor Fox mentioned, we found β-carotene and astaxanthin, but there was another point which I would like to take this opportunity to mention and that is in the starfish there has been dispute as to where the photosensitive parts are in this organism. Some say it is the eyespots, others say it is the skin. Still others have sought to compromise by saying that the eyespots permit more precise orientation but the animal's skin is light sensitive. Now the interesting thing was that Vevers and I found β-carotene and astaxanthin in the eyespots of *Marthasterias* and β-carotene and astaxanthin in the skin as well. Now this was a most suggestive observation that the animal probably has skin sensitivity. Now you may ask: how do you know that the starfish has a genuinely sensitive eyespot? There's one outstanding proof of that, I think, and that was of Hartline, who, in a paper, just gave a tracing, giving no experimental details at all, but showing that when you illuminated the eyespot of what he calls *Asterias* (whatever that may be, no species defined) you get a potential difference, a slow potential difference; in other words here is beautiful electrophysiological evidence. I do not know how he did it, because the eyespots are exceedingly minute. So starfishes seem to stand fair and square at the crossroads. Maybe it is a vitamin A rhodopsin type of thing; maybe it is specialized skin sensitivity, based upon β-carotene.

One last point. I have just come upon a paper by Rosenberg in which he has shown that β-carotene is photoconductive to a remarkable degree. And he asked the same fundamental question that has been asked time and time again: is there any need for a pigmented body to break down during vision? Why aren't you considering photoconduction?

3

Native and Extractable Forms of Chlorophyll in Various Algal Groups

M. B. ALLEN[1]

Laboratory of Comparative Biology, Kaiser Foundation Research Institute, Richmond, California

C. S. FRENCH AND J. S. BROWN

Department of Plant Biology, Carnegie Institution of Washington, Stanford, California

The fact that chlorophyll is one of the most important photoreactive pigments known scarcely requires emphasis. This pigment is present in one or more forms in all organisms that carry out the fundamental reactions of transformation of light energy into chemical energy, without which life on this earth would be severely limited, if not impossible.

The numerous varieties of photosynthetic organisms that exist have been very conservative in the pigments elaborated in their photoreactive organelles. Only a few kinds of chlorophyll are found in the living world. The most adventurous organisms appear to have been the bacteria. The photosynthetic purple bacteria contain a characteristic chlorophyll and the green bacteria at least two (cf. Paper No. 5 by Stanier and Paper No. 11 by Holt in this volume). All the phyla of organisms that carry out green plant photosynthesis, i.e., reduction of carbon dioxide to sugars accompanied by the evolution of oxygen, contain the same green pigment, chlorophyll *a*, sometimes accompanied by one of three additional chlorophylls, *b*, *c*, and *d*. The chemical structures of chlorophylls *a* and *b* are well known; that of *c* remains to be determined. Chlorophyll *d* has been shown to be identical to an oxidation product of chlorophyll *a* having a known structure (Holt and Morley, 1959). The chlorophylls present in an organism are usually identified by chromatographic separation and determination of their absorption spectra. Absorption spectra of the chlorophylls found in oxygen-liberating organisms, which are the only ones that will be discussed in this paper, are shown in Fig. 1, and the

---

[1] Supported in part by Contract AT(04-3)-232 with the United States Atomic Energy Commission and Research Grant NSF-G5584 from the National Science Foundation.

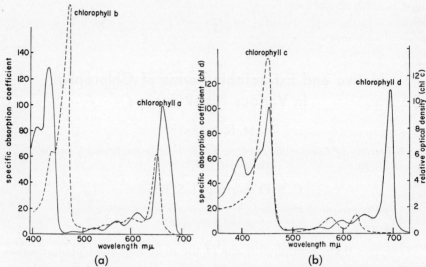

Fig. 1. Absorption spectra of chlorophylls *a*, *b*, *c*, and *d*, in ether (after Smith and Benitez, 1955).

distribution of these chlorophylls among the various groups of photosynthetic organisms is indicated in Table I.

The green algae, like the higher plants, contain chlorophylls *a* and *b*. The ratio of these chlorophylls varies somewhat from organism to organism; the euglenids, for example, contain markedly less chlorophyll *b* than the other green forms. An *a/b* ratio of approximately 3 is found, however, in a large number of the algae of this phylum. The fact that

TABLE I
DISTRIBUTION OF CHLOROPHYLLS IN PHOTOSYNTHETIC ORGANISMS

Group of organisms	Chlorophyll			
	a	*b*	*c*	*d*
Cyanophyta	+	—	—	—
Rhodophyta	+	—	—	±
Chrysophyta				
Chrysophyceae	+	—	±	—
Xanthophyceae	+	—	—	—
Bacillariophyceae	+	—	+	—
Pyrrophyta	+	—	+	—
Cryptophyceae	+	—	+	—
Chloromonadophyceae	+(?)	?	?	?
Phaeophyta	+	—	+	—
Chlorophyta	+	+	—	—
Euglenophyta	+	+	—	—

chlorophyll *b* must therefore be classed as one of the major chloroplast pigments places it in a different position from the other "accessory" chlorophylls, which occur in much smaller amounts in the organisms that contain them.

The blue-green algae contain chlorophyll *a* as their only green pigment, as do the Xanthophyceae (Heterokontae) and many of the red algae. The other red algae may or may not contain chlorophyll *d*, usually in trace amounts. The distribution of chlorophyll *d* is at present a somewhat enigmatic question. A survey of a large number of red algae by Strain (1958) has shown this pigment to occur sporadically, with little regard for other criteria by which the red algae are classified, except that it appears to be totally lacking in the "lower" Rhodophyta. Although one or two members of this group of algae have been reported to contain chlorophyll *d* in quantity (Manning and Strain, 1943; Strain, 1958), its level appears to be erratic, since re-examination of other specimens of these algae by other workers has failed to reveal such high levels. Modern techniques for measuring the absorption spectra of chlorophylls in living material have so far not provided evidence for chlorophyll *d* in the intact organism, in spite of the fact that its characteristic absorption at long wavelengths should make it easy to observe. If it does exist *in vivo* and is not formed during extraction (Holt and Morley, 1959), chlorophyll *d* is present in such small quantities that it is difficult to envisage any functional role for it.

Chlorophyll *c* is present in a diverse assemblage of organisms, ranging from giant kelps to tiny flagellates. In spite of their obvious differences there are other reasons for considering these organisms, which include the diatoms, dinoflagellates, brown algae, cryptomonads, and chrysomonads, to be more or less related. The evidence for the presence of chlorophyll *c* in two of the flagellate groups has only recently been obtained and appears worth reviewing briefly.

These two groups, the cryptomonad and chrysomonad flagellates, are widely distributed groups of organisms of considerable physiological and biochemical interest about which little has been known until recently, with the exception of morphological characteristics. The chrysomonads, as their name implies, are characteristically yellow-brown in color, whereas the cryptomonads are variously colored — red, brown, purple, blue-green, or green. The color of these organisms is due to the presence in them of chlorophylls and of various accessory pigments — carotenoids and phycobilins.

All the cryptomonads so far examined contain chlorophyll *c* in addition to chlorophyll *a*. The first crytomonad in which it was found was

Rhodomonas lens (Haxo and Fork, 1959). Later work, both by Haxo and in our laboratories has shown it to occur as well in other cryptomonads, including *Hemiselmis virescens* and *Cryptomonas ovata* var. *palustris*. Derivative absorption spectra that provided preliminary evidence for the presence of chlorophyll *c* in the latter two organisms are shown in Fig. 2 (a and b), while spectra of the chromatographically

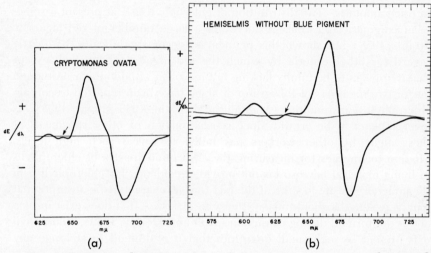

FIG. 2. Derivative absorption spectra of *Cryptomonas ovata* var. *palustris* and *Hemiselmis virescens*. Absorption tentatively ascribed to chlorophyll *c* is indicated by the arrows. The spectrum of *C. ovata* was measured with intact cells; that of *H. virescens*, with cells which had been disrupted by freezing and thawing to release the blue biliprotein found in this organism, centrifuged, washed with 0.01 M phosphate buffer, pH 7, and resuspended in the buffer. If the blue pigment is left in the cells it interferes with determination of the chlorophyll absorption spectrum.

separated pigments, confirming the presence of chlorophyll *c* in *Hemiselmis virescens*, are given in Fig. 3.

The situation in the chrysomonads appears to be more complex. Two members of the Ochromonadales, *Ochromonas danica* and *O. malhamensis*, have been found to contain no green pigments other than chlorophyll *a*. This is shown for *O. danica* in Fig. 4, in which the derivative spectrum of an acetone extract of this organism is compared with that of pure chlorophyll *a* in acetone. The absence of other green pigments has been confirmed by extraction and chromatography of the pigments of *O. danica*. Similar results have been obtained with *O. malhamensis*. In another chrysomonad, *Prymnesium parvum*, there is, however, preliminary evidence of the presence of chlorophyll *c*, as shown in Fig. 5. The chrysomonads are a highly varied group of organisms, and more representatives need to be examined in order to make clear the picture of chlorophyll distribution in these flagellates.

Chlorophyll c, although still definitely a minor pigment, requires much more study. A major obstacle to the chemical study of chlorophyll

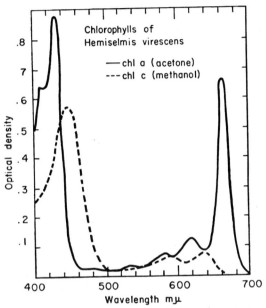

FIG. 3. Absorption spectra of chromatographically separated chlorophylls from *Hemiselmis virescens*. Chlorophyll c was obtained from a 70% methanol extract of lyophilized cells, transferred to petroleum ether containing 20% ether, and chromatographed on powdered sugar columns using n-propanol–petroleum ether mixtures as the developing solvent (cf. Smith and Benitez, 1955; Strain, 1958). It was necessary to repeat the chromatography several times to remove traces of chlorophyll a derivatives. Chlorophyll a was obtained by extraction of the residue from chlorophyll c extraction with 90% methanol (a little chlorophyll a was extracted with 70% methanol), transferred to petroleum ether, and chromatographed using 0.5% n-propanol in petroleum ether as the developing solvent.

c is the difficulty of obtaining it in pure form (Smith and Benitez, 1955). It appears probable that chlorophyll c does not contain phytol, with which the other chlorophylls are esterified (Granick, 1949). Its identification in mixtures is difficult because of its low absorption in the red end of the spectrum, which is the only region in which chlorophyll absorption spectra can be observed in unpurified preparations without interference from carotenoids. This low absorption in the red compared to that in the Soret band in the blue suggests a porphin rather than a chlorin structure for chlorophyll *c* (Granick, 1949; Rabinowitch, 1956). This pigment is readily soluble in aqueous alcohol and insoluble in nonpolar solvents such as petroleum ether. These properties aid in its identification, since they help in the separation of small quantities of chlorophyll *c* from other chlorophylls.

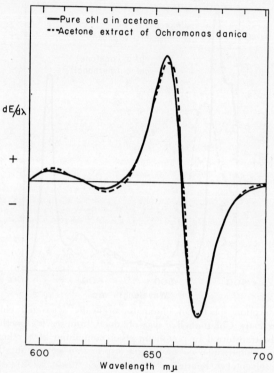

Fig. 4. Comparison of the absorption spectrum of an acetone extract of *Ochromonas danica* with that of pure chlorophyll *a* in acetone. These curves were obtained with the derivative spectrophotometer (French, 1957), which permits the detection of small differences in absorption.

The only real *terra incognita* now remaining in our knowledge of the distribution of chlorophylls in photosynthetic organisms is the group of chloromonad flagellates. These organisms are rare and difficult to culture (although "blooms" of chloromonads sometimes occur in nature); description of their chlorophylls must wait for the development of methods for obtaining vigorous cultures.

It should be noted that there are some exceptions to the general pattern of distribution of the chlorophylls. Mutants of *Chlamydomonas* are known in which chlorophyll *b* is reduced to a very low level (Chance and Sager, 1957), and mutants of *Chlorella pyrenoidosa* in which this pigment is completely lacking have been obtained (Allen, 1958). One naturally occurring alga that has been considered a member of the chlorophyta, *Nannochloris oculata*, is similarly lacking in chlorophyll *b*.

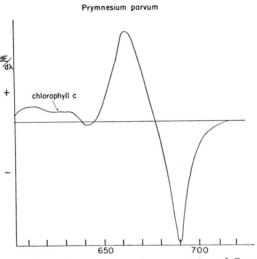

Fig. 5. Derivative absorption spectrum of a suspension of *Prymnesium parvum*. The absorption tentatively considered to indicate chlorophyll *c* is indicated by the arrow.

It is possible, however, that this alga properly belongs among the Xanthophyceae (Lewin, personal communication), in which case its pigmentation is normal. Then there is the extraordinary alga *Cyanidium caldarium*, which resembles *Chlorella* in its morphology and several features of its physiology, but which contains no chlorophyll other than chlorophyll *a*, and in addition forms a typical *c*-phycocyanin (Hirose, 1950; Allen, 1952, 1954, 1959). To add to the anomalies of its pigmentation, it has recently been found to contain zeaxanthin as its principal xanthophyll (cf. paper by Goodwin, this volume). At various times this alga has been placed among the cyanophytes (Tilden, 1898; Copeland, 1936), the chlorophytes (Hirose, 1950; Allen, 1954), the rhodophytes (Geitler, 1959; Hirose, 1958), and the cryptomonads (Bourelly, 1954; Fogg, 1956). It is anomalous in any of these positions. At present it seems best to consider it a transitional form. Study of a wider variety of organisms may be expected to reveal other such forms that do not fit into established categories for chlorophyll distribution or other characteristics.

The description of the chlorophylls of an organism can no longer be considered complete when an account of the pigments obtained on extraction is given. It has long been known that the absorption spectrum of chlorophyll in the living organism is different from that in extracts, and hence that chlorophyll *in vivo* is physically or chemically different from chlorophyll *in vitro*. For example, the red absorption maximum of chlorophyll *a* in either solution is at 662–663 mμ, while in living cells it averages around 680 mμ — values ranging from 665 to 690 mμ have been observed (cf. summary by Rabinowitch, 1951). The shift of the chlorophyll *b* absorption spectrum is more difficult to measure, as it is overlapped by chlorophyll *a*, but appears to be smaller. Rabinowitch (1951) suggests 647–648 mμ as the most probable value for the red absorption peak *in vivo*, whereas in either solution the maximum lies at 643–644 mμ (Smith and Benitez, 1955). Nothing is known of the shifts in spectrum, if any, of the minor chlorophylls, *c* and *d*.

Closer examination of the chlorophyll absorption spectrum *in vivo* in a large number of organisms has shown that not only is it different from that in solution, but also it varies from one organism to another, and may even vary in the same organism, depending on its physiological state. The suggestion that chlorophyll in living plants is variable was first made by Lubimenko (1927), and was supported by Albers and Knorr (1937) and Seybold and Egle (1940). Others felt, however, that the problems of obtaining data on living systems free of artifacts caused by reflection and scattering were such as to make interpretation of these data difficult, and considered that the results did not necessarily lead to the conclusion that the absorption spectrum of chlorophyll *in vivo* is variable (cf. Rabinowitch, 1951, 1956).

An example of the type of difference observed in the absorption spectra *in vivo* is shown in Fig. 6. Although these curves were obtained by conventional methods, recognition of such differences has been greatly aided by several technical advances made in recent years. (*1*) The opal glass method (Shibata *et al.*, 1954), which has provided a convenient way to reduce scattering artifacts. Latimer (1959) has shown that use of this method eliminates distortion of absorption peaks caused by selective scattering in the neighborhood of absorption masima. (*2*) An electronic device for curve analysis (French *et al.*, 1954) has facilitated the calculation of the absorption curves of pigments responsible for differences in spectra. (*3*) The derivative spectrophotometer (French, 1957) has made it possible to recognize spectral details that otherwise would have gone undetected. A comparison of absorption curves of *Ochromonas danica* obtained with the derivative spectrophotometer and with the usual integral instrument is shown in Fig. 7. This organism has an unusually complex chlorophyll *a* spectrum *in vivo*. The deviation of this spectrum from a simple peak is clearly evident in the integral curve, and is striking in the derivative. Derivative spectra illustrating some of the many variations that have been observed in the shape of the red peak

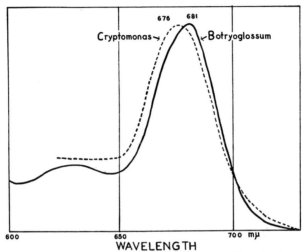

Fig. 6. Comparison of the red absorption band of chlorophyll *a* in two algae (French, 1958). The difference both in shape and in peak wavelength should be noted.

of chlorophyll *a* are collected in Fig. 8. (It is almost impossible to obtain interpretable measurements in the blue region of the spectrum because of overlapping carotenoid absorption.)

The evidence for variations in the absorption spectrum of chlorophyll *a in vivo* is by now quite extensive. The question then arises how these variations are to be interpreted. The variability can be accounted for if it is assumed that there are several spectroscopically distinct forms of chlorophyll *a* in living photosynthetic organisms and that the proportions of these spectroscopic entities vary from organism to organism and may also vary in a single organism according to the treatment to which it has been subjected. Is it possible to find evidence for the physical reality of such forms of chlorophyll?

One of the first, and strongest, pieces of evidence indicating that this question can be answered in the affirmative came from the work of Haxo and Blinks (1950). In this work, it was found that light absorbed by chlorophyll was not fully effective in inducing photosynthesis in several red algae; there was a difference between the absorption spectrum of the living material and the action spectrum for photosynthesis. Using these data, French and Young (1956) calculated the absorption spectrum of the "inactive" pigment and found that it appeared to be a part of chlorophyll a, but the wavelength of the red absorption peak was longer than that characteristic of the total chlorophyll in the alga.

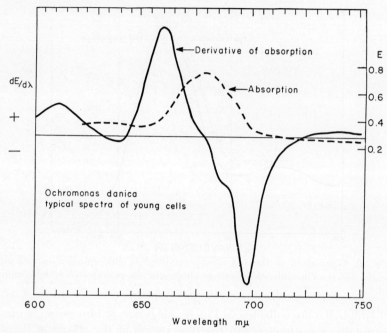

Fig. 7. Comparison of integral and derivative absorption spectra of the red band of chlorophyll a in the chrysomonad *Ochromonas danica*.

Further evidence for the existence of several forms of chlorophyll in plants has come from the work of Krasnovsky and his associates (Krasnovsky and Kosobutskaya, 1955; Vorobeva and Krasnovsky, 1956; Krasnovsky et al., 1957). These workers have found that the absorption spectrum of chloroplast suspensions shifts toward longer wavelengths after exposure to strong light. This has been interpreted as a selective bleaching of a 670 mµ component compared to a 682 mµ form of chlorophyll. The two components were also found to behave differently toward extraction by solvents. Vishniac (cf. Paper No. 24, this volume) has also found that the readily extractable chlorophyll and that which is tightly bound in the chloroplast behave differently in photochemical reactions.

Halldal (1958) found that cells of the blue-green alga *Anacystis nidulans*[2] grown under different conditions of temperature and illumination had markedly different shapes of the red chlorophyll peak. Subtraction of one absorption curve from the other yielded the spectrum of a form of chlorophyll with maximum absorption at 682 mµ, as shown in Fig. 9.

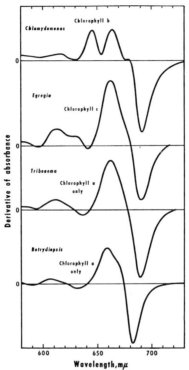

FIG. 8. Variations in the red absorption band of chlorophyll observed in different algae. The measurements on *Egregia* were made in collaboration with Dr. Francis Haxo.

The absorption spectrum of chlorophyll *a* in a mutant of *Chlorella pyrenoidosa* that lacks chlorophyll *b* was found to have a different shape when the alga was grown in the light from that observed when the alga was grown in the dark. Subtraction of the absorption curves of light-grown and dark-grown cells yielded the spectrum of a form of chlorophyll *a* with maximum at 684 mµ, as shown in Fig. 10.

A striking example of the appearance of a chlorophyll absorption

---

[2] It is unfortunate that this name has become widespread in the literature, as it is undoubtedly not the correct name for this alga.

band under certain conditions was recently observed by French and Elliot (1958). Young cultures of *Euglena gracilis* showed a chlorophyll absorption spectrum similar to that observed in other green algae. As the cultures aged, absorption at longer wavelengths became more and

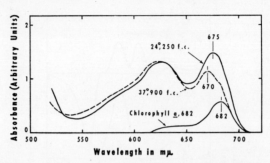

Fig. 9. Derivation of the spectrum of a form of chlorophyll *a* with maximum absorption at 682 mµ from the absorption curves of *Anacystis nidulans* grown under different conditions of light and temperature (Halldal, 1958).

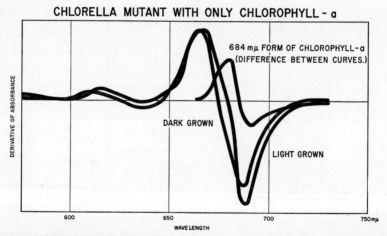

Fig. 10. Derivation of the spectrum of a form of chlorophyll *a* with maximum absorption at 684 mµ from the absorption curves of mutant G-44 of *Chlorella pyrenoidosa* grown in light and darkness.

more pronounced, as shown in Fig. 11. Analysis of the curves revealed that the long wavelength absorption could be accounted for by the appearance of the form of chlorophyll *a* with absorption maximum at 695 mµ, as illustrated in Fig. 12. Further analysis has resulted in resolution of the *Euglena* spectrum into three components with absorption maxima at 673, 683, and 695 mµ. The differing relative proportions of these as the cultures age are a function of the light intensity received by a cell as it grows. As a culture ages, the average light received is cut down. A young culture growing at low light intensity (200 foot-candles) begins forming the 695 mµ form of chlorophyll *a* within 2 days of inoculation; from then on the proportion of this form increases rapidly.

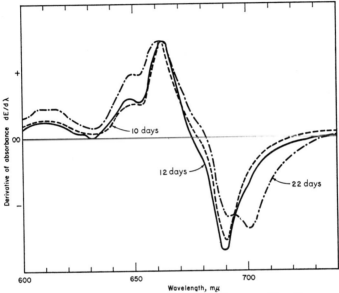

Fig. 11. Changes in absorption spectrum of *Euglena gracilis* with age of culture (French and Elliot, 1958).

Similarly striking changes in chlorophyll absorption have been observed in the chrysomonad *Ochromonas danica*, which has been previously mentioned as an illustration of a complex chlorophyll absorption spectrum *in vivo*. The curves given in Fig. 7 represent the spectrum characteristic of young growing cultures of this alga. In contrast to *Euglena*, considerable chlorophyll absorption at long wavelengths is here found in young cultures. As the cultures age, the absorption spectrum goes through a complex series of changes, finally arriving at a simple curve with absorption maximum at 675 mµ instead of the 680 mµ found in the young cultures, as shown in Fig. 13. Similar changes in absorption spectrum, shown in Fig. 14, can be observed if the cells are heated mildly or broken by passage through a needle valve. This effect of cell breakage on absorption spectrum is in contrast to that observed in *Chlorella* by Brown and French (1959). In this organism cell breakage had little effect on the chlorophyll absorption spectrum. Breaking *Euglena* with

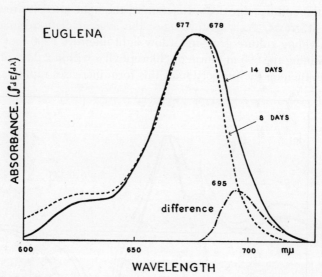

Fig. 12. Derivation of the spectrum of a form of chlorophyll *a* with absorption maximum at 695 mμ from the difference in absorption of old and young cultures of *Euglena*.

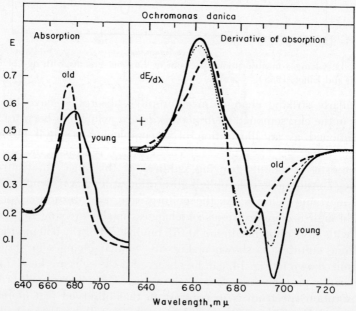

Fig. 13. Effect of age of culture on the absorption spectrum of *Ochromonas danica*. Integral spectra are given on the left; derivative spectra, on the right.

the needle valve apparatus reduces the proportion of the 695 mµ form, but does not eliminate it entirely. Breaking does not appear to affect the 673 and 683 mµ peaks.

By analysis of absorption curves of *O. danica* in different states it has been possible to resolve its complex absorption spectrum into three

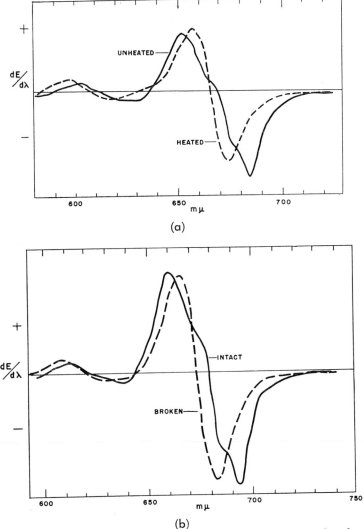

FIG. 14. Effect of (*a*) mild heating and (*b*) cell breakage on the absorption spectrum of *Ochromonas danica*. In (*a*) the cells were heated 20 minutes at 40°; in (*b*) they were broken by passage through a needle valve at room temperature.

components, with maxima at 670, 682, and 693 mµ respectively, as shown in Fig. 15. Similar analysis of the spectrum of *Chlorella pyrenoidosa* (wild type) and, as mentioned above, of *Euglena* has yielded components with absorption maxima at 672-3, 693-4, and 694-5 mµ (Brown and French, 1959). The similarity of these sets of components and their re-

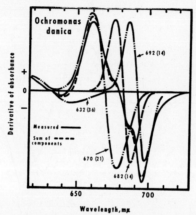

Fig. 15. Resolution of the derivative absorption spectrum of *Ochromonas danica* into components.

semblance to those obtained by analysis of data from other organisms is striking. Since the precision of the curve analysis is of the same order of magnitude as the differences obtained in the wavelength maxima of the chlorophyll components from different organisms, it is not yet possible to state whether the small differences observed are real or whether the chlorophyll *a* spectrum *in vivo* in a wide variety of organisms is composed of a combination of the same three forms.

It should be emphasized that these "forms" of chlorophyll *a* are so far known only as spectroscopic entities. It is not known whether they represent different chlorophyll-protein complexes, different states of aggregation of chlorophyll, or some other, as yet unguessed, phenomenon. The skewed absorption curves that are obtained have much the same shape as the spectrum of a colloidal suspension of chlorophyll *a* in water (French, 1958) or the spectra of adsorbed chlorophyll (Trurnit and Colmano, 1959). Since terminology for referring to the different spectral forms is necessary, it seems desirable to follow the precedent established for cytochromes and for bacteriochlorophyll and refer to them as chlorophyll *a*-682, chlorophyll *a*-695 (abbreviated as C_a-682, C_a-695, etc.).

Evidence that the spectroscopically different chlorophyll *a* forms are functional entities in photosynthetic organisms is provided by some experiments that have been carried out with *Ochromonas danica*. The work of the late Robert Emerson and his associates (Emerson *et al.*, 1957) has indicated that for efficient photosynthesis it is necessary to activate some accessory pigment, chlorophyll *b* or a biliprotein, in addition to chlorophyll *a*. In accordance with this work, mutants of *Chlorella pyrenoidosa* that lacked chlorophyll *b* were found to be photosynthetically ineffective (Allen, 1958). Scattered bits of information on naturally occurring algae without accessory pigments, such as *Nannochloris oculata* and the Xanthophyceae, have suggested that these organisms also are poor at photosynthesis.[3] Yet *O. danica*, although it contains no chlorophyll other than chlorophyll *a*, is an effective photosynthetic organism. Its light saturation curve is normal and maximal rates of oxygen evolution of 4000 μl. O_2/hour/mg. chlorophyll are obtained.

If photosynthesis and chlorophyll absorption spectra are measured during the growth of *O. danica*, it is found that during active growth the

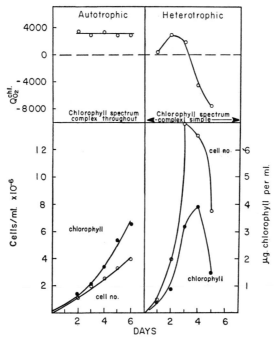

FIG. 16. Growth, photosynthetic activity, and chlorophyll absorption spectrum in autotrophically and heterotrophically grown *Ochromonas danica*.

---

[3] Recent measurements indicate, however, that some Xanthophyceae may be as active photosynthetically as Chlorophyceae.

red absorption band of chlorophyll has the complex structure shown in Fig. 7. When maximum growth is reached, the red absorption band changes to the simple form previously illustrated, and, concomitant with this loss of long wavelength absorption, the cells lose their ability to photosynthesize. Illumination of cells in this condition results only in photo-oxidation. These changes are seen most markedly in cultures grown with sugar, as illustrated in Fig. 16.

A similar result is obtained with cells in which the chlorophyll spectrum has been altered by mild heating. This treatment does not kill the cells; transfers from cultures thus treated are viable. But, as with the old cultures, their ability to photosynthesize has been lost; illumination results only in increased oxygen uptake and bleaching of chlorophyll. It appears that the presence of the long-wavelength-absorbing chlorophyll *a* is necessary for photosynthesis in *O. danica*, and it seems likely that the different forms of chlorophyll *a* may act as accessory pigments to each other, so that efficient photosynthesis can take place even though chlorophyll *a* is the only pigment activated. It may be significant that the absorption spectrum of chlorophyll *a* in *Ochromonas malhamensis*, which is a very feeble photosynthesizer (Myers and Graham, 1956), does not show the complex structure found in *O. danica*.

Acknowledgment

We are indebted to Mrs. Ruth F. Elliott for growing many of the algae.

References

Albers, V. M., and Knorr, H. V. (1937). *Plant Physiol.* **12**, 583-588.
Allen, M. B. (1952). *Arch. Mikrobiol.* **17**, 34-46.
Allen, M. B. (1954). *Rappt. et Commun. 8ᵉ Congr. Intern. Botan. Sect.* **7**, p. 41.
Allen, M. B. (1958). *Brookhaven Symposia in Biol. No.* **11**, 339-342.
Allen, M. B. (1959). *Arch. Mikrobiol.* **32**, 270-277.
Bourelly, J. (1954). Personal communication.
Brown, J. S., and French, C. S. (1959). *Plant Physiol.* **34**, 305-309.
Chance, B., and Sager, R. (1957). *Plant Physiol.* **32**, 548-560.
Copeland, J. E. (1936). *Ann. N. Y. Acad. Sci.* **36**, 1-229.
Emerson, R., Chalmers, R., and Cederstrand, C. (1957). *Proc. Natl. Acad. Sci. U. S.* **43**, 133-143.
Fogg, G. E. (1956). *Bacteriol. Revs.* **20**, 148-164.
French, C. S. (1957). *Proc. of I. S. A. Instrumentation and Control Symposium Sponsored by Northern Calif. Sect. Instrument Soc. of Am.* pp. 83-94.
French, C. S., and Elliot, R. F. (1958). *Carnegie Inst. Wash. Year Book* **57**, 278-286.
French, C. S. (1958). *Proc. 19th Ann. Biol. Colloq. Oregon State Coll. 1958* pp. 52-64.
French, C. S., and Young, V. K. (1956). *Radiation Biol.* **3**, 343-381.

French, C. S., Towner, G. H., Bellis, D. R., Cook, R. M., Fair, W. R., and Holt, W. W. (1954). *Rev. Sci. Instr.* **25**, 765-775.
Geitler, L. (1959). *Österr. Bot. Z.* **108**, 100-105.
Granick, S. (1949). *J. Biol. Chem.* **179**, 505-511.
Halldal, P. (1958). *Physiol. Plantarum* **11**, 401-420.
Haxo, F. T., and Blinks, L. R. (1950). *J. Gen. Physiol.* **33**, 389-422.
Haxo, F. T., and Fork, D. C. (1959). *Nature* **184**, 1050-1052.
Hirose, H. (1950). *Botan. Mag.* (*Tokyo*) **63**, 107-112.
Hirose, H. (1958). *Botan. Mag.* (*Tokyo*) **71**, 347-352.
Holt, A. S., and Morley, H. V. (1959). *Can. J. Chem.* **37**, 507-514.
Krasnovsky, A. A., and Kosobutskaya, L. M. (1955). *Doklady Akad. Nauk S. S. S. R.* **104**, 440-443.
Krasnovsky, A. A., Vorobeva, L. M., and Pakashina, E. V. (1957). *Fiziol. Rastenii Akad. Nauk S. S. S. R.* **4**, 124-133.
Latimer, P. (1959). *Plant Physiol.* **34**, 193-199.
Lubimenko, V. N. (1927). *Rev. gén. botan.* **39**, 547-559.
Manning, W. M., and Strain, H. H. (1943). *J. Biol. Chem.* **151**, 1-19.
Myers, J., and Graham, J.-R. (1956). *J. Cellular Comp. Physiol.* **47**, 397-414.
Rabinowitch, E. I. (1951). "Photosynthesis and Related Processes," Vol. 2, Pt. 1, p. 623. Interscience, New York.
Rabinowitch, E. I. (1956). "Photosynthesis and Related Processes," Vol. 2, Pt. 2, p. 1752-1754. Interscience, New York.
Seybold, A., and Egle, K. (1940). *Botan. Arch.* **41**, 578-603.
Shibata, K., Benson, A. A., and Calvin, M. (1954). *Biochim. Biophys. Acta* **15**, 461-470.
Smith, J. H. C., and Benitez, A. (1955). In "Modern Methods of Plant Analysis" (K. Paech and M. V. Tracey, eds.), pp. 142-196. Springer, Berlin.
Strain, H. H. (1958). "Chloroplast Pigments and Chromatographic Analysis," p. 62. Penn. State Univ., University Park, Pennsylvania.
Tilden, J. E. (1898). *Botan. Gaz.* **25**, 89-96.
Trurnit, H. J., and Colmano, G. (1959). *Biochim. et Biophys. Acta* **31**, 434-447.
Vorobeva, L. M., and Krasnovsky, A. A. (1956). *Biokhimiya* **21**, 126-136.

Discussion

Fox: One of the questions that occurred to me when I looked at the slides that Dr. Allen was showing was whether or not, when you kill a cell either by heat or by physical degradation, some of the chlorophyll can be dissolved in certain organic solvents and shift the spectrum—it goes into some of the neutral fats, for example, which are present in *Chlorella* or other algae—and whether you simply get a shift from that.

Allen: This is a possibility. However, even in those shifts that I showed, the spectrum has not changed to that of chlorophyll *a* in a fat solvent. The changes in maximum are from over 680 mμ to about 675 mμ, compared to 663 mμ in organic solvents.

Mackinney: There seems to be a different situation in the purple bacteria in that you can simplify the spectrum, at least in the case of mutants the 890 band peak disappears, and you get photosynthesis continuing. There is one other thing which worries me a little bit. If you take rhodopsin you will remember that the completely randomized suspension has a different peak from what it has apparently *in situ* and

I am wondering whether there is a difference in the packing. In that case you would not need to postulate different chromatophores. You would merely postulate that there were changes in the orientation, in the packing of these as the cell grew older in order to get the change.

ALLEN: We certainly need to find out about this. At present we have no idea whether this is a physical change due to changes in packing or whether we are changing from one form of combination to another.

4

Pteridine Pigments in Microorganisms and Higher Plants

FREDERICK T. WOLF

Department of Biology, Vanderbilt University, Nashville, Tennessee, and Biology Division, Oak Ridge National Laboratory,[1] Oak Ridge, Tennessee

Introduction

The pteridines or pterins constitute a group of naturally occurring pigmented substances. The prototype of these compounds, now known as xanthopterin, was first isolated from butterfly wings by Hopkins (1891), but its chemical structure was clarified only through its synthesis almost half a century later (Purrmann, 1940). Almost all the known pteridines are tetra-aza derivatives of naphthalene or may be considered to consist of fused pyrimidine and pyrazine rings. Two number systems for these compounds are encountered in the literature. The one used in this report was originally proposed by Kuhn and Cook (1937) and is currently preferred by the vast majority of workers in this field in English-speaking countries.

Almost all the naturally occurring pteridines have an amino group in the 2 position and a hydroxyl group in the 4 position. The various pteridines differ from one another in the nature of the substituent groups attached at positions 6 and 7. The pteridines are structurally related to riboflavin and to pteroylglutamic or folic acid.

Following established usage in this field, pteroylglutamic acid and related compounds are hereafter designated as "conjugated pteridines," to contrast them with the "unconjugated pteridines," which are our principal concern.

In general, the pteridines may be characterized as yellow pigments, amorphous to microcrystalline, with high melting points, sparingly soluble in water but more soluble in certain organic solvents. They are very stable in solution even under extreme pH conditions. They are fluorescent. They are generally separated from tissues by selective adsorption and elution techniques. They are distinguished principally by their ultraviolet-absorption spectra, which exhibit shifts in maximum with

---

[1] Operated by Union Carbide Corporation for the U. S. Atomic Energy Commission.

change in pH, and by such techniques as paper chromatography. Physiologically, at least some of them are of interest in connection with problems of growth in both normal and malignant tissues. Their pigmented nature, with light absorption in the upper ultraviolet and the lower end of the visible spectrum, suggests their possible role as photoreceptors, either in photosynthesis or in other photochemical processes.

The pteridines are exceedingly widespread in animals. Numerous review articles concerning these pigments have appeared (Gates, 1947; Albert, 1952, 1954; Berezovskii, 1953; Wolstenholme and Cameron, 1954; Karrer and Viscontini, 1955; Augustinsson, 1956; Ziegler-Gunder, 1956) to which the reader is referred for many details that, for reasons of space, cannot be mentioned here.

In contrast to the situation in animals, comparatively little is known about pteridine pigments in plants. It is the purpose of this presentation to survey the present status of knowledge concerning unconjugated pteridines in plants. It is convenient to discuss these compounds separately since they occur in bacteria, fungi, algae, and in higher plants.

Bio-assay of pteridines using the trypanosome, *Crithidia fasciculata*, has been very useful in the elucidation of the structure of pteridines (Nathan and Cowperthwaite, 1955; Nathan et al., 1956, 1958). For growth in synthetic media, this organism requires both pteroylglutamic acid and an unconjugated pteridine. The unconjugated pteridine must have a 2-amino group, a 4-hydroxy group, and a 6 substituent, which may be either hydroxymethyl or a longer aliphatic chain with two vicinal hydroxy groups (Nathan and Funk, 1959).

Bacteria

The earliest reported instance of a pteridine pigment in bacteria that has come to my attention is in the studies of Crowe and Walker (1944) with the diphtheria organism, *Corynebacterium diptheriae*. By extraction and column chromatography, these workers isolated a number of pteridine pigments from the culture filtrates of this bacterium. Several of these were not studied further, but one was clearly identified as xanthopterin (2-amino-4,6-dihydroxypteridine). These same workers (Crowe and Walker, 1949) isolated from the tuberculosis bacillus, *Mycobacterium tuberculosis*, the red pigment erythropterin (2-amino-4,6-dihydroxy-7-trihydroxypropylenepteridine) and in a further study (1954) showed that this bacterium produced xanthopterin as well.

Gaffkya homari is a coccoid bacterium that requires for growth either *p*-aminobenzoic acid or purines. According to the investigation of Nathan and Funk (1959), production of an unconjugated pteridine by the organism varies with the growth factor used to satisfy this requirement.

G. *homari* has a fairly high level of "*Crithidia* factor" activity when grown either on hypoxanthine or guanine. Only about half as much pteridine is produced by cells grown on *p*-aminobenzoic acid, and adenine-grown cultures of *G. homari* contain only small amounts of the active pteridine (Nathan and Funk, 1959).

Wacker *et al.* (1959) examined cell-free extracts of *Enterococcus, Escherichia coli,* and *Sarcina lutea* for *Crithidia* factor activity. Chromatography of these extracts and subsequent bio-assay showed three active spots, none of which was identical with biopterin (2-amino-4-hydroxy-6-dihydroxypropylpteridine) itself. None of these substances was further characterized.

In studies of the fluorescent pigments of *Azotobacter* species, Johnstone *et al.* (1959) showed that these yellow substances are clearly differentiated from riboflavin and suggested that they might well be pteridine derivatives. Additional information is needed to establish this.

In summary, among the bacteria two distinct pteridines have been definitely identified and characterized in the tuberculosis bacillus and one in the diphtheria organism (Table I). The existence of other pteridines among additional species has been established by bio-assay, but they have not as yet been identified. Still other bacterial pigments are thought to be pteridines.

TABLE I
PTERIDINES PRODUCED BY BACTERIA

Organism	Pteridine	References
Corynebacterium diphtheriae	2-Amino-4,6-dihydroxy-pteridine	Crowe and Walker (1944)
Mycobacterium tuberculosis	2-Amino-4,6-dihydroxy-7-trihydroxypropylene-pteridine	Crowe and Walker (1949)
	2-Amino-4,6-dihydroxy-pteridine	Crowe and Walker (1954)
Gaffkya homari	Pteridine with *Crithidia* factor activity	Nathan and Funk (1959)
Enterococcus sp.	Pteridine with *Crithidia* factor activity	Wacker *et al.* (1959)
Escherichia coli	Pteridine with *Crithidia* factor activity	Wacker *et al.* (1959)
Sarcina lutea	Pteridine with *Crithidia* factor activity	Wacker *et al.* (1959)

Fungi

From the culture filtrates of *Rhizopus nigricans*, which is used in the production of fumaric acid, Rickes et al. (1947) isolated, by adsorption on charcoal and fuller's earth followed by chromatography on alumina, a new compound that they called rhizopterin. The structure of this substance was determined by D. E. Wolf et al. (1947), who showed that rhizopterin is N^{10}-formylpteroic acid (Table II). Thus it is a conjugated pteridine related to the folic acid series of compounds and is not a compound of the type with which we are concerned.

TABLE II
PTERIDINES PRODUCED BY FUNGI

Organism	Pteridine	References
Rhizopus nigricans	N^{10}-Formylpteroic acid	Rickes et al. (1947)
Aspergillus oryzae	2-Amino-4,7-dihydroxy-pteridine-6-acetic acid (?)	Kaneko (1957a,b)
A. repens	Pteridine with *Crithidia* factor activity	Wolf (1957a)
Eremothecium ashbyii	2,4-Dihydroxy-6,7-dimethyl-pteridine	Masuda (1956)
	2,4-Dihydroxypteridine	Korte et al. (1958)
	2,4-Dihydroxy-8-trihydroxybutyl pteridine, or 2,8-dihydro-2-hydroxy-8-trihydroxybutylpteridine	Forrest and McNutt (1958)
Microsporum canis	Pteridine with *Crithidia* factor activity	Wolf (1957b); Wolf et al. (1958)
M. gypseum	Pteridine with *Crithidia* factor activity	Wolf (1957b); Wolf et al. (1958)
Physarum polycephalum	Pteridine with *Crithidia* factor activity	Wolf (1959)

Apparently the first report of the production of an unconjugated pteridine by a fungus is that of Kaneko (1957a,b) with *Aspergillus oryzae*, the organism used commercially in the production of the enzyme, Takadiastase. From culture filtrates of this organism, Kaneko isolated two pteridines, but neither has been chemically characterized with certainty. One of these, however, in chromatographic behavior showed great similarity to 2-amino-4,7-dihydroxypteridine-6-acetic acid. In earlier work, the writer studied the fluorescent pigment produced by *A. repens*, an osmophilic fungus that grows only on media containing high sugar concentrations (Kaneko, 1957a). Klöcker (1916) reported the fluorescent nature of a pigment produced by this fungus. The pigment was isolated by adsorption on fuller's earth, elution with methanol, and chromatography on a magnesia column to effect separation from a non-fluorescent pigment, auroglaucin. The pigment thus obtained was distinct from those found in *A. oryzae* by Kaneko. The chemistry of the pteridine from *A. repens* is not completely known, but it is a 2-amino-4-hydroxy-6-substituted pteridine that has *Crithidia* factor activity (Wolf, unpublished).

The yeast, *Eremothecium ashbyii*, is commercially used in the biosynthetic production of riboflavin. In addition to riboflavin, however, unconjugated pteridines are formed by this organism. Masuda (1956) isolated from *E. ashbyii* two pteridines, and called them "G substance" and "V substance"; the former was further characterized as a substituted derivative of 2,4-dihydroxy-6,7-dimethypteridine. Korte *et al.* (1958) reported formation of 2,4-dihydroxypteridine by this organism. Forrest and McNutt (1958) isolated from concentrates of the fermentation products of this organism a 2,4-dihydroxypteridine or a 2,8-dihydro-2-hydroxypteridine bearing a trihydroxybutyl side chain in the 8 position. The 2,4-dihydroxypteridines produced by *E. ashbyii* are of particular interest from the standpoint of comparative biochemistry since all other naturally occurring pteridines are 2-amino-4-hydroxy derivatives.

In a continuation of their earlier study, McNutt and Forrest (1958) biosynthesized pteridines and riboflavin with *E. ashbyii* in a medium containing adenine randomly labeled with C^{14}. The pteridine and riboflavin were isolated and had very similar specific activities. Although the pteridine could not be implicated as an intermediate in the formation of riboflavin from adenine, the finding of these workers supports the conclusion that the pteridine and the riboflavin both arise from a common precursor. These tracer experiments indicate the direct incorporation of the pyrimidine ring of adenine into the pteridine. In other tracer experiments with a photosynthetic alga, the radioactivity of $C^{14}O_2$ was unfortunately incorporated into the glucose moiety of a pteridine glucoside rather than into the pteridine itself (Van Baalen *et al.*, 1957).

In the fungus disease of man called *Tinea capitis*, Margarot and Deveze (1925) discovered that infected hairs were fluorescent under ultraviolet light. Several workers have investigated the nature of the fluorescent pigment. In studies of *Microsporum canis* and *M. gypseum* cultures grown both on human hair and in a synthetic medium (Wolf, 1957b), the fluorescent pigment was isolated by extraction with hot water, adsorption on charcoal, and elution with methanol. Data concerning absorption spectra, in acid and alkaline solution, fluorescence spectra, and paper chromatography of the compound were presented, indicating that this pigment is a pteridine. This finding was challenged by Chattaway and Barlow (1958). The objection of the English workers was answered (Wolf et al., 1958) by infrared spectra and bio-assay with *Crithidia*. The bioassay experiments indicated that the *Microsporum* pigment possessed *Crithidia* factor activity and is, therefore, a 2-amino-4-hydroxy-6-substituted pteridine.

The myxomycete, *Physarum polycephalum*, has as its vegetative stage a plasmodium that is yellow and fluorescent. Gray (1938, 1939, 1953) showed the necessity of light for fruiting—the transformation of the vegetative stage or plasmodium into the reproductive stage or sporangia. Blue light and acid conditions were necessary. Gray extracted the yellow pigment from the plasmodium and concluded that it functioned as a photoreceptor for the fruiting response. The chemical nature of the plasmodial pigment had not been previously known.

The pigment from plasmodia was extracted with methanol, and the methanol solution was chromatographed on an alumina column (Wolf, 1959). Two yellow fluorescent pigments were separated, each a pteridine. One of the *Physarum* pteridine pigments has an absorption peak in neutral or alkaline solution at 380 mμ, which is displaced in acid solution to 420 mμ. The properties of this pteridine are in complete agreement with what would be expected of the photoreceptor for the fruiting response according to Gray's earlier findings. This pigment has *Crithidia* factor activity and is thus an unconjugated, trisubstituted pteridine.

The occurrence of pteridine pigments in *Physarum* is particularly interesting because not only does one of them serve as the photoreceptor pigment for reproduction, but as a group they are present in large quantity making up the bulk of the pigment in an organism capable of growing to large dimensions. In other organisms, they are usually present only in very small quantities and are frequently masked by other more abundant pigments.

Algae

In studies on the blue-green alga, *Anacystis nidulans*, Forrest et al. (1957) subjected cell suspensions for short periods to a temperature of 4°C. A yellow material with fluorescent properties was observed to leach from the cells. This material consisted of two or more pteridines, the principal constituent of which was isolated in a yield equivalent to 0.05–0.1% of the dry weight of the cells. Three other blue-green algae also contained pteridines in relatively large concentrations. In *Anacystis*, cold-shock treatment with release of pteridines was accompanied by loss in photosynthetic activity, which implied a possible role of pteridines in the photosynthetic process.

The principal pteridine of *Anacystis*, designated as "compound C," was shown to be very similar to, although not identical with, biopterin (Van Baalen et al., 1957). During 5 minutes of photosynthesis by *A. nidulans* in the presence of $C^{14}O_2$, the pteridine was labeled. On acid hydrolysis of the labeled pteridine, unlabeled biopterin was found as well as a labeled but unidentified fragment. Forrest et al. (1958) identified the labeled fragment as glucose. Compound C was thus shown to be a glucoside of biopterin. Some uncertainty exists, however, concerning the exact position of the attachment of the glucose to the dihydroxypropyl side chain.

Van Baalen and Forrest (1959) isolated from *A. nidulans* and from another blue-green alga a second pteridine that they identified as 2,6-diamino-4-hydroxypteridine (Table III).

TABLE III
PTERIDINES PRODUCED BY ALGAE

Organism	Pteridine	References
Anacystis nidulans	Glucoside of 2-amino-4-hydroxy-6-dihydroxy-propylpteridine	Van Baalen et al. (1957); Forrest et al. (1958)
	2,6-Diamino-4-hydroxy-pteridine	Van Baalen and Forrest (1959)
	2-Amino-4-hydroxy-6-propionyl-5,8-dihydro-pteridine	Van Baalen and Myers (1959)
Ochromonas malhamensis	2-Amino-4-hydroxy-6-dihydroxypropyl-pteridine	Nathan and Cowperthwaite (1955)
O. danica	2-Amino-4-hydroxy-6-dihydroxypropyl-pteridine	Nathan and Funk (1959)

Compound A isolated from *A. nidulans* by Forrest et al. (1957) was at first considered (Forrest et al., 1959) to be a 6-substituted pteridine bearing both a hydroxy group and a —$COCH_2CH_3$ group in the 6 position. According to the most recent information (Van Baalen and Myers, 1959), Compound A has the structure 2-amino-4-hydroxy-6-propionyl-5,8-dihydropteridine.

In summary, research by these investigators with the blue-green alga, *A. nidulans*, has shown the presence of three distinct pteridines each of which has been chemically characterized. The implication is made of a possible participation of pteridines in photosynthesis.

The chrysomonad, *Ochromonas malhamensis*, produces two pteridines: a conjugated pteridine chromatographically identical with "citrovorum factor" and the unconjugated pteridine, biopterin (Nathan and Cowperthwaite, 1955). Pteridines are also produced by the closely related *O. danica*, and the studies of Nathan and Funk (1959) indicate that these are identical with those formed by *O. malhamensis*. Thus unconjugated pteridines have been found among the chrysomonads as well as the blue-green algae.

Fuller *et al.* (1958a) subjected cells of *Euglena gracilis* to sonic disruption, centrifugation, adsorption on charcoal, and elution. A conjugated pteridine was found on the chromatograms. In experiments with *E. gracilis*, *Chlorella pyrenoidosa*, *O. malhamensis*, and *O. danica*, these investigators report the occurrence of radioactive pteridines on paper chromatograms of cell extracts after photosynthesis in the presence of $C^{14}O_2$. Hydroxylamine or phenylhydrazine was used as a trapping agent in their fixation experiments. Both these compounds (Fuller *et al.*, 1958b) complex with an impurity in Whatman filter paper to form a yellow fluorescent compound. Thus the "pteridines" reported in fixation experiments in their earlier paper are artifacts of the trapping procedures used. Metzner (1957, 1958), from experiments with $C^{14}O_2$ fixation in *Scenedesmus* and *Chlorella*, reported the presence of an early CO_2 fixation product with the characteristics of a conjugated pteridine. Because hydroxylamine was used as a trapping agent in Metzner's experiments, his finding must also be regarded as an artifact.

Higher Plants

In contrast to the situation in microorganisms, both heterotrophic and autotrophic, very little is yet known about unconjugated pteridine pigments in higher plants. Workers studying plastid pigments and using two-dimensional paper chromatography found in higher plants pigments other than chlorophyll *a*, chlorophyll *b*, carotenes, and xanthophylls. Bauer (1952) and Lind *et al.* (1953) reported the presence on such chromatograms of colorless to pale yellow fluorescent unidentified spots. It is not known if these spots are pteridines.

The studies of Fuller *et al.* (1958a) concerned with pteridine pigments in algae included the duckweeds, *Spirodela oligorrhiza* and *Wolffia punctata*. For the reason stated earlier, these investigations permit no conclusion as to the presence of pteridines in the intact tissues of the organisms studied. Fuller *et al.* (1958a) isolated chloroplasts from spinach by differential centrifugation, extracted them with methanol, and chromatographed the extract. Small quantities of pteridines were found, but only preliminary studies have been made and they are not yet identified. This investigation clearly indicates the presence of pteridines in the chloroplasts of higher plants.

Wacker et al. (1959) studied spinach and grass as well as heterotrophic organisms. Cell-free extracts were chromatographed on paper, and the resulting spots were bio-assayed for *Crithidia* factor activity. Three active pteridines were found, none of which was identical with biopterin.

The general occurrence of folic acid in leaves (originally isolated from spinach) is so well known as not to require repetition. Iwai and Nakagawa (1958a) examined spinach, soybean, white clover, and rutabaga leaves for the presence of conjugated pteridines by a technique involving chromatography and bioautography with *Streptococcus faecalis*. The pteridine derivatives found were leucovorin (N^5-formyl-5,6,7,8-tetrahydrofolic acid), N^{10}-formylfolic acid, and another substance thought to be a derivative of teropterin. Folic acid (pteroylglutamic acid) was not detected. Thus, the naturally occurring members of the folic acid group in higher plants are mostly formylated or formylated and reduced compounds. In a further study, Iwai and Nakagawa (1958b) isolated from soybean leaves N^{10}-formylfolic acid, N^{10}-formylpteroic acid (rhizopterin), and a glutamyl conjugate of N^{10}-formylfolic acid. It was suggested that leucovorin, which was not found, might have been converted to N^{10}-formylfolic acid in the course of the isolation procedure.

One might summarize the present status of the question of unconjugated pteridines in higher plants by saying that pteridines have been shown to occur in chloroplasts, but it is not known whether these are conjugated. Unconjugated pteridines with *Crithidia* factor activity have been found in spinach and grass.

The isolation of unconjugated pteridines from higher green plants is far more difficult than isolation from microorganisms, because the former contain a far greater diversity of pigments. In a heterotrophic microorganism, advantage may be taken of the fluorescence of the pteridine, by using it as a "tracer" during chemical fractionation and separation from nonfluorescent materials. In chlorophyllous plants, chlorophyll *a*, chlorophyll *b*, several carotenes, and several xanthophylls, all of which are fluorescent, must be contended with, and the expectation of the ubiquitous presence of conjugated pteridines adds to the difficulties. We have attempted to develop a procedure for the isolation of unconjugated pteridines from higher plants by combining the extraction procedures customarily used with chlorophylls and carotenoid pigments with an adsorption-elution technique used for pteridines. The eluting agent was a mixture of equal parts of ammonia and acetone, as recommended by Fuller et al. (1958a). The procedure is as follows.

A sample of several grams of green leaf tissue is cut into small pieces then ground in acetone in a Waring blendor. The extracted cell debris is removed by suction filtration, and the acetone solution of the pigments is partitioned with petroleum ether in a separatory funnel. The pigments are transferred to the petroleum ether layer when water is added. After the petroleum ether solution of the pigments is washed several times in water, it is removed and dried with anhydrous sodium sulfate. Fuller's earth, added to the petroleum ether solution, adsorbs some of the pigments and is removed by filtration. Elution is by a solution of equal volumes of concentrated ammonium hydroxide and acetone. The solution after filtration is yellowish green and contains small quantities of chlorophyll or chlorophyll denaturation products. These green pigments are removed by passage through an alumina column and a yellow solution results.

A number of variations in this basic isolation procedure have been examined. Ether may be substituted for petroleum ether, the drying with sodium sulfate may be omitted, or the pigments may be adsorbed directly from the acetone solution. Charcoal is unsatisfactory as an adsorbent. Attempts have been made to reduce the proportion of ammonium hydroxide in the solvent mixture used in elution, but when it is much below 50% subsequent passage of the solution through an alumina column fails to remove chlorophyll or its denaturation products. Magnesia is not an acceptable substitute for alumina, even though the low rate of flow through magnesia is greatly increased by admixture with diatomaceous earth.

We are not certain that the procedure outlined above is completely effective in isolating unconjugated pteridines. Bio-assay of our preparations has not yet been done. Whereas many of the physical data obtained with our preparations are in good agreement with the findings on purified pteridines isolated from microorganisms, the ultraviolet-absorption spectra are far from satisfactory. One of our principal concerns has been the possible presence of conjugated pteridines. We have attempted removal of such pteridines by the ion-exchange technique of Heinrich et al. (1953, 1959), using Dowex 1 resin and eluting with a series of increasing concentrations of HCl. This method was not effective with our material. Perhaps the method of Rauen and Waldmann (1950) for removal of contaminating unconjugated pteridines from folic acid may be effective. This procedure involves adsorption on a Darco 60-cellulose column and elution with base.

Because of the uncertainty as to the purity of our preparations, the findings must be regarded as somewhat tentative; however, a few points seem clearly established.

Although Fuller et al. (1958a) showed the presence of pteridines in chloroplasts, it was not known whether these pigments are restricted to chloroplasts or whether they may occur elsewhere in the cell. For our purpose, pteridines have been defined as the material that, after the fractionation procedure described, accounts for absorption at 300 mμ. Chloroplasts of spinach were isolated by differential centrifugation in ice-cold 0.5 M sucrose solution according to the procedure of Granick (1938). In preliminary trials, considerable quantities of pteridines were found in the supernatant. We have attempted to measure the ratio of chlorophylls to pteridines in extracts of whole tissues and in extracts of isolated chloroplasts. Since chlorophylls are restricted entirely to chloroplasts, it might be anticipated that comparison of these ratios could lead to a possible conclusion as to pteridine distribution. Details of the procedure are as follows.

Spinach leaves were extracted with acetone in a Waring blendor, and the extract was divided into two portions. Pteridines were isolated from one aliquot according to the procedure described and made up to a known volume with methanol. The optical density was determined at 300 mμ. Total chlorophylls were determined in the remaining aliquot of the acetone extract, according to the method of Comar and Zscheile (1942). Chloroplasts were isolated from another portion of spinach leaves according to the method of Granick (1938), and the chloroplast preparation was extracted with acetone. The acetone extract was then divided in half, one half was processed for pteridines and the other was used to determine the total chlorophyll content.

Comparison of the ratios obtained in several experiments of this type indicated that, in the most favorable case, the protocol of which is presented in Table IV, 87% of the pteridines could be accounted for in the chloroplasts. Taking leakage into account, we conclude that, in all probability, the chloroplasts contain the entire complement of pteridines and these substances must henceforth be included, together with the chlorophylls and carotenoids, among the plastid pigments.

When Alaska peas are grown for about 1 week under conditions of total darkness, a pteridine may be isolated from them by the procedure described. Comparison of these preparations with those obtained from Alaska peas grown under the usual conditions of alternating light and darkness indicates that the two differ spectrophotometrically and chromatographically. A pteridine is formed by peas in darkness, and this pteridine is distinct from that formed by the same plants in light.

TABLE IV
Chlorophyll (Ch) and Pteridine (Pt) Content of Spinach Tissues and Isolated Spinach Chloroplasts

Whole tissue, extracted with 200 ml. of acetone

   A. 25-ml. aliquot partitioned into 100 ml. of ether

        chlorophyll a + chlorophyll b = 9.287 mg./liter

   B. 100-ml. aliquot worked up for pteridines in 200 ml. of methanol

        Optical density λ 300 = 0.220

Chloroplasts isolated by differential centrifugation at 60 g and 600 g in 0.5 M sucrose; chloroplasts extracted with 200 ml. of acetone

   C. 25-ml. aliquot partitioned into 100 ml. of ether

        chlorophyll a + chlorophyll b = 2.392 mg./liter

   D. 100-ml. aliquot worked up for pteridines in 200 ml. of methanol

        Optical density λ 300 = 0.050

$$\frac{[Ch]}{O.D.\ Pt}\ (cell) \neq \frac{[Ch]}{O.D.\ Pt}\ (chloroplasts)$$

$$\frac{9.287}{0.220} \neq \frac{2.392}{0.050}$$

$$42.21 \neq 48.74$$

$$100 \neq 113$$

Preparations from both dark-grown and light-grown peas were reduced chemically *in vitro* by means of zinc and HCl according to the procedure of Shiota (1959). A third compound was found that differed from either the "light" or "dark" pteridines. Preparations of the reduction product of the "light" and "dark" pteridines were identical spectrophotometrically. Therefore, the "light" and the "dark" pteridine must differ in oxidation reduction but in no other way.

We have followed the time course of the transformation of the "dark" into the "light" pteridine when dark-grown plants are subjected to light. About 16 hours is apparently required for substantial conversion to the "light" pteridine. Conversely, when light-grown plants are placed in darkness, the "light" pteridine persists in the tissues already formed for at least several days; and at the same time the "dark" pteridine may be extracted from the distal portions of the plant representing growth that occurred after the onset of darkness. These findings are consistent with the ideas that the conversion of "dark" pteridine to "light" pteridine is essentially irreversible *in vivo* and that the pteridines are relatively immobile in the plant.

It was concluded earlier that the "dark" and "light" pteridines differ in state of oxidation-reduction. We believe that the conversion of "dark" to "light" pteridine represents an oxidation. The reasoning is as follows: the "light" pteridine preparations are lemon-yellow, whereas the "dark" pteridine is a very pale yellow, and the reduction product of either is essentially colorless. If this view is correct, the "dark" would be the more reduced derivative and would bear the same relation to the "light" pteridine as leucovorin does to pteroylglutamic acid.

It is anticipated that the details of this work will be published. It is felt that this presentation will have fulfilled its purpose if it should stimulate work by others on the manifold problems presented by the pteridine pigments in plants, the function of which is so little known at the present time.

ACKNOWLEDGMENTS

It is a pleasure to acknowledge the financial support of this work by Vanderbilt University through its Committee on the Natural Sciences. Mr. Young Tai Kim has served patiently and capably as my research assistant. A portion of the work was done and this paper was prepared while I was a summer participant in the Biology Division, Oak Ridge National Laboratory, Oak Ridge, Tennessee.

REFERENCES

Albert, A. (1952). *Quart. Revs. (London)* **6**, 197-237.
Albert, A. (1954). *Fortschr. Chem. org. Naturstoffe* **11**, 350-403.
Augustinsson, K.-B. (1956). *Svensk Kem. Tidskr.* **68**, 271-281.
Bauer, L. (1952). *Naturwissenschaften* **39**, 88.
Berezovskii, V. M. (1953). *Uspekhi Khim.* **22**, 191-232.
Chattaway, F. W., and Barlow, A. J. E. (1958). *Nature* **181**, 281.
Comar, C. L., and Zscheile, F. P. (1942). *Plant Physiol.* **17**, 198-209.
Crowe, M. O'L., and Walker, A. (1944). *J. Opt. Soc. Am.* **34**, 135-140.
Crowe, M. O'L., and Walker, A. (1949). *Science* **110**, 166-167.
Crowe, M. O'L., and Walker, A. (1954). *Brit. J. Exptl. Pathol.* **35**, 18-27.
Forrest, H. S., and McNutt, W. S. (1958). *J. Am. Chem. Soc.* **80**, 739-743.
Forrest, H. S., Van Baalen, C., and Myers, J. (1957). *Science* **125**, 699-700.
Forrest, H. S., Van Baalen, C., and Myers, J. (1958). *Arch. Biochem. Biophys.* **78**, 95-99.
Forrest, H. S., Hatfield, D., and Van Baalen, C. (1959). *Nature* **183**, 1269-1270.
Fuller, R. C., Anderson, I. C., and Nathan, H. A. (1958a). *Proc. Natl. Acad. Sci. U. S.* **44**, 239-244.
Fuller, R. C., Anderson, I. C., and Nathan, H. A. (1958b). *Proc. Natl. Acad. Sci. U. S.* **44**, 518-519.
Gates, M. (1947). *Chem. Revs.* **41**, 63-95.
Granick, S. (1938). *Am. J. Botany* **25**, 558-561.
Gray, W. D. (1938). *Am. J. Botany* **25**, 511-522.
Gray, W. D. (1939). *Am. J. Botany* **26**, 709-714.
Gray, W. D. (1953). *Mycologia* **45**, 817-824.
Heinrich, M. R., Dewey, V. C., and Kidder, G. W. (1953). *J. Am. Chem. Soc.* **75**, 5425.

Heinrich, M. R., Dewey, V. C., and Kidder, G. W. (1959). *J. Chromatography* **2**, 296-303.
Hopkins, F. G. (1891). *Nature* **45**, 197.
Iwai, K., and Nakagawa, S. (1958a). *Mem. Research Inst. Food Sci. Kyoto Univ.* **15**, 40-48.
Iwai, K., and Nakagawa, S. (1958b). *Mem. Research Inst. Food Sci. Kyoto Univ.* **15**, 49-60.
Johnstone, D. B., Pfeffer, M., and Blanchard, G. C. (1959). *Can. J. Microbiol.* **5**, 299-304.
Kaneko, Y. (1957a). *J. Agr. Chem. Soc. Japan* **31**, 118-121.
Kaneko, Y. (1957b). *J. Agr. Chem. Soc. Japan* **31**, 122-126.
Karrer, P., and Viscontini, M. (1955). *Soc. Biol. Chemists India* **1955**, pp. 151-152.
Klöcker, A. (1916). *Centr. Bakteriol. Parasitenk. II, Abt.* **46**, 225.
Korte, F., Aldag, H. U., and Schicke, H. F. (1958). *Z. Naturforsch.* **13b**, 463-464.
Kuhn, R., and Cook, A. H. (1937). *Ber. deut. chem. Ges.* **70**, 761-768.
Lind, E. F., Lane, H. C., and Gleason, L. S. (1953). *Plant Physiol.* **28**, 325-328.
McNutt, W. S., and Forrest, H. S. (1958). *J. Am. Chem. Soc.* **80**, 951-952.
Margarot, J., and Deveze, P. (1925). *Bull. Soc. sci. méd. biol. Montpellier* **6**, 375.
Masuda, T. (1956). *Pharm. Bull. (Tokyo)* **4**, 71, 375.
Metzner, H., Simon, H., Metzner, B., and Calvin, M. (1957). *Proc. Natl. Acad. Sci. U. S.* **43**, 892-895.
Metzner, H., Metzner, B., and Calvin, M. (1958). *Proc. Natl. Acad. Sci. U. S.* **44**, 205-211.
Nathan, H. A., and Cowperthwaite, J. (1955). *J. Protozool.* **2**, 37-42.
Nathan, H. A., and Funk, H. B. (1959). *Am. J. Clin. Nutrition* **7**, 375-384.
Nathan, H. A., Hutner, S. H., and Levin, H. L. (1956). *Nature* **178**, 741.
Nathan, H. A., Hutner, S. H., and Levin, H. L. (1958). *J. Protozool.* **5**, 134-138.
Purrmann, R. (1940). *Ann. Chem. Liebigs* **546**, 98-102.
Rauen, H. M., and Waldmann, H. (1950). *Z. physiol. Chem.* **286**, 180-190.
Rickes, E. L., Chaiet, L., and Keresztesy, J. S. (1947). *J. Am. Chem. Soc.* **69**, 2749-2751.
Shiota, T. (1959). *Arch. Biochem. Biophys.* **80**, 155-161.
Van Baalen, C., and Forrest, H. S. (1959). *J. Am. Chem. Soc.* **81**, 1770.
Van Baalen, C., and Myers, J. (1959). *Proc. Am. Soc. Plant Physiol.* **34**, V-VI.
Van Baalen, C., Forrest, H. S., and Myers, J. (1957). *Proc. Natl. Acad. Sci. U. S.* **43**, 701-705.
Wacker, A., Lochmann, E.-R., and Kirschfeld, S. (1959). *Z. Naturforsch.* **14b**, 150-151.
Wolf, D. E., Anderson, R. C., Kaczka, E. A., Harris, S. A., Arth, G. E., Southwick, B. L., Mozingo, R., and Folkers, K. (1947). *J. Am. Chem. Soc.* **69**, 2753-2759.
Wolf, F. T. (1957a). *Physiol. Plantarum* **10**, 825-831.
Wolf, F. T. (1957b). *Nature* **180**, 860-861.
Wolf, F. T. (1959). *In* "Photoperiodism and Related Phenomena in Plants and Animals" (R. B. Withrow, ed.), pp. 321-326. American Association for the Advancement of Science, Washington, D. C.
Wolf, F. T., Jones, E. A., and Nathan, H. A. (1958). *Nature* **182**, 475-476.
Wolstenholme, G. E. W., and Cameron, M. P., eds. (1954). "Ciba Symposium on Chemistry and Biology of Pteridines." Little, Brown, Boston.
Ziegler-Günder, I. (1956). *Biol. Revs. Cambridge Phil. Soc.* **31**, 313-348.

Discussion

BANNISTER: How do the concentrations of the pteridines compare with those of chlorophyll and carotenoids in the chloroplasts?

WOLF: I do not know definitely because of doubts of the purity of my preparations. I suspect that the quantity is considerably less than that of the chlorophylls or carotenoids, which I think is one of the reasons for these things having been relatively overlooked so long.

5

On the Existence of Two Chlorophylls in Green Bacteria

R. Y. STANIER

Department of Bacteriology, University of California, Berkeley, California

This report summarizes a study of the chlorophylls of the green bacteria carried out recently in collaboration with Dr. J. H. C. Smith of the Carnegie Institute of Washington at Stanford; a full account of our work will be published elsewhere (Stanier and Smith, 1960).

In all organisms which perform green plant photosynthesis, one chemical species of chlorophyll, chlorophyll *a*, is an invariable component of the photosynthetic apparatus. In the photosynthetic bacteria this pigment does not occur, being replaced by other species of chlorophyll. The purple bacteria so far examined contain a single type of chlorophyll, identified by Fischer *et al.* (1938) as 2-desvinyl-2-acetyl, dihydrochlorophyll *a*, and designated by these workers as *bacteriochlorophyll*. The spectra of crude pigment extracts prepared from green bacteria have long indicated that they contain a different type of chlorophyll, originally termed *bacterioviridin* (Metzner, 1922), and subsequently given the more appropriate name of *chlorobium chlorophyll* (Larsen, 1953).

The difficulties of isolating and maintaining pure cultures of green bacteria have hampered work on their pigments, and the first absorption spectrum of carotenoid-free chlorobium chlorophyll was only recently published (Goodwin, 1955). This pigment was prepared from a strain of *Chlorobium thiosulfatophilum* isolated in England by Dr. June Lascelles. Kaplan and Silverman (1959) later described the spectral properties of what they presumed to be the same pigment, also isolated from a strain of *C. thiosulfatophilum* (strain L) which, although furnished to them by Professor Larsen of Trondheim, had in fact been originally isolated by Dr. Lascelles. However, the pigments described by Goodwin (1955) and by Kaplan and Silverman (1959) show markedly different spectral properties; the positions of their absorption maxima in ether differ by as much as 10 mµ, a discrepancy far too great to be explained by differences in instrumental calibration. Curiously enough, Kaplan and Silverman (1959) made no mention of the discrepancies between their data and the earlier data of Goodwin.

With the generous help of Professor Larsen and Dr. Lascelles, we have succeeded in clarifying this situation. Two strains of *C. thiosulfatophilum*, isolated some years ago in England by Dr. Lascelles, both conformed to the description of this species as established by Larsen (1953) and were presumed by Dr. Lascelles to be identical in all major respects. A restudy has shown, however, that they differ from one another with respect to the type of chlorophyll that they contain. The studies of Goodwin (1955) must have been conducted on strain PM; as already mentioned, the studies of Kaplan and Silverman were conducted on strain L. In order to distinguish these two chlorophylls we propose the designations chlorobium chlorophyll-660, for the pigment of strain PM, first studied by Goodwin (1955); and chlorobium chlorophyll-650, for the pigment of strain L, first studied by Silverman and Kaplan (1959). The numerical suffixes refer to the positions (in mµ) of the main red peaks, measured in ether solutions.

We have isolated these chlorophylls from dried cells of Lascelles' two strains of *C. thiosulfatophilum* and determined some of their properties. Quantitative absorption curves for the two pigments in ether are shown in Fig. 1. Both these chlorophylls show a general similarity to chlorophyll *a* in spectral structure, in relative peak height, and in absolute light absorption at the major peaks. The spectral differences among the three

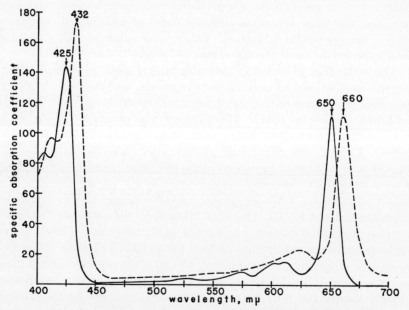

Fig. 1. The absorption spectra of the two chlorobium chlorophylls in ether.

pigments reside in the exact positions of the maxima, and in the degree of resolution of the series of minor peaks that lie between the Soret peak and the red peak. At first sight, therefore, one might be inclined to believe that the chlorobium chlorophylls differ from chlorophyll *a* only in minor chemical respects. However, this assumption is not borne out by the result of the phase test; despite many attempts, we have not succeeded in obtaining a positive test with either of the chlorobium chlorophylls. Furthermore, the extensive chemical studies of Holt on chlorobium chlorophyll-660, reported elsewhere in this volume* suggest that there are very far-reaching differences between it and chlorophyll *a*.

The position of the red peak in the absorption spectrum of the cells of different strains of green bacteria is subject to considerable variation; it can lie between 725 and 750 mμ. In reviewing these observations, Larsen (1953) pointed out that the main infrared peak of bacteriochlorophyll in the cells of purple bacteria likewise varies markedly in position from species to species. He suggested that in both groups of photosynthetic bacteria the phenomenon might be similarly explained, namely, by the association of a single type of chlorophyll with different protein complexes in the cell. We now have good evidence that, insofar as the green bacteria are concerned, this explanation is not correct; the

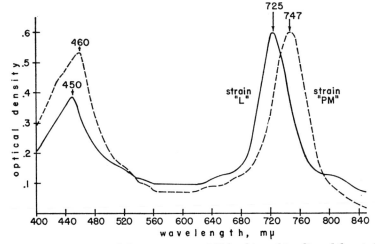

Fig. 2. *In vivo* spectra of the two strains of *Chlorobium thiosulfatophilum* isolated by Dr. June Lascelles. Strain "PM" contains uniquely chlorobium chlorophyll-660; strain "L" contains uniquely chlorobium chlorophyll-650. Data obtained from Dr. Lascelles.

---

* See Paper No. 11, in this volume, by A. S. Holt and H. V. Morley.

qualitative differences in chlorophyll content between Lascelles' two strains of *C. thiosulfatophilum* are clearly correlated with a difference between the whole cells in spectral character (Fig. 2). Strain PM, which contains chlorobium chlorophyll-660, shows a peak *in vivo* at 747 mµ; strain L, which contains chlorobium chlorophyll-650, at 725 mµ. Recently, a series of strains of *C. limicola* and *C. thiosulfatophilum* have been isolated in Berkeley by Mr. W. Sadler. All these strains contain only chlorobium chlorophyll-660, and all of them have essentially the same absorption spectrum *in vivo* as Lascelles' strain PM, with a red peak at 747 ± 2 mµ.

The fact that not one additional strain of *Chlorobium* possessing chlorophyll-650 has been found during the isolations performed by Mr. Sadler suggests that this chlorobium chlorophyll may be relatively rare among the green bacteria. However, our understanding of the ecology of green bacteria is at present so sketchy that it would be premature to draw any general conclusions about chlorophyll distributions in the group on the basis of current information.

REFERENCES

Fischer, H., Lambrecht, R., and Mittenzwei, H. (1938). *Z. physiol. Chem.* **253**, 1-39.
Goodwin, T. W. (1955). *Biochim. et Biophys. Acta* **18**, 309-310.
Kaplan, I. R., and Silberman, H. (1959). *Arch. Biochem. Biophys.* **80**, 114-124.
Larsen, H. (1953). *Kgl. Norske Videnskab. Selskabs Skrifter No.* **1**, 1-199.
Metzner, P. (1922). *Ber. deut. botan. Ges.* **40**, 125-129.
Stanier, R. Y., and Smith, J. H. C. (1960). *Biochim. et Biophys. Acta* (in press).

DISCUSSION

SCHER: I would like to mention that on checking the culture of *Chlorobium* in van Niel's collection, an *in vivo* peak at 742 mµ was found, while enrichment cultures that I have made have peaks either at 747 or 732 mµ.

VISHNIAC: Perhaps I should issue a warning at this point. In Dr. Fuller's laboratory we recently have received a number of contaminated strains of greens both from June Lascelles and from Larsen. If the cultures were transferred frequently they appeared to be impeccably pure but if they were allowed to stand for a long time there was a growth of *Chromatium* in them, which we had not introduced. As a matter of fact, in a recent letter from Norway, Dr. Larsen complains that some of his older cultures seem to turn somewhat reddish-brownish and he cannot understand why.

6

A New Leaf Pigment

HELEN M. HABERMANN[*]

*Department of Plant Biology, Carnegie Institution of Washington,
Stanford, California*

Seedlings of the white mutant of *Helianthus annuus* are frequently a bright bluish green when grown in the dark or in light of low intensity (Wallace and Schwarting, 1954). Such pigmentation can also be observed in portions of the cotyledons which remain covered by the seed coat, even though seedlings are germinated in the greenhouse and are thus exposed to direct sunlight. This pigment is rapidly bleached on exposure to bright light. Up to this time, it has been assumed that the green coloration results from the photoconversion of protochlorophyll to an unstable form of chlorophyll *a* which is bleached on continued exposure to light. The presence of protochlorophyll has been demonstrated by the absorption spectra of acetone extracts of dark-grown seedlings, and the conversion of protochlorophyll to chlorophyll has been observed by taking derivative absorption spectra of intact cotyledons before and after exposure to light (unpublished data). The apparent presence of large amounts of protochlorophyll and the absence of any detectable carotenoids (Wallace and Habermann, 1959) led us to believe that the white sunflower mutant might be an excellent source of the protochlorophyll holochrome. When dark-grown white mutant seedlings were extracted using methods outlined by Smith (1958), large amounts of an unknown, green, water-soluble pigment were obtained, rather than the expected protochlorophyll holochrome. This pigment has since been found not only in the cotyledons of mutant seedlings but also in the cotyledons and mature leaves of normal sunflowers, and in the leaves of chard and skunk cabbage. Species other than *Helianthus* were chosen at random for extraction and it thus seems possible that the pigment is widespread in occurrence.

The pigment has not yet been characterized chemically; neither has it been named, nor are its functions known. Briefly, its properties are as

---

[*] *Present address:* Department of Biological Sciences, Goucher College, Towson, Baltimore, Maryland.

follows. In neutral or alkaline solution it is green when oxidized. It can be reduced by hydrosulfite, ascorbic acid, or hydroquinone with an accompanying change in color to yellow-orange. It is reoxidized to the green form by addition of H_2O_2 or is autoxidized on exposure to air. Addition of acid results in an immediate color change to red. At all stages of purification it appears to be tightly bound to a protein.

Absorption Spectra

Figure 1 shows the absorption spectra of a partially purified preparation of the pigment in its oxidized and reduced forms. Also shown is the

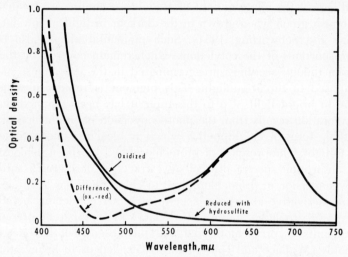

FIG. 1. Absorption spectra in the visible of oxidized and reduced forms of the new leaf pigment. Broken line is the difference spectrum (oxidized minus reduced forms) and shows the contribution of the oxidized form to total absorption.

difference spectrum (oxidized minus reduced pigment) indicating the extent to which the oxidized form contributes to the total absorption. There is only one peak in the visible portion of the spectrum (at 673 mµ) with a shoulder between 620 and 630 mµ. There is a second peak in the ultraviolet at 260 mµ and a shoulder at 320 mµ (see Fig. 2). It is well known that a peak at 260 mµ is characteristic of nucleic acid absorption spectra. According to data published by Warburg and Christian (1942), the ratio of absorption at 280 mµ to that at 260 mµ would indicate that approximately 10% nucleic acid was present in the preparation used to measure the absorption spectrum shown in Fig. 2.

Methods of Preparation

The properties of the new pigment can perhaps be best discussed by describing the procedures which have been developed for its extraction and purification. Figure 3 summarizes all the preparative steps. Initial extraction and centrifugation has thus far been done in the cold and in the dark, but such precautions may not be necessary. Ten gram portions of chilled leaves plus approximately 30 cc glycine-NaOH buffer [pH 9.5

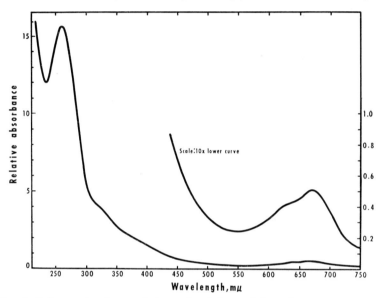

FIG. 2. Relative absorbance of the oxidized form of the unknown pigment in the ultraviolet and visible range of the spectrum.

(50 ml 0.2 M glycine plus 22.4 ml 0.2 M NaOH in a total volume of 200 ml)] were ground thoroughly with a small amount of sand in a chilled mortar; the brei was filtered through a single layer of tightly woven linen cloth to remove sand and larger particles of debris. The suspension was then centrifuged for one hour in the cold at 10,000 $\times$ g. This centrifugation removed the large cellular particles (cell wall fragments, starch grains, chloroplasts). The resulting supernatant was yellow to green in color. Next, acetone was added to a concentration of 45 to 50% by volume. The mixture was centrifuged for 10 to 15 minutes in the cold at 10,000 $\times$ g. The precipitate obtained in this centrifuging contained the remaining chloroplast fragments and precipitated protein. The supernatant was a dark bluish green and contained a high concentration of the new pigment plus traces of chlorophyll. The traces of chlorophyll

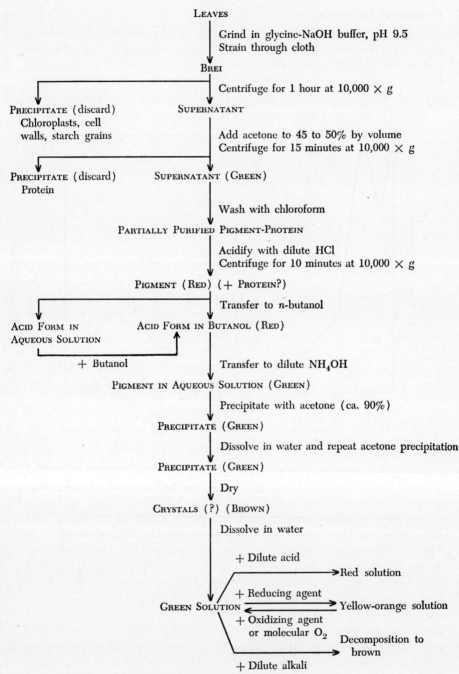

FIG. 3. Summary of preparative steps.

were removed by washing the solution with several portions of chloroform. This step concentrated the pigment because a part of the acetone moved into the chloroform layer. The pigment at this stage of purification has the absorption spectrum shown in Figs. 1 and 2. Such preparations form precipitates immediately in the presence of dilute solutions of $BaCl_2$, $HgCl_2$, and methanol, in concentrations of acetone greater than 50%, and in saturated solutions of $(NH_4)_2SO_4$. Precipitation occurs slowly in the presence of 50% saturated $(NH_4)_2SO_4$. No precipitation occurs in the presence of dilute alkali. Such evidence plus the high relative absorbance of the pigment preparation in the ultraviolet points to the presence of a protein. The pigment-protein complex is relatively stable. Preparations in this stage of purification have been stored for several days at room temperature and for several weeks under refrigeration with little apparent decomposition.

Further steps in purification have been based on the marked resemblance of the unknown to hallachrome, a pigment isolated from the marine polychaete, *Halla parthenopea*, by Mazza and Stolfi (1930) and characterized by Friedheim (1933). The acid form of hallachrome was known to be soluble in butanol and the unknown was found to behave in the same manner. Thus, the partially purified pigment was acidified with dilute HCl to the point where no further color change occurred (the solution was now bright red) and a precipitate formed. The precipitated protein was removed by centrifugation and the clear, red-colored supernatant was transferred to a separatory funnel containing an approximately equal volume of *n*-butanol. By using several additional portions of fresh butanol, most of the pigment was transferred into the organic solvent. The pigment was extracted from the butanol into a small volume of dilute NH_4OH or alkaline buffer. The pigment was finally precipitated by adding acetone to approximately 90% by volume. The precipitate was resuspended in water and final precipitation and washing were repeated several times.

The final precipitate, when dissolved in water, had the characteristics of the partially purified pigment, i.e., it was green in the oxidized form, yellow to orange when reduced, and red in acid solution. The purified preparation was not stable in the presence of alkali, however, and rapidly turned brown in basic solutions. The absorption spectrum of the final preparation in aqueous solution is shown in Fig. 4. It seems likely that the preparative steps in which the pigment was transferred through several aqueous and organic solvents and the final precipitations would have removed most of the impurities present in the initial extract. However, such purification had certainly not sharpened the absorption spectrum of the final pigment preparation. During the final steps in purification the absorption maximum in the red shifted toward shorter wavelengths and the peak at 260 mµ was reduced to only a shoulder. The reasons for this change in pattern of absorption are not known. The change may be a result of the concentration of one component in a mixture, a change from one form to another, or a degradation of the pigment. The shoulder at 320 mµ remained distinct and the ratio of peak height at

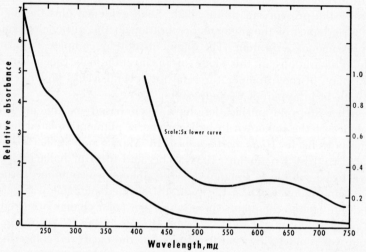

Fig. 4. Relative absorbance in the ultraviolet and visible portions of the spectrum of the oxidized pigment in aqueous solution. The unknown has passed through all steps in purification summarized in Fig. 3.

260 mµ in the ultraviolet to that in the region of maximum absorption in the red did not change significantly. It has therefore been impossible to determine whether or not the protein associated with this pigment is an artifact of preparation. It is obvious, however, that once joined together, the pigment and its protein are extremely difficult to separate.

The preliminary attempts at chromatographic purification of the pigment have been unsuccessful because the appropriate adsorbants have not yet been found. A tricalcium phosphate column could not be developed because, once adsorbed, the pigment could not be removed. Cellulose columns were equally unsuccessful because the pigment moved with the solvent front. When the pigment in acid form was passed through a cellulose column there was slight adsorption. A blue pigment was eluted from such columns with dilute NH_4OH. This blue pigment was yellow-orange when reduced and lavender in acid solution. This indicates either that the original solution contained at least two pigments or that there was a change in the pigment on the column. The absorption spectrum of the blue form had two distinct absorption maxima in the red portion of the spectrum, one at the usual position of 673 mµ and a second at the position of the shoulder (630 mµ) in the partially purified preparations.

Nature of the Pigment and Possible Functions

This new colored substance may well be a decomposition product of one or more of the naturally occurring leaf components. It resembles the naturally occurring quinones, hallachrome and pyocyanine. It is possible that it is a precursor of the melanins normally formed when plant tissues are ground and left exposed to the air because extractions carried out at pH values lower than approximately 8.5 or at room temperature have yielded discolored products. Perhaps this pigment has been preserved in a form very little changed from its state *in vivo*. The most probable form in which such a pigment would exist within the reduced environment of intact living cells would be the yellow (reduced) form and it is certainly possible that a colorless form exists. There is no direct evidence for the presence of this pigment *in its extracted form* in living leaves — in normally pigmented leaves the carotenoids and chlorophylls would completely mask it. The formation of considerable amounts of protochlorophyll in dark-grown mutant seedlings and the rapid bleaching of the blue-green pigment formed by the white mutant tend to confuse the issue further.

Anything which can be said about the chemical nature of the pigment at this time is no more than speculation. However, its marked similarity to known and chemically characterized pigments from bacterial and animal sources may provide some useful clues for its identification. Table I summarizes some of the properties of the unknown, pyocyanine and hallachrome. Both pyocyanine and hallachrome undergo color changes with changing levels of oxidation-reduction or pH, and both are quinones.

There is thus far very little experimental evidence to elucidate the role of this pigment in plant metabolism. It is of interest to note, however, that pyocyanine and hallachrome were shown by Friedheim (1933) to be powerful stimulants of respiration. When these compounds were added to suspensions of red blood cells, sea urchin eggs, or bacteria, rates of oxygen consumption were increased by several hundred per cent. The ease with which pigments of this kind are oxidized or reduced makes them likely candidates for participation in electron or hydrogen transport systems. It has been possible to demonstrate spectrophotometrically that the oxidized form of pigment "x" is capable of oxidizing reduced cytochrome c.

Hill and Walker (1959) recently reported the stimulation of rates of photosynthetic phosphorylation by pyocyanine and concluded that the previously known activity of phenazine methosulfate in phosphorylating

TABLE I
Comparison of Properties of Leaf Pigment "x," Hallachrome, and Pyocyanine[a]

Pigment	Source	Color	
		Oxidized	Reduced
Leaf pigment "x" ?	Helianthus annuus	Alkaline solution (green) Acid solution (red)	Yellow-orange
Hallachrome	Halla parthenopea	pH>8.5 (green) pH<8.5 (red)	Brown ? (semi-quinone)
Pyocyanine	Bacillus pyocyaneus	pH>4.9 (blue) pH<4.9 (red)	Green (semi-quinone)

[a] Information on hallachrome and pyocyanine from Hewitt (1933).

systems was due to its rapid conversion to pyocyanine in the light. These observations suggest that the new pigment may be a naturally occurring counterpart of pyocyanine in phosphorylating systems *in vivo*. Crane (1959) and co-workers have isolated lipid-soluble quinones from photosynthetic tissues and Bishop (1959) has demonstrated that certain p-quinone derivatives (among them coenzyme Q_{255} isolated by Crane) are capable of restoring the photochemical activity of petroleum ether extracted chloroplasts. The many observations of the stimulating effects of quinones on the partial reactions of photosynthesis make the possibility of involvement of pigment "x" in photosynthesis an intriguing speculation.

Acknowledgments

The work reported here was done at the Carnegie Institution of Washington, Department of Plant Biology, Stanford, California, during the summer of 1959. The author wishes to thank Drs. C. S. French, J. H. C. Smith, and other members of the Carnegie staff for their valuable help and advice. This work was supported by a grant from The National Science Foundation (NSF-G8972).

References

Bishop, N. I. (1959). *Proc. 9th Botan. Congr. Montreal* p. 34.
Crane, F. L. (1959). *Plant Physiol.* 34, 128-131.
Friedheim, E. A. H. (1933). *Biochem. Z.* **259**, 257-268.
Hewitt, L. F. (1937). "Oxidation-reduction Potentials in Bacteriology and Biochemistry," pp. 42-46. P. S. King and Son, London.
Hill, R., and Walker, A. (1959). *Plant Physiol.* **34**, 240-245.
Mazza, F. P., and Stolfi, G. (1930). *Boll. soc. ital. biol. sper.* **5**, 74-77.
Smith, J. H. C. (1958). *Carnegie Inst. Wash. Year Book* **57**, 287-290.
Wallace, R. H., and Habermann, H. M. (1959). *Am. J. Botany* **46**, 157-162.
Wallace, R. H., and Schwarting, A. E. (1954). *Plant Physiol.* **29**, 431-436.
Warburg, O., and Christian, W. (1942). *Biochem. Z.* **310**, 384-421.

Discussion

MILLOTT: It very early occurred to me that the properties of this pigment are strikingly similar to those of a napthoquinone. The word quinone has appeared in your own lecture from time to time. I tried to suppress making this somewhat revolutionary suggestion, but I wonder if I might just list various places where there are precise similarities. The absorption peaks of a napthoquinone like echinochrome are at 261 mµ, 340 mµ, and then in the case of echinochrome A, 463 mµ. That third peak can vary greatly, particularly if you are dealing with salts. Point number two: napthoquinone, shaken with sulfite, decolorizes. Shake it with air, it comes back to its own color. Point number three: it is an acid-base indicator, yellow in alkaline solution, red in acid. Point number four: it doesn't fluoresce. Point number five: when you treat it with alkali, very often it decomposes, and it gives, on the chromatogram, very disturbing, distracting things, pigments which resemble melanins — black, brown, and so on — very difficult to elute. Then with salts you can get a variety of colors. You can get conjugation with proteins. So all together, if I may say so, the similarities seem rather striking there.

HABERMANN: Yes, this is true.

MOSS: I'd like to add one point to the comment that has just been made. Our experience with the napthoquinone type of compound has been that if it is in the oxidized quinone form, it does not fluoresce. But if you reduce it, either with a hydrosulfide or with sodium borohydride, it will be intensely fluorescent. This is true of any type of Vitamin K type of compound, so long as it has a napthoquinone structure.

HABERMANN: We were looking for fluorescence in both the oxidized and reduced states and we thought we had found a great deal of fluorescence until we washed the chlorophyll contaminants out, and then it fluoresced neither in the oxidized nor the reduced state.

MILLOTT: Of course this depends on solvent also.

HENDRICKS: There was a report by Hill, roughly in 1926, on the occurrence of a blue compound of this general characteristic in *Mercurialis annis* and herbarium specimens supposedly carry the blue color. Moreover, it has been obtained in solution and it has this characteristic of ready oxidation and reduction. We once got some *Mercurialis* — it grows wild in the Eastern part of the United States — but none of the stuff we picked up showed it, so we finally got some supplied from the botanical garden in Kew that showed the expected characteristics.

ARNON: What is the approximate concentration of this pigment?

HABERMANN: This is awfully hard to say because we do not have any accurate molecular weight determination and the preparation from leaf material is rather sloppy, but it has a remarkably high concentration. You can start with about ten grams of leaves, and end up with a hundred or two hundred cubic centimeters of concentrated pigment preparation. So you get an awful lot of green material, which is not chlorophyll.

ARNON: Could you get it out of chloroplasts?

HABERMANN: Well, I have no idea. We throw most of the chloroplasts away, but of course it is very possible that this is leached out of some of the broken chloroplasts.

VISHNIAC: Perhaps it might be appropriate here to mention the occurrence of some other pigments which may or may not be similar but which are attached so consistently to certain enzymatic activities that one might ascribe a possible role to them. One was found by Racker in the purification of ribulose diphosphate carboxylase. It has a yellow pigment with it which cannot be separated from the enzyme without destroying the enzymatic activity. The other was found by San Pietro in the purification of the photosynthetic pyridine nucleotide reductase. It is quite intensely red and it has not been possible so far to separate this color from the enzyme without destroying the enzymatic activity.

7

Fossil Pigments

J. R. VALLENTYNE

Department of Zoology, Cornell University, Ithaca, New York

Introduction

A calculation based on three "order of magnitude" geochemical estimates will serve to introduce the viewpoint to be developed here. The estimates are as follows.

(a) The total weight of organic compounds annually synthesized by photosynthetic plants is about 1.6×10^{17} gm.*

(b) Photosynthetic organisms have existed on the earth for at least 1.6×10^9 years, and possibly for somewhat over 2×10^9 years (Barghoorn, 1957).

(c) The total mass of organic compounds now present in organic sediments and sedimentary rocks, petroleum, living organisms, etc., is of the order of 6×10^{21} gm. (Wickman, 1956).

Assuming that the modal rate of photosynthetic carbon fixation during the past 2×10^9 years has been one-fifth of the present rate, it follows that the total mass of organic matter produced by green plants during the biological history of the earth has been about 6×10^{25} gm. This is an immense (theoretical) weight. It corresponds to 1/100 the mass of the entire earth (6×10^{27} gm.). Only 6×10^{21} gm., or roughly 0.01% of this amount, still persists in organic combination. This residuum represents the net accumulation (over a long time) of the small annual losses of organic carbon from the biosphere brought about by the burial of organic matter in marine sediments. The chemical nature of this fossil organic matter has long been a question of great geochemical interest. Ideally one would like to know all the molecular transformations that have occurred in passing from once-living organisms to the organic substances in the oldest sedimentary rocks; however, this end has been difficult to approach. Only part of the difficulty can be attributed to the

---

* This represents the organic matter equivalent of the annual rate of photosynthetic carbon fixation on land (Schroeder, 1919) and a mean of the values listed for the oceans by Riley (1944) and Steemann-Nielsen and Jensen (1957).

chemical complexity of the fossil material. Much of it has resulted from a failure to apply appropriate methods of analysis. The existing data on the composition of fossil organic matter have largely been derived from elemental or gross molecular (i.e., class or group) analysis. These data are, of course, essential for an over-all understanding of the problem; but they are not well suited for use in deductive arguments. A minority of workers have pursued the problem by analysis for particular compounds. This approach suffers from the disadvantage of incompleteness, but the data are full of meaning in deductive terms. This paper will deal solely with the second (i.e., molecular) approach, primarily from a biological viewpoint. The writer considers that the transformation from living to fossil occurs at the time of an organism's death.

The biochemical approach to the study of fossil organic matter is relatively new. It dates from approximately 40 years ago when it was clearly demonstrated that a well known biological compound, cellulose, had been almost perfectly preserved in deposits known to be 20 or so million years old (Gothan, 1922; Wisbar, 1923; Schultz and Hamackova, 1924). Prior to that time no one had dared to consider such preservation possible. Geologists were probably overimpressed with the instability of biological molecules, and biologists with the activity of sedimentary bacteria and evidence of high temperatures deep within the earth's crust.

Opinions have since changed. It is now known that a variety of biological substances have been preserved in a relatively intact manner for periods of several hundred millions of years in extreme cases, e.g., proteinaceous compounds (Abelson, 1954, 1957; Jones and Vallentyne, 1960), carbohydrates (summarized by Vallentyne, 1960), porphyrins (summarized by Dunning, 1960), and many other organic compounds. There is even a report of porphyrins in pre-Cambrian deposits (Radchenko and Sheshina, 1956). We will deal here only with the case for fossil pigments. For other aspects of the subject the reader is referred to a forthcoming volume edited by Irving A. Breger (1960).

The discoveries of these "phenomenal" cases of molecular preservation have usually been attended with considerable surprise and interest; but we must ask if the data and the approach are really important. The writer is convinced that they are. A number of intriguing problems are now open to analysis and experiment. One of the most important of these is the *direct* study of biochemical evolution, i.e., through the study of fossil *ancestors* rather than living *cousins*. Though such studies will be fraught with innumerable interpretative difficulties, there is little doubt that they will provide the only sound basis for discussing questions of biochemical evolution. A study of the molecular nature of pre-Cambrian organic carbon may ultimately provide the best information on the nature of pre-Cambrian life. As a second example we can consider the use of organic compounds as "built-in" low-temperature geothermometers. All carbonate shells and organic sediments that have been studied to date contain proteinaceous materials (Abelson, 1954; Kleerekoper, 1957). Knowing the age of a deposit and the kinds and amounts of amino acids present, the temperature history can be partly reconstructed (Jones and Vallentyne, 1960). The method applies to a temperature range (below 250°C) that has otherwise been difficult to approach geologically. As a third example we might consider the origin and history of petroleum. A solution to this problem must ultimately come from data on the molecular composition of recent and fossil organic matter as Smith (1954) and Meinschein (1959) have already shown.

Let us now turn to the immediate problem: the biogeochemistry of certain photoreactive pigments. The treatment is limited to chlorophylls, carotenoids, and their breakdown products, for these are the only groups of biological pigments for which adequate data exist. The subject will be approached through a discussion of the qualitative and quantitative changes accompanying the destruction of chlorophyll and carotenoid pigments in aquatic, or once-aquatic, environments. Virtually nothing is known of the fate of these pigments in terrestrial environments, except that destruction there appears to be much more complete.

The Fate of the Chlorophylls

PLANKTON AND SESTON*

A truly fundamental study of the green pigments present in freshwater and marine seston has yet to be undertaken. So far as the author is aware, no one has ever attempted to demonstrate on a rigorous basis that chlorophyll *a* actually does occur in freshwater or marine seston. This is not belaboring the point at all. It indicates our present state of ignorance.

The use of pigment extraction methods for the quantitative estima-

---

* The terms *plankton* and *seston* have often been used indiscriminately in chemical work. Seston is the total particulate matter in a sample of water, whereas plankton refers only to the living individuals in the particulate matter. With the aid of a microscope it is possible for the biologist to study the plankton as such; but the chemist almost invariably deals with a sample of seston, i.e., a mixture of plankton, organic detritus, and inorganic detritus. This is an important point to bear in mind, for sestonic organic matter may, under certain conditions, show more chemical resemblance to the organic matter of surface sediments than to that characteristic healthy planktonic organisms.

tion of phytoplankton abundance was first developed by Harvey (1934). Numerous modifications of the basic method have since been proposed that permit, as Harvey's original method did not, a separate determination of chlorophylls and carotenoids. The most refined of these is the spectrophotometric method developed by Richards and Thompson (1952) and later modified by Creitz and Richards (1955). This method permits separate estimations of chlorophylls a, b, and c, but the validity of the results depends on the assumption that these are the only pigments that selectively absorb light at three wavelengths in the red end of the spectrum in 90% acetone extracts of seston. There is little doubt that chlorophylls a, b, and c are present in samples of seston taken from surface waters, but there is some reason to suspect (see below) that additional pigments may be present, particularly in nutrient-deficient or light-deficient waters.

Most of the data obtained by quantitative "phytoplankton chlorophyll" determinations are of little use to us in the present context since the data are usually expressed as "chlorophyll" per unit volume of water rather than per unit dry weight of seston or sestonic organic matter. Gillbricht (1952) and Banse (1957) have determined both the amount of "chlorophyll" and the dry weight of seston in unit volumes of water, thus permitting a calculation of the amounts of "chlorophyll" per unit dry weight of seston. Some average values found by Banse (1957) for seston collected from the Baltic Sea are listed in Table I. It will be noted that the organic matter of deep-water seston contains appreciable quantities of "chlorophyll." Much of this "chlorophyll" must have occurred in the form of detritus (or tripton) rather than in living phytoplanktonic cells. An earlier study of Wisconsin lake waters (Kozminski, 1938) revealed the same phenomenon. By using a combination of microscopic and chemical methods Gillbricht (1952) estimated that chlorophyll constituted 4–12% of the ash-free dry weight of plankton in the Bay of Kiel. The upper limit seems rather high, for it exceeds the usual value given for the chlorophyll content of spinach leaf chloroplasts (8% of the dry weight). The corresponding value for "chlorophyll-containing detritus" was 3% of the dry weight. He also found, on the basis of studies made throughout the year, that the phytoplankton usually constituted only a minor fraction of the total weight of seston even in near-surface waters during times of great algal productivity. The maximum phytoplankton content of the seston did not exceed 25% at any time of the year. A study of the nature of the detrital "chlorophyll" that accompanies the phytoplankton chlorophyll may well show a closer resemblance to sedimentary chlorophyll degradation products (see below) than to the

TABLE I

Average Values for "Chlorophyll" and Seston in Four Water Layers (0–220 Meters) in the Baltic Sea[a]

Water layer	"Chlorophyll" (μg./liter)	Seston (mg./liter)	Per cent organic matter in seston	Per cent "chlorophyll" in	
				dry seston	sestonic organic matter
Warm surface waters	0.94	0.24	ca. 33	0.4	ca. 1.2
Thermocline region	1.41	0.40	ca. 25	0.35	ca. 1.4
Upper bottom water	0.31	0.12	ca. 25	0.26	ca. 1.0
Deep bottom water	0.17	0.22	ca. 10	0.08	ca. 0.8

[a] Data as described by Banse (1957) for stations east of Bornholm.

chlorophylls of healthy phytoplanktonic cells. Gorham (1960) has reported suggestive evidence of chlorophyll *a* breakdown in seston collected from Esthwaite Water, England. There was a shift of the 432 mμ chlorophyll *a* band to lower wavelengths in hypolimnetic seston.

On the basis of reasonable assumptions involving two different approaches Orr *et al.* (1958) calculated that only about 0.2% of the chlorophyll rings synthesized in the waters above the Santa Barbara basin (off the coast of Southern California) actually survived to reach the underlying sediments at a depth of 600 meters. The extent of decomposition in deeper basins, and presumably in the open ocean, was disproportionally higher. The decomposition of seston "chlorophyll" has been studied in the laboratory by Skopintsev and Bruk (1940). They found that after 53 days of aerobic laboratory decomposition at 6° and 16°C only 48 and 14%, respectively, of the initial amounts of "chlorophyll" had remained with intact dihydroporphyrin ring systems.

Recent and Late-Pleistocene Sediments

It has long been realized that green, chlorophyll-like pigments occur in both freshwater and marine sediments, but it is only within the past few years that serious attempts have been made to fractionate and identify the pigments. One of the first studies of sedimentary chlorophyll degradation products (SCDP) was made by Trask and Wu (1930). Later attempts to fractionate and identify SCDP were made by Fox and Anderson (1941), Beatty (1941), Fox *et al.* (1944), Koltz (see Phinney 1946), Vallentyne (1955), and Andersen and Gundersen (1955). In the light of more modern work these studies can only be regarded as primitive steps in the right direction. Orr and Grady (1957) and other workers have noted the similarity of SCDP absorption spectra to that of pheophytin *a*; but an unequivocal demonstration of the presence of pheophytin *a* (or for that matter any other known dihydroporphyrin) has yet to be published. Fortunately, some recent unpublished information is available for discussion (Corcoran, 1957; S. R. Brown, personal communication).

Corcoran (1957) obtained chromatographic, spectral, and chemical evidence for the presence of pheophytins *a* and *b*, and pheophorbides *a* and *b* in marine sediments. These were the dominant pigments in the marine sediments that he analyzed. Quantitative evidence was presented suggesting that a conversion of pheophytins to pheophorbides occurred in the sediment. Chlorophylls *a* and *b* were found in one sample of surface sediment from the Gulf of Mexico, but these primary pigments were otherwise found to be lacking. Chromatographic fractions resembling pheophytin *d* were found in some other marine sediments.

Brown (personal communication) has made a thorough study of the complex mixture of SCDP present in the surface and subsurface sediments of Little Round Lake, Ontario. The sediments of this lake are notoriously rich in SCDP and carotenoids. Using diverse chromatographic methods Brown reports that the following compounds have been isolated: pheophytins a and b, chlorophyllides a and b, pheophorbides a and b, methylpheophorbide a, and some undegraded chlorophyll b. His identifications were based on chromatographic, spectral, and chemical data in comparison with known standards. A fraction with an absorption spectrum similar to that of chlorophyll c was also obtained, but its exact identity, like that of its suspected counterpart, remains uncertain. The existence of chlorophyllides a and b in the sediments of this lake shows that at least some of the chlorophyll molecules lose the phytyl group before the magnesium. This reaction is doubtless mediated by the enzyme *chlorophyllase*. There was a tendency for pheophytins to be replaced by pheophorbides in the deeper sediments of a core taken from the lake. Allomerization products of virtually all the above compounds were found in a middle section of the core, in addition to the more "normal" pigments. Brown states that the allomerized compounds were not formed during extraction procedures, but were characteristic of the pigments in the sediment. Brown also found some bacteriochlorophyll-like fractions in his study.

The general picture emerging from the above studies, subject to future modification, is as follows.

(a) Chlorophyll a, the most abundant green pigment in plants, is of *extremely* rare occurrence in sediments.

(b) Chlorophyll b, which tends to lose its magnesium ion less readily than chlorophyll a, is found in at least some near-surface sediments.

(c) With the exception of chlorophyll b and bacteriochlorophyll, all the sedimentary green pigments appear to be transformation products of known chlorophylls.

(d) Two degradational pathways appear common, the first more so than the second:

$$\text{chlorophyll } a \to \text{pheophytin } a \to \text{pheophorbide } a;$$
and
$$\text{chlorophyll } a \to \text{chlorophyllide } a \to \text{pheophorbide } a.$$

The molecular transformations referred to above are depicted in the left hand part of Figure 1. Many details remain to be filled in with regard to a complete understanding of the sedimentary picture, but it is most encouraging to find that one can now speak with some degree of assurance about the types of green pigments in sediments.

Vallentyne (1955) and Orr and Grady (1957) have developed routine methods for the quantitative determination of SCDP *in toto*. Vallentyne (1955) chose to express SCDP concentrations in terms of an arbitrary system of SCDP units. One unit was defined as the amount of green pigment giving a light absorbance reading of 0.100 (1 cm. path) at the wavelength of maximum absorption in the red end of the spectrum (usually 667 mµ) when dissolved in 10 ml. of 90% acetone. This procedure is somewhat more cautious than that employed by Orr and Grady (1957) who expressed concentrations as pheophytin *a* weight equivalents. One SCDP unit is approximately equivalent to 27 µg. of pheophytin *a*. Until the exact identities, relative amounts and extinction coefficients of the sedimentary compounds are known in detail, it will be impossible to obtain any accurate weight estimate from spectrophotometric data. So long as this limitation is remembered, one method of expression is as good as the other.

Vallentyne and Craston (1957) determined the amounts of SCDP (*in toto*) in the surface sediments of some relatively shallow Connecticut lakes, up to 18 meters in maximum depth. There was no marked tendency for the SCDP concentrations per unit ignitable matter to decrease as a function of depth below the lake surface; if anything, there was a slight tendency toward the reverse. The concentrations of SCDP in the organic matter of marine sediments do tend to decrease with increasing depth of the water column (Orr *et al.*, 1958). This is doubtless a reflection of extensive decomposition in the water column as well as of the high phytoplankton production rates characteristic of coastal waters. Vallentyne and Swabey (1955) and Fogg and Belcher (1960) have made detailed analyses of the amounts of SCDP (*in toto*) as a function of depth in sediment cores known to cover a time interval of somewhat over 12,000 years. In both cases there was a general tendency for the ignitable matter of the deepest sediments to be low in SCDP relative to the surface sediments, but green pigments were present at all levels of the cores that contained organic matter. The above authors and Orr and associates (1958) as well, found marked decreases in the amounts of SCDP per unit ignitable matter with increasing depth in near-surface sediments. The above facts suggest that there is a slow rupture of dihydroporphyrin ring systems in wet sediments, this process being somewhat more intensified near the sediment surface than at deeper levels. The nature of the decomposition processes is not known.

A recent report of vanadium-porphyrins in the basal silts of a postglacial lake sediment deserves special mention. In a core taken from North Cooking Lake, Alberta, Hodgson et al. (1960) found green hydroporphyrin pigments in the uppermost gyttja layers; but in the basal silts, separated from the gyttja by an intervening layer of clay, only vanadium-porphyrins were present. The concentration was low (2 p.p.m. dry weight). Extracts of the silt yielded a chromatographic fraction with absorption maxima at 410, 536, and 576 mµ in chloroform, with very little background absorption. From the properties of the material in this fraction and the positions of the absorption maxima there seems little doubt that a vanadium-porphyrin complex was present. It is of interest to note that a chromatographic fraction was earlier obtained from one level of a marine sediment core with properties similar to a vanadium-porphyrin complex (Fox, 1944; Fox et al., 1944). These are the only data suggesting that metalloporphyrins do occur in late Pleistocene sediments; however, future work may well reveal a wider distribution. The paucity of data does not necessarily imply rarity of occurrence.

Sedimentary Rocks and Petroleum

Dihydroporphyrins (unidentified) occur only in traces in pre-Pleistocene sedimentary rocks (Treibs, 1936) and are apparently lacking in petroleum. They are replaced in these older deposits by metalloporphyrins (Ni and V) and to a lesser extent by free porphyrins. Most workers assume that the hydroporphyrins of recent sediments have been transformed into the metalloporphyrins of older deposits; but the sequence and nature of the reactions as well as their times of occurrence have yet to be elucidated.

Metalloporphyrins and free porphyrins were first isolated from sedimentary rocks and petroleum by Treibs (1934a, b, 1935a, b, 1936) and Dhéré and Hradil (1934). These compounds are now known to occur in petroleum, oil shales, coals, guano, phosphorites, coprolites, and gilsonite. Treibs' contributions to the study of fossil porphyrins form a classic example of molecular biogeochemistry at its best. He isolated the commonest fossil porphyrins in pure form, proved their structure, indicated their relation to plant and animal porphine derivatives, and showed both by experiment and deduction how his data supported the idea that petroleum originated under reducing and relatively low-temperature conditions from the organic matter of plants and animals. It is beyond our scope to consider all the data that have since been accumulated on fossil porphyrins by other workers, particularly since an excellent review has recently been published by Dunning (1960). We need only be concerned with the chemical nature of the pigments and their relation to the hydroporphyrins of recent sediments.

The two porphyrins whose structures are shown in the right-hand end of Fig. 1, occur frequently in fossil materials: desoxophylloerythrin and particularly its decarboxylation product, desoxophylloerythroetioporphyrin. They occur most commonly in combination with either nickel or vanadium. The relation of these compounds to chlorophyll a is evident from the presence of the five-membered ring linking pyrrole group III with the methene-C between pyrrole groups III and IV. Three other porphyrins, more closely related to hematins than to chlorophylls, were identified by Treibs (1934b, 1935b) in various fossil materials: mesoporphyrin, mesoetioporphyrin, and deuteroetioporphyrin; but the correctness of these identifications is still in doubt (Dunning, 1960). The chlorophyll-derived and hematin-derived fossil porphyrins can readily be distinguished on the basis of spectral data. The former have a "phyllotype" spectrum in which one absorption band (band III in the porphyrin spectral terminology) is lower in height than either of the two adjacent spectral bands (II and IV). The latter have an "etio-type" spectrum in which band III is intermediate in height between bands II and IV. This difference in spectral characteristics is associated with substitution on the methene-C linking pyrrole groups III and IV of the chlorophyll-derived porphyrins. Fossil porphyrins of the "phyllo spectral type" have almost certainly been derived from the chlorophylls of plant tissues, whereas those of the "etio spectral type" have probably arisen from the iron-porphyrin compounds of both plant and animal tissues.

The extraction of porphyrins from fossil materials is usually accomplished with reagents such as HBr-acetic acid that remove the metals from metalloporphyrins, thus permitting the investigator to utilize the ampholytic properties of uncomplexed porphyrins for further extraction and isolation procedures. The metals commonly chelated by the fossil porphyrins are nickel and vanadium, the latter existing as a vanadyl, $-V=O$, group (Erdman et al., 1956). Iron-porphyrins may also occur in some organic shales (Treibs, 1934b; Moore and Dunning, 1955), but proof is not complete.

Vanadium-porphyrins are commonly concentrated in the asphalt-resin (heavy) portions of crude petroleum, and impoverished in the lighter fractions. According to Radchenko and Sheshina (1955, 1956) nickel-porphyrins are present in light crude oil fractions, and are not enriched in asphalt-resin fractions. Due to the presence of the oxygen atom in the vanadyl moiety of vanadium-porphyrins, vanadium complexes are more polar than the corresponding nickel complexes. This property can be used to advantage in chromatographic separations (Blumer, 1956). All the porphyrins in petroleum appear to be com-

7. FOSSIL PIGMENTS

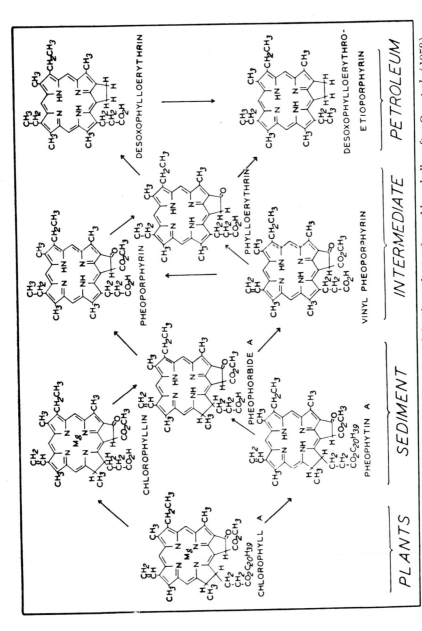

FIG. 1. Possible steps involved in the formation of fossil porphyrins from chlorophyll *a* after Orr *et al.* (1958).

plexed with metals. Free porphyrins are either very rare or lacking. Organic shales (Blumer, 1950) and some coals (Treibs, 1935b) do contain free porphyrins, but the amounts are usually small relative to those of the metal complexes. This is particularly true of shales.

Many early and middle Paleozoic shales and crude oils have been shown to contain metalloporphyrin complexes. Nickel-porphyrins have even been reported from pre-Cambrian deposits (Radchenko and Sheshina, 1956). It behooves the investigator under these circumstances to show that metalloporphyrin complexes do have sufficient thermal stability to persist under geological conditions for such long periods of time. This has been done. Montgomery (unpublished) and Hodgson and Baker (1957) found that fossil vanadiumporphyrins decompose according to first-order reaction kinetics when heated in the presence of excess petroleum. The activation energy required for ring rupture is of the order of 52–53 kg.-cal. per mole. Extrapolation suggests that vanadium-porphyrins would survive complete destruction for somewhat over 10^9 years if continuously preserved in a medium of petroleum at a temperature of 100°C.

It has occasionally been suggested that the fossil metalloporphyrin complexes might be remains of similar complexes present in formerly living organisms. Glebovskaya and Volkenshtein (1948) proposed this possibility for vanadium-porphyrins, and Radchenko and Sheshina (1956) have made a somewhat similar suggestion in relation to nickel-porphyrins. Most workers are not in agreement with these contentions. The following facts merit close attention.

(a) Pheophytin a is converted to desoxophylloerythroetioporphyrin (with an intermediate stage of phylloerythrin) when pheophytin a is heated for several hours in a medium of petroleum at 300–360°C (Treibs, 1934b).

(b) The nickel complex of pheophytin a is converted to a nickel-porphyrin complex when the former is heated in 35% KOH in ethylene glycol for 20 minutes at 200–250°C (Hodgson et al., 1960).

(c) On heating organic sediments (containing dihydroporphyrins) with excess nickel ammonium sulfate in sealed tubes at 102–140°C, the free dihydroporphyrin concentration decreases with a concomitant increase in the concentration of nickel-dihydroporphyrins (Hodgson et al., 1960). Extrapolation to low temperatures suggests that this reaction could proceed at measurable rates in nickel-rich sediments *in situ*.

(d) The metal ions in vanadium-porphyrins (Treibs, 1935a; Erdman et al., 1956), nickel-porphyrins (Dunning et al., 1953) and in the nickel complex of pheophytin *a* (Hodgson *et al.*, 1960) have all been complexed under relatively mild conditions in the laboratory.

(*e*) In all samples of crude petroleum that have been examined so far the concentrations of nickel and vanadium exceed the amounts necessary to complex all the porphyrins present (Dunning, 1960).

(*f*) The *most abundant* dihydroporphyrins in recent sediments have not retained the magnesium ions characteristic of their parent compounds, the chlorophylls. The high ratio of nickel and vanadium to magnesium in crude petroleum also supports the notion that magnesium is lost from the hydroporphyrins prior to the time of accumulation of petroleum (Baker and Hodgson, 1959).

A mature consideration of the above facts indicates that if chlorophyll *a* had occurred in early Paleozoic plants it would have followed the same geologic pathways of decomposition as are now observed for chlorophyll *a* in seston and recent sediments. This would have eventually led to the formation of the metalloporphyrins that are now observed in sedimentary rocks and petroleum. One cannot conclude on the basis of present data that chlorophyll *a* actually did occur in early Paleozoic plants. I am only saying that existing data do not disagree with the idea that chlorophyll *a* or similar magnesium-containing dihydroporphyrins were then in existence. The burden of proof to the contrary rests with those who hold such a belief. The occurrence of nickel-porphyrins in Pliocene petroleum and vanadium-porphyrins in postglacial silts suggests that if these compounds are relatively unmodified biological constituents they should be found in living organisms. No thoroughly documented case of this sort exists. The normal biological occurrence of iron and cobalt in porphyrin and porphyrin-like ring systems, respectively, suggests that vanadium and nickel may be similarly complexed; but if so, the amounts must be small and the biological occurrences comparatively rare.

We may briefly summarize what has been said above about the fate of the chlorophylls in the geologic column. There is a relatively rapid breakdown of chlorophylls *a* and *b* following the death of cells containing these pigments, eventually resulting in a complete conversion to other compounds. The rupture of the ring structure proceeds rapidly in the water column, more slowly in wet sediments, and still more slowly in sedimentary rocks and petroleum. The main decompositional sequence of *unruptured* chlorophyll ring systems in waters and wet sediments is

$$\text{chlorophyll} \rightarrow \text{pheophytin} \rightarrow \text{pheophorbide}$$

but also known is the sequence

$$\text{chlorophyll} \rightarrow \text{chlorophyllide} \rightarrow \text{pheophorbide}$$

The reactions leading to the formation of the metalloporphyrins and free porphyrins characteristic of sedimentary rocks and petroleum have not been elucidated in detail nor has the time of occurrence been fixed; but it seems reasonable to assume that the metals have secondarily entered either a dihydroporphyrin or porphyrin ring system. An outline of the possible decompositional sequences (without reference to the secondary metals) is given in Fig. 1. The concentrations of different porphine-type compounds in various geological materials are summarized in Table II.

The Fate of the Carotenoids

The geologic occurrences of carotenoids differ from those of porphine derivatives in two notable respects: (a) carotenoids that occur in living organisms have also been isolated in pure form from freshwater and marine sediments; and (b) carotenoids, as such, have not been reported to occur in sedimentary rocks or petroleum. The second point of difference may simply be due to the lack of adequate studies.

Data on the types of carotenoids in various algal groups are summarized by Goodwin in this volume.* We may take it that the concentrations generally range from 20 to 200 mg. per 100 gm. dry weight. The usual ratios of hypophasic to epiphasic carotenoids in algae (xanthophylls/carotenes, or the h/e ratio) are 4:1 to 9:1. All the carotenoids known to occur in phytoplankters and zooplankters can be expected to be found in collections of seston. It has never been determined if seston contains some carotenoid degradation products in addition, but such a finding would not be unexpected. Carotenoids that have been identified in sestonic extracts are as follows: β-carotene (Heilbron and Lythgoe, 1936; Tischer, 1938; Pinckard et al., 1953); echinenone, also called myxoxanthin (Heilbron and Lythgoe, 1936; Tischer, 1938); myxoxanthophyll (Heilbron and Lythgoe, 1936; Tischer, 1938); lutein (Heilbron and Lythgoe, 1936); flavacin (Tischer, 1938); and peridinin (Kylin, 1927; Pinckard et al., 1953). The few h/e ratios that have been determined for marine seston lie in the range of 7:1 to 14:1 (Fox, 1944). Studies on the occurrences of astacene and other carotenoids in marine zooplankters have been reviewed by Kon (1954).

Carotenoids were first reported to occur in marine sediments by Trask and Wu (1930). Later and more informative studies have since been made by Baudisch and von Euler (1934), Lederer (1938), Karrer and Koenig (1940), Fox et al. (1944), Savinov et al. (1950), Andersen and Gundersen (1955), Vallentyne (1956, 1957), and Züllig (1956, 1960). A survey of the above works suggests that something in excess of

---

* See Paper No. 1, in this volume, by T. W. Goodwin.

TABLE II

A Summary of the Types and Concentrations of Chlorophylls and Their Transformation Products in Plankton, Seston, Organic Detritus, Freshwater and Marine Sediments, Organic Shales and Petroleum

Sample	Constituents	Concentration		References
		p.p.m. dry weight	p.p.m. organic matter	
Planktonic algae	Chlorophylls a, b, c	3000–67,000	4000–120,000	Rabinowitch (1945); Gillbricht (1952)
Marine seston	??	200–10,000	8000–40,000	Gillbricht (1952); Banse (1957)
"Chlorophyll-containing detritus"	??		??–34,000	Gillbricht (1952)
Recent organic sediments				
(a) freshwater (2–18 meters)	Various hydroporphyrins	220–1700[a]	1150–4500[a]	Vallentyne and Craston (1957)
(b) marine: shallow basins (up to 120 meters)	Various hydroporphyrins	1–13[a]	130–2100[a]	Orr et al. (1958)
California basins (148–2060 meters)	Various hydroporphyrins	2–103[a]	81–1710[a]	Orr et al. (1958)
open sea (3760–4800 meters)	Various hydroporphyrins	0.02–1.6[a]	3–44[a]	Orr et al. (1958)
Organic shales	Mostly metalloporphyrins	tr.–4200	—	Treibs (1934a, b, 1935a); Blumer (1950)
Petroleum	Mostly metalloporphyrins	tr.–1600	tr.–1600	Radchenko and Sheshina (1955)

[a] Pheophytin a equivalents. tr. = traces.

20 carotenoids occur in freshwater and marine sediments. Only six of these have been rigorously identified: β-carotene (Lederer, 1938; Savinov *et al.*, 1950; Vallentyne, 1956, 1957); α-carotene (Vallentyne, 1956, 1957); echinenone (Lederer, 1938); rhodoviolascin (Karrer and Koenig, 1940); lutein (Fogg and Belcher, 1960); and myxoxanthophyll (Züllig, 1960). There is suggestive evidence that lycopene may occur in lake sediments and that the following carotenoids may occur in marine sediments (Fox *et al.*, 1944): rhodopurpurin, flavorhodin, torulene, fucoxanthin, petaloxanthin, zeaxanthin (or diatoxanthin), sulcatoxanthin, antheroxanthin, glycymerin, and leprotene. Other fractions with carotenoid-like spectra have been listed by various authors.

The most conspicuous characteristic of sedimentary carotenoids is the prevalence of h/e ratios in the vicinity of unity, or even lower (Fox, 1944; Muraveisky and Chertok, 1938; Fogg and Belcher, 1960). These data clearly indicate a preferential destruction of xanthophylls, both in the water column and in wet sediments. This may be partly attributed to the greater susceptibility to attack by molecular oxygen that is shown by the xanthophylls. Fogg and Belcher (1960) reported fluctuations in the h/e ratio at different levels in a sediment core taken from Esthwaite Water, England. They calculated (on the assumption that there had been no destruction of SCDP in the sediments of this lake) that the time required for the destruction of 50% of the lutein initially present was of the order of 16,000 years. This estimate would have to be lowered if decomposition of SCDP had occurred during the time interval studied (roughly 12,000 years).

The concentrations of carotenoids in the organic matter of freshwater and marine sediments appear to be in about the same range as those typical of green leaves and algae. This indicates that carotenoids are broken down at approximately the same rate as the total organic matter. The rate of xanthophyll destruction is somewhat faster, and that of carotenes somewhat slower.

A number of freshwater and marine sediment cores have been analyzed for carotenoids as a function of depth (Fox and Anderson, 1941; Fox *et al.*, 1944; Züllig, 1956; Vallentyne, 1956; Fogg and Belcher, 1960). The data so obtained suggest that carotenoids are relatively stable in wet sediments. The oldest carotenoid-containing sediments yet discovered are estimated to be about 100,000 years old (Andersen and Gundersen, 1955). In the sediments of one bog lake Vallentyne (1956) found much higher concentrations of the α- and β-carotene fraction in 6,000–8,000 year old sediments than in sediments deposited either later or earlier. There appeared to be a correspondence between the amounts of carotenes per unit weight of organic matter and the estimates of former productivity in the lake basin as determined by more conventional methods.

One notable fact about sedimentary carotenoids is the apparent lack of cis-isomers, a point first drawn attention to by Vallentyne (1956). As a test case Vallentyne (1957) examined a sample of 20,000 year old sediment from Searles Lake, California, for cis-isomers of β-carotene. Trace amounts of neo-β-carotene-B and neo-β-carotene-U were detected in carefully treated extracts; but the ratio of cis-isomers to trans isomers was found to be much lower than the ratios in equilibrium solutions in organic solvents. It is evident that something in sediments either hinders steric transformation or else shifts the equilibrium position in favor of the all-trans form.

One of the nicest cases showing the utility of a biochemical approach to paleobiology has recently been provided by Züllig (1960). Züllig examined the near-surface sediments of several Swiss lakes for concentrations of myxoxanthophyll as a function of depth. Each of these lakes has had a history of increasing pollution during the present Century. This has led to the development of high populations of the blue-green alga, Oscillatoria rubescens. Züllig was interested in knowing if the concentrations of myxoxanthophyll in the sediments of these lakes could be used to determine past changes in the abundance of blue-green algae. (So far as is known, myxoxanthophyll is a carotenoid peculiar to the blue-green algae.) The results are depicted in Fig. 2. They clearly show the validity of the approach. One might think that a simpler method would be to examine the sediments for microscopic remains of blue-green algae, but such an approach is futile. The cells are not preserved therein.

It has been pointed out by Lochte and Littman (1955) that one of the known constituents of petroleum, 2,2,6-trimethylcyclohexane-1-carboxylic acid, might be a remnant of a carotenoid-type molecule, perhaps derived from a sedimentary compound such as β-carotene by hydrogenation of the double bond in one of the ring systems and oxidation of the carbon atom linking the ring to the rest of the molecule. The case is well put. A search for 2,2,6-trimethylcyclohexane-1-carboxylic acid in sediments and sedimentary rocks would seem to be warranted.

In concluding this section on the biogeochemistry of carotenoids, one point should be strongly made: some synthesis of bacterial and fungal carotenoids undoubtedly occurs in near-surface sediments. At deeper sedimentary levels where biological activity becomes negligible, there is probably little microbial synthesis. The point is that it cannot be taken for granted that all of the sedimentary carotenoid molecules have been synthesized in the overlying water column. Carotenoids thus differ in this respect from sedimentary dihydroporphyrins, where the assumption of no postdepositional synthesis is more likely to be true.

We may briefly summarize the carotenoid data as follows. At least 20 different carotenoids occur in wet sediments up to 100,000 years old. Six of these have been rigorously identified and shown to have counterparts in living cells. The concentrations of carotenoids in sedimentary organic matter are of the same order of magnitude as those in living algae, but

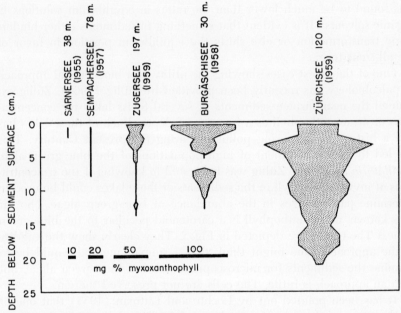

FIG. 2. Concentrations of myxoxanthophyll (mg. % dry weight) in the surface sediments of some Swiss lakes, redrawn from Züllig (1960). The years refer to the time of sampling.

there is a selective destruction of xanthophylls during the decomposition of phytoplankton as evidenced by lower h/e ratios in sediments than in phytoplankton. Carotenoids have not been reported to occur in sedimentary rocks and petroleum, but one petroleum acid may have originated by the decomposition of carotenoids possessing ionone ring systems.

A Concluding Remark

One question remains that has not yet been considered. Why is it that hydroporphyrins and carotenoids are preserved in sediments whereas many other pigments of living organisms are evidently not? A final answer to this question must await further research, for the other pigments have not been adequately looked for in sediments. Fox (1944) and others have pointed out that three factors tend to favor the preservation of hydroporphyrins and carotenoids in wet sediments: (*a*) the lack of oxygen; (*b*) the lack of light; and (*c*) the low temperatures. To these a fourth factor must be added: the water-insolubility of the compounds in their natural states. This is a rule that has apparently been completely overlooked in studies of sedimentary organic matter. An accumulation of water-soluble compounds in sediments cannot be expected except in the case of *in situ* decomposition processes in buried sediments. The bulk of sedimentary organic matter is (almost by definition) water-insoluble in its natural state. Soluble compounds do not reach the sediments.

There may be a second reason as to why "stability" (or, more appropriately, persistence) is associated with water-insolubility. We do not know why proteins, hemicelluloses, and other seemingly nutritious compounds are not completely decomposed in surface sediments by the microorganisms that are so abundant there. There is a suggestion that resistance to bacterial attack may be partly associated with water-insolubility. The mechanism is not as yet thoroughly understood; but it may be comparable to the difficulty of lipid digestion in the vertebrate gut without the aid of dispersing agents. One of the most notable cases in point is given by Whittaker and Vallentyne (1957). These workers showed that mono- and disaccharides exist in relatively high concentrations in the surface sediments of some lakes. When the state of the sugars in the sediment was examined, it was found that they did not occur in the interstitial water as was first expected. They were bound to some (unknown) water-insoluble constituents of the sediments by a relatively loose bond that could be broken with cold 70% ethanol. The existence of this bond evidently acted not only to decrease solution processes, but also to hinder bacterial attack. A similar state of affairs may exist for dihydroporphyrins, carotenoids, and other organic compounds in sediments; however, further work will be required to prove its occurrence.

Acknowledgment

I am indebted to H. N. Dunning for allowing me to read the manuscript of his comprehensive review of natural organic pigments (Dunning, 1960) while the present paper was in preparation.

References

Abelson, P. H. (1954). *Carnegie Inst. Wash. Year Book* **53**, 97-101.
Abelson, P. H. (1957). *Geol. Soc. Am. Mem.* No. **67**, 87-92.
Andersen, S. T., and Gundersen, K. (1955). *Experientia* **11**, 345-348.
Baker, B. L., and Hodgson, G. W. (1959). *Bull. Am. Assoc. Petrol. Geologists* **43**, 472-476.

Banse, K. (1957). *Kiel. Meeresforsch.* **13**, 186-201.
Barghoorn, E. S. (1957). *Geol. Soc. Am. Mem.* No. **67**, 75-86.
Baudisch, O., and Von Euler, H. (1934). *Arkiv Kemi Mineral. Geol.* **11A** (No. 21), 1-10.
Beatty, R. A. (1941). *J. Exptl. Biol.* **18**, 144-152.
Blumer, M. (1950). *Helv. Chim. Acta* **33**, 1627-1637.
Blumer, M. (1956). *Anal. Chem.* **28**, 1640-1644.
Breger, I. A., ed. (1960). "Organic Geochemistry." Pergamon, New York.
Corcoran, E. F. (1957). Ph.D. Thesis, University of California, Los Angeles.
Creitz, G. I., and Richards, F. A. (1955). *J. Marine Research (Sears Foundation)* **14**, 211-216.
Dhéré, C., and Hradil, G. (1934). *Schweiz. mineral. petrog. Mitt.* **14**, 279-295.
Dunning, H. N. (1960). *In* "Organic Geochemistry" (I. A. Breger, ed.) Pergamon, New York. (In press.)
Dunning, H. N., Moore, J. W., and Denekas, M. O. (1953). *Ind. Eng. Chem.* **45**, 1759-1765.
Erdman, J. G., Ramsey, V. G., Kalenda, N. W., and Hansen, W. E. (1956). *J. Am. Chem. Soc.* **78**, 5844-5847.
Fogg, G. E., and Belcher, J. H. (1960). *New Phytologist* (in press).
Fox, D. L. (1944). *Science* **100**, 111-113.
Fox, D. L., and Anderson, L. J. (1941). *Proc. Natl. Acad. Sci. U. S.* **23**, 295-301.
Fox, D. L., Updegraff, D. M., and Novelli, D. G. (1944). *Arch. Biochem.* **5**, 1-23.
Gillbricht, M. (1952). *Kiel. Meeresforsch.* **8**, 173-191.
Glebovskaya, E. A., and Volkenshtein, M. V. (1948). *J. Gen. Chem. U. S. S. R. (Eng. Transl.)* **18**, 1440-1451.
Gorham, E. (1960). *Limnol. Oceanog.* **5**, 29-33.
Gothan, W. (1922). *Braunkohle* **21**, 400-401.
Harvey, H. W. (1934). *J. Marine Biol. Assoc. United Kingdom* **19**, 761-773.
Heilbron, I. M., and Lythgoe, B. (1936). *J. Chem. Soc.*, pp. 1367-1380.
Hodgson, G. W., and Baker, B. L. (1957). *Bull. Am. Assoc. Petrol. Geologists* **41**, 2413-2426.
Hodgson, G. W., Hitchon, B., Elofson, R. M., Baker, B. L., and Peake, E. (1960). *Geochim. et Cosmochim. Acta* (in press).
Jones, J. D., and Vallentyne, J. R. (1960). (In preparation).
Karrer, P., and Koenig, H. (1940). *Helv. Chim. Acta* **23**, 460-463.
Kleerekoper, H. (1957). Thèse présentée à la Faculté des Sciences de l'Université de Paris, pp. 1-205.
Kon, S. K. (1954). *Bull. soc. chim. biol.* **36**, 209-225.
Kozminski, Z. (1938). *Trans. Wisconsin Acad. Sci.* **31**, 411-438.
Kylin, H. (1927). *Z. physiol. Chem.* **166**, 39-77.
Lederer, E. (1938). *Bull. soc. chim. biol.* **20**, 611-634.
Lochte, H. L., and Littman, E. R. (1955). "The Petroleum Acids and Bases." Chemical Publ. Co., New York.
Meinschein, W. G. (1959). *Bull. Am. Assoc. Petrol. Geologists* **43**, 925-943.
Moore, J. W., and Dunning, H. N. (1955). *Ind. Eng. Chem.* **47**, 1440-1444.
Muraveisky, S., and Chertok, I. (1938). *Compt. rend. acad. sci. U. R. S. S.* **19**, 521-523.
Orr, W. L., and Grady, J. R. (1957). *Deep-Sea Research* **4**, 263-271.
Orr, W. L., Emery, K. O., and Grady, J. R. (1958). *Bull. Am. Assoc. Petrol. Geologists* **42**, 925-962.

Phinney, H. K. (1946). *Am. Midland Naturalist* **35**, 453-459.
Pinckard, J. H., Kittredge, J. S., Fox, D. L., Haxo, F. T., and Zechmeister, L. (1953). *Arch. Biochem. Biophys.* **44**, 189-199.
Rabinowitch, E. I. (1945). "Photosynthesis and Related Processes," Vol. I. Interscience, New York.
Radchenko, O. A., and Sheshina, L. S. (1955). *Doklady Akad. Nauk S. S. S. R.* **105**, 1285-1288.
Radchenko, O. A., and Sheshina, L. S. (1956). *Doklady Akad. Nauk S. S. S. R.* **109**, 614-616.
Richards, F. A., and Thompson, T. G. (1952). *J. Marine Research (Sears Foundation)* **11**, 156-172.
Riley, G. A. (1944). *Am. Scientist* **32**, 129-134.
Savinov, B. G., Mikhailovnina, A. A., and Shapiro, S. A. (1950). *Compt. rend. acad. sci. U. R. S. S.* **72**, 1087-1089.
Schroeder, H. (1919). *Naturwissenschaften* **7**, 8-12 and 23-29.
Schultz, F., and Hamackova, J. (1924). *Bull. soc. chim. France* **35**, 183-187.
Skopintsev, B. A., and Bruk, E. S. (1940). *Mikrobiologiya* **9**, 595-606.
Smith, P. V. (1954). *Bull. Am. Assoc. Petrol. Geologists* **38**, 377-404.
Steemann-Nielsen, E., and Jensen, E. A. (1957). *Galathea Rept.* **1**, 49-136.
Tischer, J. (1938). *Z. physiol. chem.* **251**, 109-128.
Trask, P. D., and Wu, C. C. (1930). *Bull. Am. Assoc. Petrol. Geologists* **14**, 1451-1463.
Treibs, A. (1934a). *Ann. Chem. Liebigs* **509**, 103-114.
Treibs, A. (1934b). *Ann. Chem. Liebigs* **510**, 42-62.
Treibs, A. (1935a). *Ann. Chem. Liebigs* **517**, 172-196.
Treibs, A. (1935b). *Ann. Chem. Liebigs* **520**, 144-150.
Treibs, A. (1936). *Angew. Chem.* **49**, 682-686.
Vallentyne, J. R. (1955). *Can. J. Botany* **33**, 304-313.
Vallentyne, J. R. (1956). *Limnol. Oceanog.* **1**, 252-262.
Vallentyne, J. R. (1957). *Arch. Biochem. Biophys.* **70**, 29-34.
Vallentyne, J. R. (1960). In "Organic Geochemistry" (I. A. Breger, ed.). Pergamon, New York. (In press).
Vallentyne, J. R., and Craston, D. F. (1957). *Can. J. Botany* **35**, 35-42.
Vallentyne, J. R., and Swabey, Y. S. (1955). *Am. J. Sci.* **253**, 313-340.
Whittaker, J. R., and Vallentyne, J. R. (1957). *Limnol. Oceanog.* **2**, 98-110.
Wickman, F. E. (1956). *Geochim. et Cosmochim. Acta* **9**, 136-153.
Wisbar, G. (1923). *Braunkohle* **22**, 42.
Züllig, H. (1956). *Schweiz. Z. Hydrol.* **18**, 5-143.
Züllig, H. (1960). *Proc. Intern. Assoc. Limnol.* **14** (in press).

Discussion

ARNON: There is present-day evidence that vanadium actually may play a part in photosynthesis. Under certain conditions there is actually a very marked effect of vanadium on photosynthesis in certain organisms, so it seems to me that one could make a case, at least a speculative one, that one must suppose a type of vanadium porphyrin in explaining these things that occurred hundreds of millions of years ago.

VALLENTYNE: May I ask if this vanadium is a vanadium porphyrin?

ARNON: Well, I do not know about the vanadium porphyrin. It has not been looked for in cases where vanadium has had an effect, but there is some evidence that vanadium affects photosynthesis.

BOGORAD: An interesting thing is, why do you not get cupric complexes? Because, I think, at least with the pure porphyrin series, the cupric ion goes in very fast and forms a stable compound. This would support Dr. Arnon's suggestion that you might possibly have had some physiologically functioning vanadium-porphyrin complexes because if the cupric ion got in, it might not be displaced so readily by vanadium.

FRENCH: Is there much copper in these oil deposits?

VALLENTYNE: I was just asking myself that question. Certainly the amounts are not equivalent to nickel, iron, and vanadium, but what they are I cannot exactly say. Would you care to comment on that, Dr. Fox?

FOX: The only thing that comes to my mind is that this is a highly reducing environment. It is possible the copper might be tied up with sulfide or something of that sort, and would be out of the picture, as far as reactivity is concerned. I think the evidence is very much on your side, however.

ARNON: Well, in terms of chemical synthesis, it is not any more difficult, say, to make a magnesium porphyrin than it is to make a vanadium porphyrin. From an evolutionary point of view, at least, it would offer no greater difficulty to visualize one than it does the other.

FOX: No, I think the only thing is that you do not have the signposts for it. The vanadium chromogens that we do know are not, I believe, vanadium porphyrins, but vanadium tetrapyrroles, and I do not think we know any copper porphyrins in nature. Hemocyanin is certainly not a porphyrin.

NEILANDS: There is a curicin from the feathers of a South African bird.

ARNON: I wonder if I may ask a question about this evolutionary point of view. Which do you place first in the scale of evolutionary development of porphyrins, the hematins or the magnesium porphyrins?

FOX: Are you asking me to make a choice?

VALLENTYNE: I do not think there is any evidence one way or the other in this sort of work.

FOX: Does the evidence of the fossils not now bear upon the question of which became the most prolific? There had to be a tremendous lot of living forms before there began to be any appreciable amount of fossils. By that time the magnesium porphyrins greatly outweighed the hemoproteins, as to this day.

VALLENTYNE: I would like to turn the tables on you and ask you a question. I think you said a minute ago that chemically it is just as easy to stick a vanadium in there as a magnesium. Is that really so?

ARNON: Well, I was sticking my head out. I see no reason why vanadium porphyrin would be impossible.

SCHER: Since vanadium is known, at least in open chain tetrapyrrole structures, we might also assume that somewhere in the microbial world there is a vanadium counterpart, in which vanadium, in a structure similar to what you might propose, might play a biological role. We have intended to look at this sort of thing with enrichment culture techniques, but so far this has not been successful. But that does not mean it does not exist, and we intend to look further.

VERNON: I think you will find that the transition elements like copper and vanadium will complex more strongly than magnesium would, and actually they are more stable than a magnesium complex, so there is no difficulty there as far as the vanadium going in. At least the copper complex is more stable than the magnesium complex of pheophytin.

PRINGSHEIM: Do any of the chemicals furnish any clue to the organism from which petroleum originated?

VALLENTYNE: Not really. Only the ratio of the porphyrins in petroleum that had the little carbocyclic ring on the third pyrrole group to the forms that are more like the protoporphyrin types of structure that do not have that, has been used to suggest that plant materials are much more important in the origin of petroleum than are animal materials, but of course you do get the protoporphyrin type of structure in both plants and animals.

Fox: You did mention in passing, did you not, that you found a recombination with iron of some of the pheophytin types?

VALLENTYNE: Yes. Recently, Moore and Dunning and I have suggestive evidence based on the abundance of porphyrins in a material from the Green River shale, which is oil shale, and the abundance of iron, nickel, and vanadium. What we have shown there is, in the first place, that the porphyrin is metal-complexed and, secondly, that there is insufficient nickel and vanadium to complex all of the porphyrin that occurs in their extract. So that presumably iron could be part of the metal complex.

Fox: If iron occurred in the natural deposit, this would make the lack of copper complexes a little more puzzling.

8

Pigments in Phototaxis

SELINA BENDIX[1]

*Laboratory of Comparative Biology, Kaiser Foundation Research Institute,
Richmond, California*

Introduction

Phototaxis is a subject which is becoming increasingly interesting to modern photobiologists who have come to realize that the physiological importance of light to photautotrophs cannot be evaluated solely by a study of photosynthesis and photoperiodism. As we shall see later, there is increasing evidence for the fundamental relationships between phototaxis and photosynthesis and the basic energy economy of the plant cell.

The first observations on phototaxis were made by Treviranus in 1817. He observed negative phototaxis (motion away from a stimulating light source) in *Draparnaldia glomerata* and *Ulothrix subtilis* on exposure to direct sunlight. Nearly fifty years later, Cohn (1865) attempted to find out what color light was most effective in producing phototaxis in swarm spores. Using colored glass filters, he found blue light to be phototactically most effective. Strasburger (1878), using a sodium light with filters and, later, sunlight dispersed by a large prism, came to the same conclusion as Cohn. Engelmann (1881, 1882a) was the first to recognize the need to take incident energy into account in determining an action spectrum for phototaxis. In studying the accumulation of *Bacterium photometricum* in a prism dispersed spectrum he obtained a band of accumulation between 800 and 900 mμ, fainter bands at 590 mμ, and between 520 and 550 mμ. He found the absorption spectrum of a thin layer of bacteria to be similar to this action spectrum and, because of its correlation with photosynthesis, decided that phototaxis was due to a chemotaxis toward oxygen (Engelmann, 1881, 1888). He also found that *Paramecium bursaria*, which contains *Chlorella*, under anaerobic conditions aggregates in a spectrum at the spots of maximum absorption of chlorophylls *a* and *b* (Engelmann, 1882b). He was also the first experi-

---

[1] Portions of this paper are from a dissertation presented in partial fulfillment of the requirements for the degree of Ph.D. in Zoology at the University of California, Berkeley, California.

mentally to demonstrate that the eyespot in *Euglena* was the phototactic photoreceptor (Engelmann, 1882b).

Action Spectra

FLAGELLATES

The first balanced energy action spectra for phototaxis were obtained by Mast. He was one of the many who have worked on *Euglena*. *Gonium* was found to be similar to *Euglena*, with a maximum at 485 mµ; *Pandorina* and *Spondylomorum* had broader based curves with maxima at 534 mµ (Mast, 1917). Mast thought that the differences between genera correlated with differences in the carotenoid pigments present. Laurens and Hooker (1918, 1920) determined the action spectrum of phototaxis in *Volvox* and found it to be similar to that of *Euglena* as determined by Mast.

Luntz (1931a, b, 1932) studied phototaxis in *Eudorina elegans*, *Volvox minor* and *Chlamydomonas*. He found a maximum at 492 mµ, no reaction in the red, and a weak reaction at 366 mµ. Colorless *Chilomonas* showed a maximum at 366 mµ, and no maxima in the visible. The thresholds for movement in *Eudorina* and *Volvox* were lower than the thresholds for orientation. The same holds true for *Micrasterias* (Bendix, 1957).

Durston (1925) found positive phototaxis in the brine flagellate *Dunaliella salina* to be brought about by blue or yellow, but not by red light. Blum and Fox (1933) used filters to determine the action spectrum for this response in red and green forms of *D. salina*. Both forms gave approximately the same response, with a maximum at 500 mµ. Slight differences in shape of the curves for the two forms were attributed to the screening effect of inactive pigments present. The phototactic response was retained in the virtual absence of molecular oxygen. Haxo and Clendenning (1953) found that *Ulva lactuca* gametes showed a maximum response at about 440 mµ and no reaction above 550 mµ.

Halldal (1958) determined phototactic action spectra for five species of Volvocales, gametes of two *Ulva* species and three members of the Dinophyceae. He concluded that two, or possibly three, pigments were involved. The members of the Volvocales and the *Ulva* gametes gave the same action spectra with a maximum at 493 mµ and a small shoulder at 435 mµ; *Gonyaulax catenella* and *Peridinium trochoideum* were similar but not identical, with a maximum at 475 mµ and a small shoulder at 435 mµ. *Prorocentrum micans* exhibited a maximum at 570 mµ. The action spectra for positive and negative phototaxis were the same in the Volvocales. Absorption spectra of living suspensions of the Volvocales and Dinophyceae did not indicate the presence of any pigment corresponding to the phototactic action spectra. The same major fat-soluble pigments were present in both *Peridinium* and *Prorocentrum*. None of the active pigments were successfully isolated.

As Table I indicates, *Euglena* has been a favorite experimental sub-

TABLE I
Action Spectra Maxima for Phototaxis in *Euglena*

Author	Year	Action Spectra
Positive Phototaxis		
Strasburger	1878	ca. 415–435
Engelmann	1882b	ca. 470–490
Loeb and Maxwell	1910	460–510
Loeb and Wasteneys	1916	450–506
Mast	1917	485
Brachet	1937	420–460
Bünning and Schneiderhöhn	1956	490–500
Gössel	1957	410 (achlorophyllous)
Wolken and Shin	1958	465, 630 photokinesis[a]
		420, 490 phototaxis[a]
Negative Phototaxis		
Bünning and Schneiderhöhn	1956	415
Gössel	1957	410 var. *bacillaris*
		410 var. *bacillaris*
		PBZG4

[a] In Wolken and Shin's terminology photokinesis equals light induced random swimming and phototaxis equals oriented motion toward or away from a light source.

ject for phototactic action spectra. Although the results of various workers are similar, they are not identical. Polarized light appears to be more efficient in inducing phototaxis than nonpolarized light, but produces some changes in the action spectrum (Wolken and Shin, 1958).

Diatoms

Verworn (1889) found short wavelengths to be most effective in eliciting phototaxis in *Navicula brevis*. Heidingsfeld (1943) confirmed this for *N. radiosa*. Nultsch (1956), in a detailed study with interference filters, found a maximum at about 550 mµ for several diatoms, including genera other than *Navicula*.

Red and Brown Algae

Phototaxis has been observed in the primitive red alga *Porphyridium cruentum* (Dangeard, 1930; Vischer, 1935; Geitler, 1944; Pringsheim and

Pringsheim, 1949) and in the spores of brown algae (Thuret, 1850; Kylin, 1918; Papenfuss, 1935), however, no action spectra are yet available for any of these forms.

Blue-Green Algae

Pieper (1913, 1915) attempted to determine the action spectrum of phototaxis in *Oscillatoria formosa* with filters. Since he made no energy measurements, his results are difficult to interpret. Red and yellow elicited a stronger response than white light. Green and white light gave a positive phototactic response at low intensities, but none at high, suggesting inhibition of the response by high intensities. On the other hand, he obtained no response in the blue, even at "high" intensity. In view of the low emission in the blue of incandescent sources, this suggests the possibility of insufficient energy to elicit a response. Dangeard (1928) found various species of *Oscillatoria* to collect in the orange, red, and infrared; again, no energy measurements. The response was constant for a given species, but varied from species to species. He interpreted this as due to interspecific pigment differences.

Drews (1957) used interference filters to observe phototaxis in several blue-green algae. He studied "topotaxis," or orientation in a light beam and movement toward or away from a source, and "phobotaxis," or the light spot trapping phenomenon [see Bendix (1960) for a description of this phenomenon]. His results are summarized in Table II. He suggests that the differences in behavior of Oscillatoriae and Nostocaceae are correlated with the difference in their mode of movement. The Nostocaceae do not rotate as they move (Ullrich, 1926).

TABLE II
Phototactic Responses in Blue-Green Algae[a]

Organism	Topotaxis[b]	Phobotaxis[b]
Oscillatoriae		
Oscillatoria mougeotii	400–730 mμ	515–770 mμ
Phormidium uncinatum	400–610 mμ	610–770 mμ
Nostocaceae		
Anabaena variabilis	orange to light red	—[c]
Cylindrospermum lichiniforme		—[c]

[a] After Drews (1957).

[b] See text for Drews' definitions of these terms.

[c] Drews could not obtain phobotaxis in these forms under any conditions which he tested.

Purple Bacteria

Buder (1919) resolved Engelmann's blue-green maximum into peaks at 530, 490, and 470 mµ (due to carotenoids), and peaks at 900, 850, 800, and 590 mµ as well as a blue peak which were ascribed to bacterio-chlorophyll. Manten (1948), studying *Rhodospirillum rubrum*, found that bacteriochlorophyll was responsible for most of the response, with a secondary contribution by carotenoids, and no participation of the dominant carotenoid, spirilloxanthin. Clayton (1953a, b, c) obtained an action spectrum which was like Manten's above 570 mµ, but different below this point, suggesting participation of spirilloxanthin, although scarcer carotenoids were still more active in proportion to their concentration. This apparent discrepancy has been resolved by Goodwin and Sissens (1955), who found that production of spirilloxanthin by *R. rubrum* cultures varies with the age of the culture, increasing with time.

The Role of the Stigma

Buder (1917) studied phototaxis in the colorless genera *Polytoma* and *Polytomella*. Only species with eyespots were found to be phototactic. Hartshorne (1953) studied phototaxis in an eyespotless mutant of *Chlamydomonas reinhardi* produced by ultraviolet irradiation. He found that lack of the eyespot was not associated with complete insensitivity to light, although the eyeless mutant reacted with much less uniformity and precision than the wild type. Perception of light is therefore not restricted to the eyespot. It seems that the relative light sensitivity of eyespot and cytoplasm varies from organism to organism.

Gössel (1957) has used interference filters to study a number of achlorophyllous *Euglena gracilis* strains. *E. gracilis* var. *bacillaris,* with both photoreceptor and stigma (see Fig. 1 for an explanation of these terms), showed negative phototaxis only, with a maximum at 410 mµ. Variety *E. bacillaris PBZ-G4,* with a photoreceptor but no stigma, showed both positive and negative phototaxis with primary maxima for both responses at 410 mµ, but with different secondary maxima. *Astasia longa,* with neither photoreceptor nor stigma, was not phototactic under any condition tested.

It is currently believed (Gössel, 1957; Bünning and Tazawa, 1957; Wolken and Shin, 1958) that the photoreceptor only is involved in negative phototaxis and that positive phototaxis is due to periodic darkening of the photoreceptor by the stigma, or, in the absence of the stigma, by cytoplasmic proteins. Gössel did not attempt to explain why the colorless *Euglena* strain with both stigma and photoreceptor should react negatively only.

The Effects of Environmental Factors

The effect of temperature on phototaxis has been studied by Strasburger (1878), Mast (1911), Bracher (1919), and Mainx (1929). Strasburger and Mast found that *Euglena* changed from positive to negative phototaxis at low temperatures (around 5°C) under uniform illumination.

Mast (1918) studied the effect of chemicals on phototaxis in the colonial form *Spondylomorum quaternarium*. Chloroform, ether, chloral hydrate, and acids made negative specimens respond positively. Positive phototaxis was unaffected by all organic and inorganic acids tested.

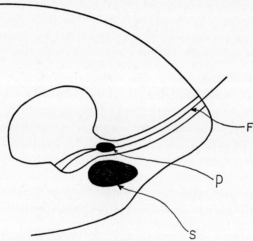

Fig. 1. Photoreceptive and related structures in *Euglena*. KEY: S = stigma, P = photoreceptor, F = flagellum.

Formalin, sugar, H_2O_2, $MgSO_4$, $CaCl_2$, KNO_3 were found to have no effects on phototaxis. Burkholder (1933) found normal phototactic movement in *Oscillatoria* in the pH range 6.4–9.5. Movement was hindered outside this range.

Mainx (1929) studied the effects of many agents on phototaxis, for example: pH, oxygen tension, temperature, anaesthetics, alcohols, osmotic pressure, previous light exposure, etc. He found that conditions which lower the threshold of positive phototaxis raise the energy level at which the transition from positive to negative takes place.

Halldal (1956, 1957) studied phototaxis in *Platymonas subcordiformis* and other motile green algae and dinoflagellates in which irregular positive and negative phototactic behavior at a given light intensity proved to be partially correlated with the concentration of calcium and magnesium ions in the medium. Movement was relatively independent of oxygen and carbon dioxide tension. In experiments with different ratios of calcium chloride to magnesium chloride in the medium, he found positive phototaxis when the Ca/Mg ratio was less than 1:6; random motion at 1:6; and negative phototaxis at Ca/Mg ratios greater than 1:6. It is possible that Klebs' (1896) observations that phototaxis in the microspores of *Ulothrix zonata* was strongly influenced by the composition of the medium were related to a similar phenomenon.

Phototaxis in Chloroplasts

Phototaxis in higher plants is generally observed as a movement of chloroplasts. The chloroplasts of certain algae also exhibit phototactic movements. Bohm (1856), who studied chloroplast movements in well over 100 species, was the first to notice that the position of chloroplasts in the cell was light dependent.

Among algae, chloroplast phototaxis has mainly been studied in *Mougeotia* (a saccoderm desmid, syn. *Mesocarpus*). Haupt (1958) has recently performed some interesting experiments on this alga. Using interference filters, he found maximum effectiveness in the 600–700 mµ region, with a secondary peak between 430 and 480 mµ. Three to six times as much energy was required for response at 430 mµ as at 679 mµ. The 480 to 580 mµ region had very little effect. If cells with the chloroplast parallel to the direction of the incident light are irradiated for 1 minute at 679 mµ, then 45 minutes later in 60–80% of the cells the chloroplast will have rotated to a position perpendicular to the incident beam. This effect can be suppressed by irradiating with 733 mµ for 3 minutes immediately after the 1 minute at 679 mµ, which cuts the reaction to 20–30% of the cells. Irradiation with 733 mµ given before 679 mµ has no effect. The sequence 679, 733, and 679 mµ gives the same results as 679 mµ alone; 733 mµ will not cause movement in the opposite direction.

Mosebach (1958) studied conditions affecting the change-over from positive to negative phototaxis in *Mougeotia*. The intensity at which this change occurs depends on the pretreatment of the experimental material; the intensity is lowered by predarkening and raised by a few hours of preillumination. The CO_2 concentration in the medium surrounding the cells affects these responses, but is not the primary cause of them. This was demonstrated by performing experiments with running water so that photosynthesis would not affect the CO_2 concentration in the medium.

Chloroplast Action Spectra

Famintzin (1867) studied phototactic chloroplast movements in the moss *Mnium*, where he found yellow light to give darkness orientation and blue light to cause daylight orientation. Frank (1872) and Stahl (1880) thought that red was necessary to produce darkness orientation. Senn (1908) could induce parastrophe only with blue light. (See Table III for definition of terms relating to chloroplast arrangement in cells.)

TABLE III
TERMS RELATING TO CHLOROPLAST POSITION IN THE CELL

Name of chloroplast condition	Position in the cell	Orientation of chloroplasts with respect to incident beam	Corresponding light condition
APOSTROPHE	Against internal cell walls	Parallel	Darkness
EPISTROPHE	Against walls perpendicular to incident light beam	Perpendicular	Dispersed light
PARASTROPHE	Against walls perpendicular to leaf surface or parallel to incident light beam	Parallel	Direct sun or very strong artificial light

Linsbauer and Abramowicz (1909) concluded that the apostrophe-epistrophe reaction was phototactic, but that the epistrophe-apostrophe change was associated with changes in the concentration of CO_2. They did not have any actual evidence for the CO_2 effect. Blue light was found to induce the apostrophe-epistrophe change most efficiently in *Funaria hygrometrica* (Voerkel, 1934). Carotenoids were implicated. Epistrophe could not be induced by red light, even at high intensities. Voerkel did not succeed in experimental induction of parastrophe and concluded that long wavelength ultraviolet was responsible for this response.

Phototactic chloroplast movements in several species have been studied by Zurzycka and Zurzycki. Work on *Lemna trisulca* (Zurzycka, 1951) gave results which agreed with those of Voerkel on the relative effects of blue and red on the epistrophe-apostrophe reaction, but Zurzycka was also able experimentally to achieve the epistrophe-parastrophe transition. The action spectrum for the latter reaction had two maxima, one above 600 mµ, the other below 500 mµ. She concluded that the apostrophe-epistrophe transition was due to carotenes or xanthophylls and that the epistrophe-parastrophe negative phototactic response was correlated with the absorption of light by chlorophyll.

The Infrared Effect

Discrepancies in the results obtained with red light by various experimenters have been resolved by Babushkin (1955b) who showed that the effect is produced not by red but by infrared, which was not adequately filtered out in some experiments. Zurzycka obtained a "red" effect using a 3% $CuSO_4$ solution as an infrared filter; Babushkin found that 6% $CuSO_4$ transmitted sufficient infrared to produce the effect, but that 6% $CuSO_4$ plus infrared opaque glass eliminated all chloroplast movement in red light. This effect was confirmed by production of chloroplast movements with an incandescent source and an RG-8 filter, which transmits only infrared.

Babushkin (1955b) obtained an action spectrum for chloroplast phototaxis by taking advantage of the fact that when chloroplasts move to the surface walls of leaf cells the leaf looks less transparent. He developed a contact print method in which a 20 × 40 mm. piece of leaf upon which a spectrum had been projected was placed on photographic paper and illuminated. The paper was developed so as to accentuate the contrast. Light intensity was regulated so that where chloroplasts were in epistrophe no light was transmitted and the paper was white and where the chloroplasts remained in apostrophe the paper was dark.

The action spectrum he obtained had 5 maxima: 420, 430, 450, 462, and 477 mμ (all ± 2–5 mμ). The broadest maximum was at 450 mμ. The action spectrum is corrected for energy differences across the spectrum. This action spectrum indicates participation of chlorophylls and several carotenoids. The main maxima are due to violaxanthin (420 mμ) and chlorophyll a (430 mμ). The entire chloroplast pigment complex seems to participate in the response, however.

Effect of Light Intensity

Since 1952 Babushkin (1955a, b, c), primarily working with tobacco, has engaged in an intensive study of the effect of light intensity on phototaxis in chloroplasts. Phototactic behavior of chloroplasts was characterized by determining the light intensity at which full epistrophe was attained (I_1) and the intensity at which the epistrophe-parastrophe transition began to take place (I_2). The method was sensitive to ± 5 ergs/cm.² second for I_1 and ± 20 ergs/cm² second for I_2. Tobacco was employed because the top row of palisade cells is easily visible without injury to the leaf. A temperature increase in the 10–25°C range was found to cause an increase in I_1 and a decrease in I_2. In natural populations, the higher the light intensity under which an individual plant grew, the greater the interval between I_1 and I_2. The more light loving a species was, the greater the interval between I_1 and I_2.

Chloroplast Phototaxis and Metabolism

The effect of phototactic chloroplast movements on photosynthesis has been investigated by Zurzycki (1955) and Babushkin (1955c). Zurzycki used a micro-technique based on a capillary tube respirometer which could be used for measurements of gaseous metabolism of leaves several millimeters square or algal filaments several millimeters long, with an accuracy of 10^{-3} µl. Chloroplast movements were studied in *Funaria, Lemna, Mougeotia,* and *Spirogyra,* the last as control material in which the chloroplasts could not move. Changes in chloroplast arrangement and photosynthesis were correlated at low intensities only. At low intensities, increase of assimilation area (apostrophe-epistrophe and parastrophe-epistrophe reactions) is related to change in photosynthetic rate. As long as the photosynthetic rate increases proportionally to the light intensity, the chloroplasts are in epistrophe. Full parastrophe is reached at the point of photosynthetic light saturation (Zurzycki, 1955).

A strong reduction or complete blocking of photosynthesis by hydroxylamine hydrochloride or sodium azide in *Lemna trisulca* inhibited or stopped entirely those phototactic movements which started with the profile arrangement (parastrope-epistrophe, apostrophe-epistrophe) and had no effect on phototactic movements beginning from the horizontal chloroplast position (epistrophe-parastrophe, epistrophe-apostrophe). A 70% reduction in the respiratory rate by berberine sulphate brought no change in the phototactic movements (Zurzycki and Zurzycka, 1955).

Babushkin (1955c) studied photosynthetic behavior manometrically and with $C^{14}O_2$. He used tobacco leaves grown under 2000 lux of fluorescent light so that the compensation point would be as low as possible. He found that the lower I_1 the higher the photosynthetic rate. When photosynthetic rate was plotted against light intensity, marked dips occurred which coincided with the apostrophe-epistrophe and epistrophe-parastrophe transitions. $C^{14}O_2$ studies indicated that the photosynthetic rate decrease actually takes place first, since it shows up on 3 minute exposures which are insufficient to cause chloroplast movement. These dips suggest the possibility of the diversion of energy from photosynthesis to the movement of chloroplasts.

As a result of his work, Babushkin has come to the following conclusions. Photosynthesis and phototaxis cease in chloroplasts with the destruction of chlorophyll. The action spectrum of phototaxis at short wavelengths indicates the participation of both chlorophyll and carotenoids. The magnitude of I_1 correlates with the shape of the photosynthetic efficiency curve. Transitions in chloroplast position in the cell correlate with minima in photosynthetic rate curves. Chloroplasts have a single pigment system operating in phototaxis and the light phase of photosynthesis. The action spectrum of phototaxis can only be explained by the assumption that the products of photoreaction are different for red and blue light. As will be seen later, Bendix (1957) also has evidence for a difference in the action of red and blue light in a completely different organism.

Phototaxis in Micrasterias

Bendix (1957) studied phototaxis in the unicellular desmid *Micrasterias rotata* var. *evoluta*, which, like other desmids, lacks organized photoreceptors. *Micrasterias* cells are highly pigmented (100 cells suffice for spectrophotometric determination of pigments) and easily observed because of their large size (275 μ in diameter) and low velocity (3 mm./hour maximum). Phototactic responses were studied in prism dispersed collimated beams or with a spectrometer plus filters to eliminate stray light. Energy measurements were made with a thermopile mounted on a horizontal moving arm controlled by a rack and pinion that permits locating the thermopile slit to an accuracy of ± 0.01 mm. over a range of 50 mm.[2]

Micrasterias cells show negligible random motion in darkness or in the light. Movement, when it occurs, is oriented by a light source. Cells placed in a beam dispersed by a prism (5 cm. visible spectrum) show two separate phototactic responses: one in the plane of the beam (PN response), and one perpendicular to the beam (L response) (Fig. 2). The L response is unidirectional, causing cells to move out of the long wavelength end of the visible spectrum, but no further than 570 mμ. As intensity increases, the PN response goes to a positive (toward the source) maximum and back to zero to a null response region, then goes to a negative (away from the source) maximum and again returns to zero.

It is interesting to speculate on the nature of selective pressures in evolutionary favor of survival of an organism possessing two different phototactic responses. Moon (1940) has found that the blue-to-red ratio

---

[2] This apparatus was designed and built by Prof. J. E. Gullberg, Zoology Dept., Univ. of Calif. The author wishes to acknowledge with gratitude Prof. Gullberg's help with optical matters involved in these experiments.

in natural sunlight may vary from 1.3 to 0.4 according to weather conditions. Assuming that this organism contains a physiologically important substance which absorbs in the far red, we may postulate the following theory: the PN response is a typical phototactic response which moves the organism into a light intensity region optimal for photosynthesis in this organism. When atmospheric conditions are such that the percentage of red is relatively high in sunlight, there is potentially damaging destruction to the essential far red absorbing substance. At the expense of optimal photosynthetic efficiency, the L response operates to move the or-

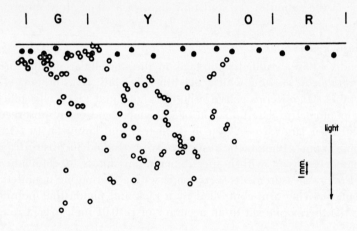

● initial cell positions
○ positions at 144 hrs. (after a number of generations)

FIG. 2. Phototactic movement of *Micrasterias* cells in a spectrum showing negative phototaxis in the plane of the stimulating source and avoidance of the red end of the spectrum.

ganism to a lower light intensity, thus keeping too much of the far red absorbing substance from being destroyed.

THE L RESPONSE

Figure 3 shows the positions and behavior of a group of cells 125 hours after being placed at the red end of a spectrum. The L response is observed in a range of intensities from an irradiance too low to support division (290 μwatts/cm.^{2}) to an irradiance high enough to irreversibly inhibit division (2400 μwatts/cm.^{2}). This response was found to be independent of the direction of the absolute energy gradient across the region in which it is observed and is not correlated with the red chlorophyll absorption peak.

Infrared produces the response, but is not most effective. The water cell had to be removed in order to obtain the response with the Wratten 88A and 87 filters (88A gives 1% transmission at 725 mµ, 87 gives 1% transmission at 745 mµ). Preliminary results indicate that the maximum of the action spectrum is between 695 and 725 mµ. There is no indication of a reversible reaction of the type studied by Haupt (1958); cells have never been observed to reverse direction of movement. No experiments have yet been done to see if the response can be selectively inhibited by certain wavelengths. In view of the widespread occurrence

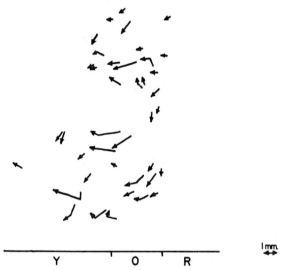

Fig. 3. Lateral phototactic movement out of the red end of the spectrum. Each arrow represents the movement of a single cell during 7.5 hours. Movements less than one cell diameter not recorded.

of the red—far red reaction, it would not be surprising to find that the L response is in some fashion connected with this reaction.[3]

The PN Response

The threshold of the positive PN response in white light is a little below 2 foot-candles (see Fig. 4), in monochromatic light it goes as low as 5×10^{-4} µwatts/cm.^{2} at 690 mµ (approximately 1.5×10^5 quanta per second per cell). In white light this response saturates between 2 and 6 foot-candles; in monochromatic light it saturates at as low an irradiance as 5×10^{-3} µwatts/cm.^{2} (690 mµ). Fig. 5 shows the response

---

[3] Cf. Paper No. 19 by Hendricks, in this volume.

curve at 440 mμ where the threshold is over ten times that at 690 mμ. A rough action spectrum for the threshold of the PN response is given in Fig. 6. There are not enough points below 600 mμ to give any definite information about this region.

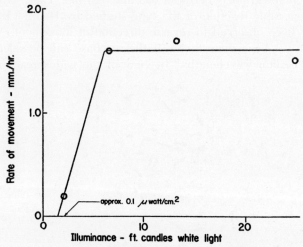

Fig. 4. Phototactic response in white light.

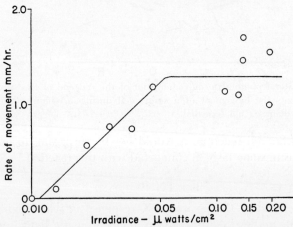

Fig. 5. Relationship between phototactic response and log irradiance at 440 mμ.

The action spectrum for the positive PN response in Fig. 7 is given as E.U.F. (energy utilization factor) values, where E.U.F. is the rate of phototaxis in mm./hour divided by the irradiance producing the response in μwatts/cm.^{2} E.U.F. is not efficiency, since it is not dimensionless. If cell mass and light-exposed surface are assumed to be constant, then E.U.F. = a constant × efficiency. More points are needed before positive conclusions can be drawn; however, present data do indicate that one or more chlorophylls are involved, and that light absorbed at the blue peak is relatively inefficient. Maxima are at approximately 690, 645, 440, and 410 mμ. The 690 mμ maximum may be due to a long wavelength absorbing form of chlorophyll a.[4] Absorption spectra of *Micrasterias* pigments indicate that the *in vivo* chlorophyll a maximum is close to 685 mμ.

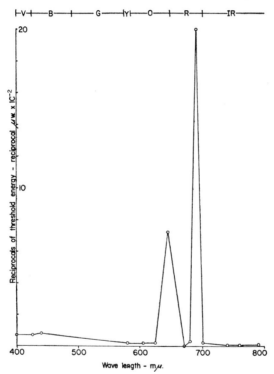

FIG. 6. Action spectrum for phototactic threshold energies.

Since equipment sensitive enough to detect light energy at the lowest intensities to which *Micrasterias* reacts was not available, two independent methods of calculating the low energy values were developed. These methods depend upon the fact that the monochromator source was found to closely approximate the radiant emittance of a black body in the wavelength region of interest. The first method employs an em-

---

[4] Cf. Paper No. 3 by Allen *et al.*, in this volume.

pirically derived relationship between the energy at 700 mµ and the voltage across the source which is extrapolated to other wavelengths. The second method was suggested by the work of Weaver and Hussong (1939) and used the formulas and tables of Planck's Function. Under the experimental conditions employed, the color temperature of the

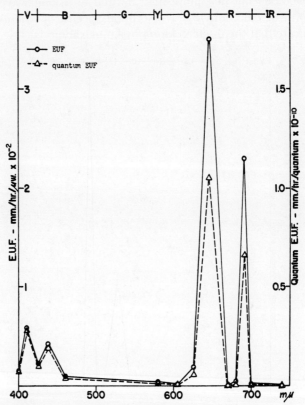

Fig. 7. Action spectrum for positive phototactic response. (See text for explanation of E.U.F. values.)

emergent monochromator beam proved to bear a linear relationship with the per cent of the rated voltage at which the source was operated. In view of the necessary assumptions and extrapolations involved in the two methods, agreement in calculated values is surprisingly good. Generous estimates of possible errors do not produce any changes in the shapes of the action spectra.

NEGATIVE PHOTOTAXIS AND THE NULL RESPONSE

It was not possible to obtain an action spectrum for the negative PN response, since monochromatic beams of sufficient intensity were not available. A complete set of values for the energy levels of the various phototactic responses from threshold of the positive PN response to loss of the negative PN response is not available for any wavelength. The best available set of data, for orange light (color visually determined), is given in Table IV. The ability to respond negatively is lost after a few

TABLE IV
PHOTOTACTIC BEHAVIOR OF *Micrasterias* AT VARIOUS ENERGY LEVELS OF ORANGE LIGHT

Threshold of positive PN response	Below 1.0 μwatts/cm.²
Threshold of null response region	5 × 10³ μwatts/cm.²
Threshold of negative PN response	6 × 10³ μwatts/cm.²
End of negative PN response	9.5 × 10³ μwatts/cm.²

hours, but the ability to respond to a positive stimulus is not lost by cells which have exhausted their negative PN response.

If we presume that the positive PN response drops back to zero due to inhibition, then the reciprocals of the low energy null response boundary values are a measure of the inhibitory effectiveness of the various

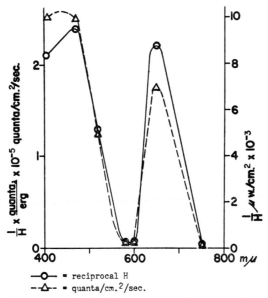

FIG. 8. Action spectrum for the low energy null response boundary. (See text for explanation.)

spectral regions. A rough action spectrum of this sort is presented in Fig. 8, and suggests the participation of chlorophyll, as there are two peaks of the same order of magnitude in the blue and red. Such results might be explained by phototactic inhibition by a photosynthetic by-product or by the presence of a product of the photodecomposition of chlorophyll.

Orientation

During positive PN movement 92% of 500 cells were observed to orient with their transverse axes within 25° or less of perpendicular to the stimulating beam (see Fig. 9). During negative PN motion 90% of the cells orient. At the threshold of positive PN movement, phototactic

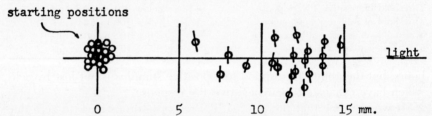

Fig. 9. Orientation of *Micrasterias* cells in monochromator beam. 500 mµ. 12.5 hour exposure. Lines through circles represent transverse axes of cells which orient perpendicular to direction of stimulating beam.

movement is observed at a slightly lower irradiance than orientation. The first cells to respond, however, are those which are accidentally well oriented.

Cells which are exposed to irradiances high enough to produce peak negative PN response orient while responding, but subsequently have their orientation mechanism destroyed. When given a low intensity stimulus, these cells now give a positive PN response without orientation. This suggests that for some reason the cells could not simply reverse direction, but had to turn around. If this is the case, the organism should have some evidence of anterior-posterior differentiation or of differential phototactic sensitivity of the two semicells. In *Micrasterias* there is no visible anterior-posterior differentiation and there is no difference in phototactic sensitivity yet detected between the new and old semicells, although such differential sensitivity does occur in the desmid *Penium curtum* (Braun, 1851).

What is the mechanism of the orientation in *Micrasterias*? The cells respond to the direction of the incident stimulus regardless of wavelength and intensity gradients perpendicular to the beam. How do the cells determine their angle with respect to the incident light? It would be interesting to investigate these phenomena further with polarized light.

REFERENCES

Babushkin, L. N. (1955a). *Doklady Akad. Nauk S.S.S.R.* **102**, 1215-1218; *Referat. Zhur. Biol.* **1956** (62609).
Babushkin, L. N. (1955b). *Doklady Akad. Nauk S.S.S.R.* **103**, 333-335; *Referat. Zhur. Biol.* **1956** (48639).
Babushkin, L. N. (1955c). *Doklady Akad. Nauk S.S.S.R.* **103**, 507-510; *Referat. Zhur. Biol.* **1956** (76249).
Bendix, S. W. (1957). Thesis, Univ. of California, Berkeley.
Bendix, S. W. (1960). *Botan. Rev.* **26**, 145-208.
Blum, H. F., and Fox, D. L. (1933). *Univ. Calif. (Berkeley) Pubs. Physiol.* **8**, 21-30.
Bohm, J. A. (1856). *Sitzber. Akad. Wiss. Wien. Math.-naturw. Kl.* **22**, 479-512.
Bracher, R. (1919). *Ann. Botan. (London)* **33**, 93-102.
Bracher, R. (1937). *J. Linnean Soc. London Botany* **51**, 23-42.
Braun, A. (1851). "Betrachtungen über die Erscheinung der Verjüngung in der Natur, insbesondere in der Lebens— und Bildungs—geschichte der Pflanze." W. Engelmann, Leipzig.
Buder, J. (1917). *Jahrb. wiss. Botan.* **58**, 105-220.
Buder, J. (1919). *Jahrb. wiss. Botan.* **58**, 525-628.
Bünning, E., and Schneiderhöhn, G. (1956). *Arch. Mikrobiol.* **24**, 80-90.
Bünning, E., and Tazawa, M. (1957). *Arch. Mikrobiol.* **27**, 306-310.
Burkholder, P. R. (1933). *J. Gen. Physiol.* **16**, 875-881.
Clayton, R. K. (1953a). *Arch. Mikrobiol.* **19**, 107-124.
Clayton, R. K. (1953b). *Arch. Mikrobiol.* **19**, 125-140.
Clayton, R. K. (1953c). *Arch. Mikrobiol.* **19**, 141-165.
Cohn, F. (1865). *Jahresber. Schles. Ges Vaterl. Kultur* **42**, 35-36.
Dangeard, P. A. (1928). *Ann. Protist. (Paris)* **1**, 3-10.
Dangeard, P. A. (1930). *Compt. rend. acad. sci.* **190**, 819-821.
Drews, G. (1957). *Ber. deut. botan. Ges.* **70**, 259-262.
Durston, A. D. (1925). Thesis, Stanford Univ., Stanford, California.
Engelmann, T. W. (1881). *Pflüger's Arch. ges. Physiol.* **26**, 537-545.
Engelmann, T. W. (1882a). *Botan. Z.* **40**, 419-426.
Engelmann, T. W. (1882b). *Pflüger's Arch. ges. Physiol.* **29**, 387-400.
Engelmann, T. W. (1888). *Pflüger's Arch. ges. Physiol.* **42**, 183-188.
Famintzin, A. (1867). *Jahrb. wiss. Botan.* **6**, 45-49.
Frank, G. (1872). *Jahrb. wiss. Botan.* **8**, 216-303.
Geitler, L. (1944). *Flora (Jena)* **137**, 300-333.
Goodwin, T. W., and Sissens, M. E. (1955). *Biochem. J.* **61**, xiii-xiv.
Gössel, I. (1957). *Arch. Mikrobiol.* **27**, 288-305.
Halldal, P. (1956). *Carnegie Inst. Wash. Year Book* **55**, 259-261.
Halldal, P. (1957). *Nature* **179**, 215-216.
Halldal, P. (1958). *Physiol. Plantarum* **11**, 118-153.
Hartshorne, J. N. (1953). *New Phytologist* **52**, 292-297.
Haupt, W. (1958). *Naturwissenschaften* **45**, 273-274.
Haxo, F. T., and Clendenning, K. A. (1953). *Biol. Bull.* **105**, 103-114.
Heidingsfeld, I. (1943). Doctoral dissertation, Breslau.

Klebs, G. (1896). "Die Bedingungen der Fortpflanzung bei einigen Algen und Pilzen." G. Fischer, Jena.
Kylin, H. (1918). *Svensk Botan. Tidskr.* **12**, 1-64.
Laurens, H., and Hooker, H. D. (1918). *Anat. Record* **14**, 97-98.
Laurens, H., and Hooker, H. D. (1920). *J. Exptl. Zool.* **30**, 345-368.
Linsbauer, K., and Abramowicz, E. (1909). *Sitzber. Akad. Wiss. Wien. math.-naturw. Kl. Abt. I* **118**, 418-421.
Loeb, J., and Maxwell, S. S. (1910). *Univ. Calif. Publs. (Berkeley) Physiol.* **3**, 195-197.
Loeb, J., and Wasteneys, H. (1916). *J. Exptl. Zool.* **20**, 217-236.
Luntz, A. (1931a). *Z. vergleich. Physiol.* **14**, 68-92.
Luntz, A. (1931b). *Z. vergleich. Physiol.* **15**, 652-678.
Luntz, A. (1932). *Z. vergleich. Physiol.* **16**, 204-217.
Mainx, F. (1929). *Arch. Protistenk.* **68**, 105-176.
Manten, A. (1948). Doctoral dissertation, Utrecht.
Mast, S. O. (1911). "Light and the Behavior of Organisms." Wiley, New York.
Mast, S. O. (1917). *J. Exptl. Zool.* **22**, 471-528.
Mast, S. O. (1918). *J. Exptl. Zool.* **26**, 503-520.
Moon, P. (1940). *J. Franklin Inst.* **230**, 583-617.
Mosebach, G. (1958). *Planta* **52**, 3-46.
Nultsch, W. (1956). *Arch. Protistenk.* **101**, 1-68.
Papenfuss, G. F. (1935). *Botan. Gaz.* **96**, 428, 431.
Pieper, A. (1913). *Ber. deut. botan. Ges.* **31**, 594-599.
Pieper, A. (1915). Doctoral dissertation, Berlin.
Pringsheim, E. G., and Pringsheim, O. (1949). *J. Ecol.* **37**, 57-64.
Senn, G. (1908). "Die Gestalts- und Lageveränderung der Pflanzen Chromatophoren." W. Engelmann, Leipzig.
Stahl, E. (1880). *Botan. Z.* **38**, 297-304.
Strasburger, E. (1878). *Jena. Z. Naturw.* **12**, 551-625.
Thuret, G. (1850). *Ann. sci. nat. Botan. et biol. végétale* [3], **14**, 246-247.
Treviranus, L. C. (1817). "Vermischte Schriften von G. K. und L. C. Treviranus," Vol. II. J. F. Röwer, Bremen.
Ullrich, H. (1926). *Planta* **2**, 295-324.
Verworn, M. (1889). "Psychophysiologische Protistenstudien." G. Fischer, Jena.
Vischer, W. (1935). *Verhandl. naturforsch. Ges. Basel* **46**; quoted in Geitler (1944).
Voerkel, H. (1934). *Planta* **21**, 156-205.
Weaver, K. S., and Hussong, H. E. (1939). *J. Opt. Soc. Am.* **29**, 16-19.
Wolken, J. J., and Shin, E. (1958). *J. Protozool.* **5**, 39-46.
Zurzycka, A. (1951). *Acta Soc. Botan. Polon.* **21**, 17-38.
Zurzycki, J. (1955). *Acta Soc. Botan. Polon.* **24**, 27-63.
Zurzycki, J., and Zurzycka, A. (1955). *Acta Soc. Botan. Polon.* **24**, 663-674.

DISCUSSION

MILLOTT: Have you met with the work of Gaston Villot, Strasbourg? He is a professor of psychology who has been studying the reversal of a tactic response. He has worked, as far as I can remember, on two animals—*Daphnia*, the cladoceran crustacean with compound eye and with a skin sense, and with rotifers. Now in the case of *Daphnia*, he says, the animal acts to light by two agencies. It orientates precisely by its compound eye but it flies away from or is attracted towards light by virtue of what he calls the "dermatoptic sense." The dermatoptic sense in the animal's flying away from light is most active in the blue. He then goes to ascribe this to what he calls the "reaction photopathologique." He thinks, you see, that the animal has developed a kind of response to light whereby light creates a destructive thing. The destructive effect is minimized if the animal goes further, then the balance tips the other way, and the animal comes back into the light. Now *Daphnia* is a complex organism with a central nervous system, almost certainly of fair complexity and extent, which he neglects. But at the same time it did occur to me that this approach might prove helpful, because there is no reason why, if there is anything in his approach, this destructive effect of light at the blue end should not affect your alga as well.

9

Is Pigmentation a Clue to Protistan Phylogeny?

ELLSWORTH C. DOUGHERTY AND MARY BELLE ALLEN[1]

Laboratory of Comparative Biology, Kaiser Foundation Research Institute, Richmond, California

Introduction

Can the photoreactive pigments of photosynthetic protists (i.e., of certain bacteria and most algae) be used as phylogenetic markers? In other words, do the chlorophylls, carotenoids, and biliproteins illuminate the evolutionary history of the Protista?[2] We believe that the answer to this question is a qualified "yes." Various of the pigments participating in photosynthesis do, on present evidence, characterize certain groups of organisms that, in the light of over-all knowledge of their morphology and biochemistry, appear taxonomically natural. But two important reservations must be made: (*a*) detailed knowledge is limited to a relatively few members of most taxonomic assemblages (even lacking for the chloromonads) and interpretations are hazardous without data from many forms; (*b*) there are certain groups (to be discussed here) in which pigment distribution is seemingly anomalous and in which one cannot yet select the most likely among possible phylogenetic interpretations.

When we speak of a "phylogenetic marker," we mean a trait, structural or functional (including biochemical), that indicates an evolutionary relationship within or between taxa of organisms—in other words, that helps reconstruct family trees.

That a given feature is a valid phylogenetic marker often cannot be decided with confidence. Knowledge of many groups—and, perhaps, particularly of the Protista—is still so fragmentary that discussion of interrelationships must inevitably involve a generous degree of speculation. In our own thinking we have applied the following principle: better to have a working hypothesis, even if based on fragile evidence, than to shrug aside a question of phylogeny as prematurely posed. Moreover,

---

[1] Supported in part by Grant G-5584 from the National Science Foundation.
[2] We take the Protista to include the bacteria, algae, protozoa, and fungi (Dougherty and Allen, 1958).

puzzling over phylogenetic problems stimulates inquiry into scientific unknowns that might not otherwise be obvious or attractive to us.

Speculation on protistan evolution by earlier thinkers, such as the brilliant Ernst Haeckel, could be bold, because the authors were ignorant of so much. As knowledge has burgeoned and scientists have confined themselves more and more narrowly within subdivisions of biology, the older concepts have become discredited or at best controversial, and few present-day biologists have undertaken serious speculation on protistan origins and affinities. For example, it is striking with respect to the protists that, in comprehensive works such as Fritsch's (1935, 1945) and Smith's (1955) volumes on algae and in the uneven but remarkable work by Copeland (1956), who has attempted a taxonomic synthesis of most of the Protista, overt evolutionary speculation is, with few exceptions, highly cautious. By contrast most of the writers on "protozoa" in Grassé's "Traité de Zoologie" (1952a, 1953) have been more imaginative, especially Grassé himself (1952b) in his introductory essay. But these French biologists, even including Grassé, have written with the outlook of protozoologists rather than of protistologists (*sensu lato*), and this orientation has limited their approach.

As biochemically oriented biologists concerned with protistan evolution, we are particularly interested in chemical features giving evolutionary insights. Ideally one might hope to find a decreasing hierarchy of such attributes, which would, on suitable interpretation, characterize assemblages of organisms from very large to small and would thus aid in the reconstruction of protistan phylogeny. Let us temporarily exclude the photoreactive pigments from consideration, since they are *sub judice* here. Recent work on certain other chemical differences appears to have provided phylogenetic markers at the highest taxonomic levels. Thus the general occurrence, in significant amounts, of α,ε-diaminopimelic acid in the monerans[3] (bacteria and blue-green algae) (Work and Dewey, 1953) and its virtual absence in higher organisms [where it has been found, it is in minute amounts only, as in *Chlorella ellipsiodea* (Fujiwara and Akabori, 1954)] seem an important chemical distinction between organisms with primitive, nonvesicular "nuclei" (*prokarya* of Dougherty, 1957a) and all higher organisms, which possess true nuclei (*eukarya*), set off as compartments within cellular compartments. Conversely, the monerans apparently lack sterols (Heilbron, 1942), although there are a number of claims that conflict, for specific organisms, with this generalization; by contrast, these compounds may well be universally present

---

[3] See Stanier and van Niel (1941) and the *Discussion* of this paper for the modern application of this term.

9. PIGMENTATION AND PROTISTAN PHYLOGENY

in higher organisms. [Heilbron (1942) claims that aerial green algae of the genus *Trentepohlia* lack sterols—the only reported exception among non-monerans.]

Students of protistan metatropology can well yearn for additional phylogenetic markers, whether morphological or biochemical. Although, with advancing knowledge of morphology, major natural groupings of the Protista now seem recognizable (except for some of the animal protists—or protozoa *sensu stricto*—and for certain "fungi"), characters for estimating relationship *between* certain major natural groupings (the phyla, for example) are more difficult to decide upon and assess. The fossil record gives us little help; at least some metaprotistan groups have seemingly existed in modern form since the pre-Cambrian (Tyler and Barghoorn, 1954; Barghoorn, 1957). So, if one seeks historical evidence on major protistan affinities, one must look in even more ancient rocks, and that appears quite unpromising. Bigelow (1958), a student of insects, recently has laid down the dictum that, without a fossil record, there is no sound basis for phylogenetic reconstruction. To accept this point of view would, for protistologists, seem tantamount to foredooming any attempt at understanding the phylogeny of most lower organisms—a pessimistic attitude with which we have scant patience. A hundred years from now, we suspect, scientists will have found ingenious substitutes for the mere lack of a fossil record.

More profound studies of protistan morphology, especially of fine structure, may well establish the significance of morphological features not now recognized as fundamental or not even at present known and thus aid in construction of an improved protistan classification and more satisfactory protistan phylogeny. But, if we may judge from the examples of α,ε-diaminopimelic acid and the sterols, the comparative biochemistry of protists may also be expected to yield further important phylogenetic clues.

We are not suggesting that biochemical evidence will necessarily prove more fundamental than morphological. In fact, one may hope that ultimately this distinction will disappear. Let us suppose that, for each species, the exact structure, including steric configuration, of the entire hereditary material, which by general agreement is DNA, can one day be determined. This would be an ultimate triumph of micromorphology and at the same time, indistinguishably, of biochemistry.

Understanding the phylogeny of the protists will inevitably require the gradual, painstaking fitting together of evidence from all sources. One can logically anticipate a progressive reciprocation, with biochemical similarities stimulating morphological studies, and *vice versa*.

So much then for the general principles of our approach. In the balance of this paper we take up in succession the chlorophylls, carotenoids, and biliproteins; for each class of pigments we consider the possible interpretations of their distribution in phylogenetic terms, with a minimum of correlation with other characters. Then follows a general discussion fitting information on all three groups of pigments to what seems to us a reasonable contemporary picture of protistan phylogeny.

Recently Strain (1958) has comprehensively summarized the chloroplast pigments of higher plants and algae. Goodwin (1959) has reviewed the carotenoid pigments of all photosynthetic organisms. These two workers are in disagreement with respect to the occurrence of a number of carotenoids. A tabulation of bacterial and algal pigments according to taxonomic groups is given in Table I; this combines the data of Strain and Goodwin.

Chlorophylls

At the present time there are four major chlorophylls known—that is, occurring singly, in certain organisms, unaccompanied by other chlorophylls. Three—bacteriochlorophyll and two kinds of chlorobiochlorophyll[4] (see Paper No. 5, by R. Y. Stanier, in this symposium)—are restricted to bacteria (the purple bacteria and the green sulfur bacteria respectively); in these, molecular oxygen is not produced in photosynthesis. The third, chlorophyll *a*, has been found in all organisms in which the complete photosynthetic process is linked with production of molecular oxygen.

Distribution of the chlorophylls is discussed elsewhere in this volume.[*] The present state of our knowledge is summarized in Table I.

If one hypothesizes that each major chlorophyll had a single phylogenetic origin, one can view photosynthetic organisms in two major groupings: the non-oxygen-evolving bacteria (with chlorobiochlorophyll and bacteriochlorophyll) on the one hand, and the oxygen-producing blue-green algae and all higher plants (with chlorophyll *a*) on the other.

Judged, for the moment, solely by major chlorophylls, the bacteria and blue-green algae can be interpreted as having diverged from their remote, common, presumably photosynthetic, moneran ancestor before the evolution of photosynthesis with oxygen as a product. On this reasoning the blue-green line later gave rise to higher organisms (presumably a single initial lineage) after the evolution of the oxygen-pro-

---

[4] A name suggested by Dougherty (1957c) for the "*Chlorobium*-chlorophyll" of the green sulfur bacteria.

[*] See Paper No. 3, by Allen *et al.*, in this volume.

ducing mechanism invariably associated with chlorophyll a in all known modern organisms.

If one considers the secondary chlorophylls each as monophyletic in origin, then the presence of chlorophyll b suggests a common origin of euglenoids and green algae, i.e., the so-called "green" algal line (from which it is generally agreed that the multicellular land plants originated), while the presence of chlorophyll c suggests a common origin for the cryptomonads, dinoflagellates, chrysomonads, diatoms, and brown algae, i.e., the so-called "brown" algal line. This leaves the heterokonts at, or near, the base of the metaprotistan radiation, unless they have lost chlorophyll c. Chlorophyll d is peculiar to the higher red algae and speaks for their relative evolutionary isolation.

Carotenoids

Carotenoids are found in all photosynthetic organisms and fall into two classes—the hydrocarbon carotenoids, or *carotenes,* and their oxidized derivatives, the *xanthophylls*. Both carotenes and xanthophylls occur as *cyclic* compounds (with one or two β-ionone end groups, e.g., α-, β-, and γ-carotenes) and as *acyclic* compounds.

The distribution of carotenoids in the various protistan groups is discussed elsewhere in this volume† and is summarized in Table I. The conflicting reports of Strain on the one hand and of Goodwin on the other for the xanthophylls of certain groups (particularly the heterokonts and euglenoids) and for the occurrence versus nonoccurrence of zeaxanthin in several groups make it prudent to speculate conservatively on the phylogenetic significance of these oxidized carotenoids generally. Nevertheless, certain generalizations on the carotenes and xanthophylls appear justified.

As with chlorophylls, the photosynthetic bacteria appear to show more fundamental diversity in their carotenoid content than do all other photosynthetic organisms. The purple bacteria are unique in apparently having only acyclic carotenoids (two carotenes and at least three xanthophylls; see Goodwin, 1955; Goodwin and Land, 1956b), several of which are possibly group-specific. By contrast, recent evidence on three species of green sulfur bacteria (Goodwin and Land, 1956a) has revealed cyclic carotenoids—γ-carotene as their only carotene, and rubixanthin (3-hydroxy-γ-carotene) as their typical xanthophyll.

β-Carotene is almost as characteristic of the protists with oxygen-evolving photosynthesis as is chlorophyll a. This again speaks for evolutionary affinities of the moneran blue-green algae with the meso- and

---

† Cf. Paper No. 1, by Goodwin, in this volume.

TABLE I
Distribution of Photoreactive Pigments in Plastids of Protistan Groups[a,b,c]

Key: $+$ = present; $-$ = absent; $\pm$ = irregularly present or absent; $+(-)$ = generally present, absent in a few forms; $-(+)$ = generally absent, present in a few forms; $?+$ = presence doubtful, or insufficiently verified; $?$ = possibly present in traces.

Photoreactive pigments	Schizophyta		Archephyta (Cyanophyceae)	Rhodophyta (Rhodophyceae)	Cryptophyta
	Eubacteriae	Chlorobacteriae			
Chlorophylls					
Bacteriochlorophyll	$+$	$-$	$-$	$-$	$-$
Chlorobiochlorophyll(s)	$-$	$+$	$-$	$-$	$-$
Chlorophyll a	$-$	$-$	$+$	$+$	$+$
Chlorophyll b	$-$	$-$	$-$	$-$	$-$
Chlorophyll c	$-$	$-$	$-$	$-$	$+$
Chlorophyll d	$-$	$-$	$-$	$\pm$	$-$
Carotenoids					
Carotenes					
Lycopene	$+$	$-$	$-$	$-$	$-$
α-Carotene	$-$	$-$	$-$	$+/\pm$	$-$
β-Carotene	$-$	$-$	$+$	$+(-)^d/+$	$+$
γ-Carotene	$-$	$+$	$-$	$-$	$-$
ε-Carotene	$-$	$-$	$-$	$-$	$?+$
Flavacene	$-$	$-$	$+$	$-$	$-$
Xanthophylls					
Acyclic xanthophylls (named)	$+$	$-$	$-$	$-$	$-$
Rubixanthin	$-$	$+$	$-$	$-$	$-$
Echinenone	$-$	$-$	$+^e$	$-$	$-$
Myxoxanthophyll	$-$	$-$	$+$	$-$	$-$
Zeaxanthin	$-$	$-$	$+/$	$-/\pm$	$?+^g$
Lutein	$-$	$-$	$?/\pm$	$+/\pm$	$-$
Violaxanthin	$-$	$-$	$-$	$?+/-$	$-$
Neoxanthin	$-$	$-$	$-$	$-$	$-$
Fucoxanthin	$-$	$-$	$-$	$-$	$-$
Diatoxanthin	$-$	$-$	$-$	$-$	$-^g$
Diadinoxanthin	$-$	$-$	$-$	$-$	$-$
Flavoxanthin	$-$	$-$	$-$	$-$	$-$
Peridinin	$-$	$-$	$-$	$-$	$-$
Dinoxanthin	$-$	$-$	$-$	$-$	$-$
Siphonaxanthin	$-$	$-$	$-$	$-$	$-$
Siphonein	$-$	$-$	$-$	$-$	$-$
Unnamed or unidentified xanthophylls	$+$	$-$	$-/+^e$	$-$	$+$
Biliproteins[h]					
Phycocyanin(s)	$-$	$-$	$+(-)$	$+(-)$	$+$
Phycoerythrin(s)	$-$	$-$	$\pm$	$+(-)$	$\pm$

[a] Based largely on the data of Goodwin (1959; Paper No. 1, this volume) and Strain (1958); where there is disagreement between them, data of both workers are given, separated by a diagonal line—to the left for the former, to the right for the latter. [b] The minor pigments neofucoxanthin A and B, neodiadinoxanthin, neodinoxanthin, and neoperidinin are not listed here (see Smith, 1955; Strain, 1951, 1958), nor is oscillaxanthin (see Goodwin, 1959). [c] No information exists on the plastid pigments of the Chloromonadophyta (Chloromonadineae). [d] Only one species without β-carotene known—*Phycodrys sinuosa* (see Goodwin, 1959). [e] Two "myxoxanthin-" (= echinenone-)like pigments (Strain, 1958). [f] Not recorded by Goodwin (1959) through misprint (Goodwin, personal communication). [g] Probably zeaxanthin, but closely similar diatoxanthin not definitely ruled out (Haxo and Fork, 1959). [h] Several kinds of both types of biliproteins are known; certain of these appear to be group specific, but further work is needed to clarify the over-all distribution.

| | Phaeophyta | | | Pyrrhophyta (Dinophyceae) | Chlorophyta | | Euglenophyta (Euglenineae) |
Heterokontae [=Xanthophyceae]	Chrysophyceae	Diatomophyceae [=Bacillariophyceae]	Phaeophyceae		Chlorophyceae	Charophyceae[i]	
—	—	—	—	—	—	—	—
—	—	—	—	—	—	—	—
+	+	+	+	+	+	+	+
—	—	—	—	—	+(—)[j]	+	+
—	±	+	+	—	—	—	—
—	—	—	—	—	—	—	—
						+	
—	—	—	—	—	—	—	—
—	—	—/±	—/±	—	±[k]	—	—
+	+	+	+	+	+	+	+
—	—	—	—	—	—	+	—
—	—	+/—	—	—	—(+)[l]/—	—	—
—	—	—	—	—	—	—	—
—	—	—	—	—	—	—	—
—	—	—	—	—	—	—	—
—	—	—	—	—	—	—	—
—	—	—	—	—	—(+)[m]/+	—	?/—
+/—	+	?/—	?+/—	—	+	+	+
+/—	—	—	+	—	+	+	—
+/—	—	—	—	—	+	+	+/—
—	+	+	+	—	—	—	—
—	—	+	?/—	—	—	—	—
—	—	+	—	—	—	—	—
—	—	—	?/—	+	—/?	—	—
—	—	—	—	+	—	—	—
—	—	—	—	+	—	—	—
—	—	—	—	—	—(+)[n]	—	—
—	—	—	—	—	—(+)[n]	—	—
—/+[o]	—	—	—	—	—/±[p]	?	—/±[q]
—	—	—	—	—	—(+)	—	—
—	—	—	—	—	—(+)	—	—

[i] One species studied only—*Chara fragilis* (see Strain, 1958). [j] Lacking in a few organisms only. [k] The major carotene of most Siphonales. [l] In Siphonales only (as trace). [m] Major carotenoid in one species—the enigmatic *Cyanidium caldarium* (see Goodwin, Paper No. 1, this volume). [n] In the Siphonales only. [o] Four unique xanthophylls claimed. [p] One unique xanthophyll claimed to be sometimes present. [q] Two unique xanthophylls claimed.

metaprotists. The occurrence of α-carotene partially replacing β-carotene in certain red algae (Strain, 1958) and entirely in the cryptomonad, *Cryptomonas ovata* (Haxo and Fork, 1959), is compatible with a relatively close phylogenetic relationship between the red algae and cryptomonads.

At least some of the xanthophylls also lend themselves to evolutionary speculations. If, for protists other than bacteria, one hypothesizes a single phylogenetic origin for certain xanthophylls, they can be used in order to make interesting inferences. The clearest example is that of fucoxanthin, which is shared by chrysomonads, diatoms, and brown algae and supports a common evolutionary origin for these members of the "brown" line of metaprotists. The common possession of diadinoxanthin by diatoms and dinoflagellates is suggestive of an affinity between these two groups. And the fact that small amounts of zeaxanthin occur in blue-green algae and that this xanthophyll is probably the dominant one of the only cryptomonad so far studied is compatible with a relatively primitive status for the cryptomonads among the metaprotists. The anomalous organism *Cyanidium caldarium*, which looks morphologically to be a unicellular green alga (Allen, 1954, 1959), has been found to have zeaxanthin as its major xanthophyll (Goodwin, Paper No. 1, this volume); from this one might indeed suspect possible cryptomonad affinities.

Group-specific xanthophylls have been found in all blue-green algae (echinenone, myxoxanthophyll), dinoflagellates (peridinin, dinoxanthin), and the siphonalean green algae (siphonaxanthin, siphonein); these compounds accordingly appear to be useful phylogenetic markers for the groups in question.

With the foregoing examples there is no conflict between the data of Strain and of Goodwin. It does not seem profitable at this time to base speculation on pigments for which disagreement exists.

Biliproteins

The biliproteins, water-soluble pigments with open-chain tetrapyrroles as the chromophoric group, appear to characterize all pigmented members of three groups—the blue-green and red algae (Strain, 1951, 1958; Smith, 1955) and the cryptomonads (Allen *et al.*, 1959; Ó hEocha and Raftery, 1959; Haxo and Fork, 1959). There are, however, distinct differences in absorption spectra and physical properties between the biliproteins of these groups.‡ Biliproteins have also been found in a few exceptional green algae (Boresch, 1922; Allen, 1954, 1959). We have speculated elsewhere (Dougherty and Allen, 1959) that the common

---

‡ Cf. Paper No. 12, by Ó hEocha, in this volume.

possession of biliproteins may indicate evolutionary affinities, but at the same time alternatively pointed out that their possession may result from the independent origin of these compounds from a microconstituent of the cells, in the same manner as hemoglobin has repeatedly arisen in widely separate groups of organisms.

Discussion

So far we have treated each group of pigments largely in isolation. Now they should be correlated with one another and with other characters fundamental to our phylogenetic thinking.

In common with most students of the Protista, we subscribe to the view that all organisms known to science in our planet's present-day biosphere trace their ancestry back to a common life form or group of life forms in the remote past (cf. Dougherty, 1955). Whether or not life on Earth had a single origin as opposed to multiple origins is not at issue in the foregoing concept. Whatever the case may have been, it seems clear to us that primitive *monerans*, organisms of a relatively highly evolved type compared with the primordial life form or forms, must have served as the stock from which all *known* contemporary organisms have descended. (One should not close one's mind to the possibility that more primitive organisms remain, but have not been detected, either because of minute size or ecological restrictions, or both.)

The basic monophylogeny of the Protista and higher organisms is given its most comprehensive support by the facts of comparative biochemistry; there is an underlying similarity of fundamental metabolism despite the many more or less superficial differences to be found between groups.

We regard the Protista as falling into three major levels of structural organization, which, in our view, represent three major evolutionary steps (Dougherty and Allen, 1958, 1959) and which we represent in Fig. 1. The most primitive level is the *moneran* and refers to the bacteria, blue-green algae, and related organisms. The second is the *mesoprotistan* and comprehends the red algae. And the third is the *metaprotistan* and includes all higher organisms except the metaphytes and metazoa. Figure 1 gives several important features on the basis of which these three groups can be distinguished. [For terminology see notes of Dougherty (1957a,b).]

We conceive of each of these levels as fundamentally monophyletic: the two major types of monerans (bacteria and blue-green algae) coming from their common, primitive moneran ancestor; the mesoprotists arising as a single line from an early moneran (blue-green algal) stock; and the metaprotists as a single line from an early mesoprotistan (red algal) stock. Any other view leads to invoking parallel evolution of so many features as to strain one's credulity. All three groups possess constellations of characters that are unlikely to have developed more than once in evolution: the monerans with their simple nucleus (prokaryon) and

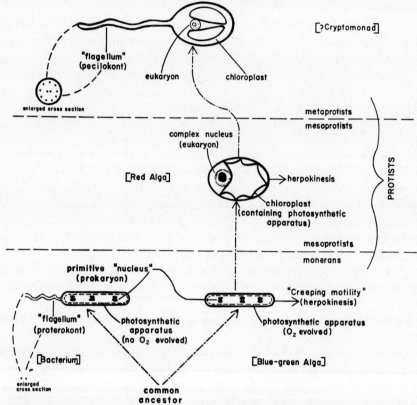

FIG. 1. Major morphological steps in protistan evolution (hypothetical). (For terminology see Dougherty, 1957a, b; Dougherty and Allen, 1958.)

primitive cytoplasmic organization; the mesoprotists (and metaprotists) with their more elaborate nucleus (eukaryon) and associated centrioles and their photosynthetic apparatus compartmentalized in the chloroplast; and the metaprotists with their complex flagellum (pecilokont) and associated system of blepharoplast-rhizostyle-stigma.

The photosynthetic pigments have an important contribution to make to the understanding of the foregoing picture.

Despite the differences among the bacterial photosynthetic systems and between them and that of the blue-green algae, it seems reasonable to believe that all chlorophyll-carotenoid systems go back to a common evolutionary origin in the ancestral monerans. The bacteria on many grounds appear collectively to be the most primitive surviving organisms. There are many examples of biochemical diversity in lower organisms in contrast to relative uniformity in their more highly evolved descendants. The diversity of the bacteria with respect to major chlorophylls and carotenoids in contrast to the relative uniformity of higher forms is reminiscent, for example, of the situation with the sterols of the Metazoa: many lower metazoa are known to contain a striking variety of sterols (Bergmann, 1949), but, in the more evolved forms, cholesterol is the only important representative of this class of compounds in metabolically active tissues.

It is interesting that purple bacteria (both sulfur and non-sulfur) are unique among photosynthetic organisms in lacking cyclic carotenoids, whereas those green bacteria so far studied, in having the cyclic compounds γ-carotene and rubixanthin, are closer to the blue-green algae and other photosynthetic plants, which have the cyclic compound β- and often α- (or rarely, only α-) carotene and two or more cyclic xanthophylls. The difference in pigmentation between green and purple bacteria has been regarded (Dougherty, 1957c) as sufficiently important to justify recognition of a separate bacterial class, Chlorobacteriae, for the green bacteria, as opposed to the class Eubacteriae, in which most other bacteria—including the purple bacteria but excluding the spirochetes (class Spirochaetae)—may be grouped.

Chlorophyll a and β- (or α-) carotene thus seem to cut across a natural relationship—that of bacteria to blue-green algae, groups resembling one another in fundamental respects that distinguish them from higher organisms. Yet one must find the origin of the meso- and metaprotists from more primitive organisms, and the coexistence of chlorophyll a and β-carotene strengthens the idea of an affinity of all higher, or eukaryous, organisms with the blue-green algae.

At this point it is appropriate to point out that the moneran pigments appear to illustrate the possibility of phylogenetic reconstructions based essentially on contemporary organisms. Reference is made earlier in this paper to the recent article by Bigelow (1958), to whose views on the unsoundness of reconstructing phylogeny without fossil evidence we take exception. But we heartily concur with certain of his general observations—the one of greatest interest to us at this point being that the rate of evolutionary change for certain characters has been fast in some phylogenetic lines and slow in others. Thus, from one moneran group, the blue-green algae, arose a line in which the basic moneran features were transformed into those of the mesoprotist; and that mesoprotistan line subsequently radiated to give the more conservative higher mesoprotists and the phylogenetically versatile metaprotists, from the latter of which, in turn, all higher organisms (metaphytes and metazoa) arose. Yet throughout this prodigious evolutionary diversification the photosynthetic apparatus, where preserved, has remained extraordinarily conservative with respect to the presence of chlorophyll a and, with rare exceptions, of β-carotene. This could well have been true for a period extending over more than half the time of life on Earth. By contrast, those lines that have remained moneran have been conservative in their basic structure and experimental in their pigment synthesis.

But why assume that the system of chlorophyll a plus β-carotene necessarily had a single evolutionary origin? In answering this we invoke the biliproteins. The common possession of this class of chromoproteins by the blue-green and red algae fits very well with the derivation of the red algal photosynthetic apparatus from that of the blue-green algae and hence of the mesoprotists from the blue-green algae. One must, however, assume either that the modern blue-green algae evolved echinenone and myxoxanthophyll *after* the red-algal line split off, or that the latter lost these pigments.

Up to this point, the known picture of chlorophyll-carotenoid distribution serves to clarify and strengthen the general picture of protistan evolution adopted here. By contrast the problem of metaprotistan origin from the mesoprotistan stock is somewhat more difficult. We have already noted that the pigmented cryptomonads resemble the blue-green and red algae in possessing biliproteins. It would be attractively simple if one could regard the cryptomonads as the most primitive metaprotistan group. This runs into at least two difficulties: (a) the cryptomonads are morphologically specialized unicellular metaprotists; and, (b) the few species of cryptomonads so far studied have recently been shown to have chlorophyll c (Haxo and Fork, 1959; Allen et al., Paper No. 3, this volume), and the only one of these studied in detail was shown to have α-carotene instead of β-carotene, and most probably zeaxanthin as the major xanthophyll.

Yet, with respect to morphology, it may be doubted that one can select another pigmented metaprotistan group as definitely more primitive structually than the cryptomonads. The chrysomonads are frequently considered primitive flagellates, yet their internal structure, so far as described, does not look to us convincingly more primitive than that of the cryptomonads.

The presence of chlorophyll c in the cryptomonads is perhaps a more difficult problem. One way out of the impasse is to postulate that this pigment was developed in the stem metaprotistan line and secondarily lost in some chrysomonads and in the heterokonts and replaced by chlorophyll b in the euglenoids and green algae. Any other interpretation would seem to call for the independent origin of chlorophyll c or the biliproteins at least twice.

The diatoms, brown algae, and dinoflagellates all possess chlorophyll c and in this are related to the cryptomonads and chrysomonads. They belong to the metaprotists commonly referred to as the "brown line," which seems a reasonably natural lineage. Xanthophyll patterns suggest a close affinity between the chrysomonads, diatoms, and brown algae and a more distant affinity between the diatoms and dinoflagellates. In this company the heterokonts are an enigmatic group.

The fact that the green algae have a xanthophyll pattern quite similar to that of the red algae suggests that the green algae may be relatively primitive and close to the red algae. It is of considerable interest therefore that a few biliprotein-containing green algae are known (Boresch, 1922; Allen, 1954, 1959), suggesting, moreover, that the cryptomonad pattern may indeed be transitional in certain respects between the red and green algae.

The origin of the metaphytes from the green algae is well supported by pigmentation (see Goodwin, Paper No. 1, this volume).

In this paper we give *what* we think to have happened to photoreactive pigments in the evolution of photosynthetic organisms, but do not attempt to provide any explanation as to *why* or *how*. The studies of the late Robert Emerson indicating that two pigments must be activated by light for efficient photosynthesis to occur indicate that the elaboration of accessory pigments would be highly adaptive (Emerson and Chalmers, 1958). An interpretation of pigment evolution in terms of the development of specific enzymes required to synthesize the chlorophylls, carotenoids (particularly the production of xanthophylls from carotenes), and biliproteins is an obvious goal, but somewhat beyond the limits of contemporary knowledge.

In Fig. 2 we summarize in a highly tentative fashion a phylogeny of the metaprotists that seems likely on the basis of the foregoing discussion. It must again be emphasized that detailed knowledge is limited to relatively few species for most of the protistan groups and that, for some, present knowledge is fragmentary (chrysomonads, cryptomonads) or even lacking (chloromonads). It seems inevitable that the major lacunae will be filled in the near future, for vigorously growing axenic cultures of a number of chrysomonads and cryptomonads are now available; also, chloromonads are known to occur as algal blooms, and a few species are available in culture, although these have grown rather poorly with techniques applied to date.

Additional species of groups already worked on in some detail are also available in axenic culture, and study of their pigments will widen the base of knowledge. We look to the near future for such work too.

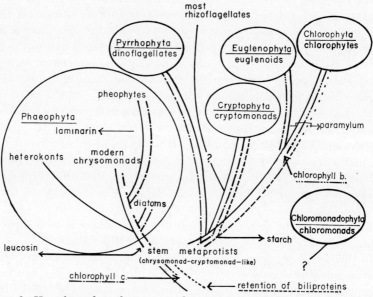

FIG. 2. Hypothetical evolutionary radiation of the algal metaprotists. (KEY: – – – = retention of biliproteins; – · – = possession of chlorophyll c; – · · · – = possession of chlorophyll b.)

In conclusion it seems fair to say that knowledge of the photoreactive pigments does appear to illuminate the phylogeny of the Protista in several important respects. Our approach in this paper has been to assume, for the chlorophylls (both primary and secondary), carotenoids, and biliproteins, single phylogenetic origins. On this basis we have presented a largely consistent evolutionary interpretation of their distribution in the Protista. It nevertheless cannot be excluded that some of these pigments may have arisen independently more than once; this would seem more likely for the secondary chlorophylls, the xanthophylls, and the biliproteins than for the primary chlorophylls and β-carotene. We should not be surprised if future work leads to modification, possibly abandonment, of some of the views expressed, and assumptions made, here. But, if this presentation serves to stimulate such studies, we shall count our primary aim fulfilled.

REFERENCES

Allen, M. B. (1954). *Rappt. Commun. 8ᵉ Congr. intern. botan. Paris. Sect.* 17, pp. 41-42.
Allen, M. B. (1959). *Arch. Mikrobiol.* **32**, 270-277.
Allen, M. B., Dougherty, E. C., and McLaughlin, J. J. A. (1959). *Nature* **184**, 1047-1049.
Barghoorn, E. S. (1957). *In* "Treatise on Marine Ecology and Paleoecology" (J. W. Hedgpeth, ed.), Vol. 2, Chapt. 4. Natl. Acad. Sci., Washington, D. C.
Bergmann, W. (1949). *J. Marine Research (Sears Foundation)* **8**, 137-176.
Bigelow, R. S. (1958). *Systemat. Zool.* **7**, 49-59.
Boresch, K. (1922). *Ber. deut. botan. Ges.* **40**, 288-292.
Copeland, H. F. (1956). "The Classification of Lower Organisms." Pacific Books, Palo Alto, California.
Dougherty, E. C. (1955). *Systemat. Zool.* **4**, 145-160, 190.
Dougherty, E. C. (1957a). *J. Protozool.* **4**, Suppl., 14.
Dougherty, E. C. (1957b). *J. Protozool.* **4**, Suppl., 14.
Dougherty, E. C. (1957c). *J. Protozool.* **4**, Suppl., 14.
Dougherty, E. C., and Allen, M. B. (1958). *Experientia* **14**, 78.
Dougherty, E. C., and Allen, M. B. (1959). *Proc. 15th Intern. Congr. Zool. London* pp. 184-186.
Emerson, R., and Chalmers, R. (1958). *Phycol. Soc. Am. News Bull.* **11**, 51-56.
Fritsch, F. E. (1935). "The Structure and Reproduction of the Algae," Vol. 1. Cambridge Univ. Press, London and New York.
Fritsch, F. E. (1945). "The Structure and Reproduction of the Algae," Vol. 2. Cambridge Univ. Press, London and New York.
Fujiwara, T., and Akabori, S. (1954). *J. Chem. Soc. Japan* **75**, 990-993.
Goodwin, T. W. (1955). *Ann. Rev. Biochem.* **24**, 497-522.
Goodwin, T. W. (1959). *Advances in Enzymol.* **21**, 295-368.
Goodwin, T. W., and Land, D. G. (1956a). *Biochem. J.* **62**, 553-556.
Goodwin, T. W., and Land, D. G. (1956b). *Arch. Mikrobiol.* **24**, 305-312.
Grassé, P.-P., ed. (1952a). "Traité de Zoologie: Anatomie, Systématique, Biologie," Vol. 1, Fasc. I (Phylogénie. Protozoaires: Généralités, Flagellés.) Masson, Paris.
Grassé, P.-P. (1952b). *In* "Traité de Zoologie: Anatomie, Systématique, Biologie," Vol. 1, Fasc. I, pp. 37-132. Masson, Paris.
Grassé, P.-P., ed. (1953). "Traité de Zoologie: Anatomie, Systématique, Biologie," Vol. 1, Fasc. II (Protozoaires: Rhizopodes, Actinopodes, Sporozoaires, Cnidosporidies.) Masson, Paris.
Haxo, F. T., and Fork, D. C. (1959). *Nature* **184**, 1051-1052.
Heilbron, I. M. (1942). *Nature* **149**, 398-400; *J. Chem. Soc.* **1942**, 79-89.
Ó hEocha, C., and Raftery, M. (1959). *Nature* **184**, 1049-1051.
Smith, G. M. (1955). "Cryptogamic Botany," Vol. 1 (Algae and Fungi.) McGraw-Hill, New York.
Stanier, R. Y., and van Niel, C. B. (1941). *J. Bacteriol.* **42**, 437-466.
Strain, H. H. (1951). *In* "Manual of Phycology" (G. M. Smith, ed.), pp. 243-262. Chronica Botanica, Waltham, Massachusetts.

Strain, H. H. (1958). "Chloroplast Pigments and Chromatographic Analysis. 32nd Annual Priestley Lectures." Penn. State Univ., University Park, Pennsylvania.
Tyler, S. A., and Barghoorn, E. S. (1954). *Science* **119**, 606-608.
Work, E., and Dewey, D. L. (1953). *J. Gen. Microbiol.* **9**, 394-409.

Discussion

WEBER: I wonder if you would consider including a possible hypothetical ancestor of the higher protists. It is to me still disturbing to have the red algae the ancestors of most other photosynthetic as well as nonphotosynthetic organisms.

DOUGHERTY: I do not want you to take too seriously the actual forms given. The thing that I find very difficult to believe is that anything as complicated as the particulate nucleus of mesoprotists and higher organisms, or even as the chloroplast, occurred as a parallel evolution more than once. There is too much involved in this.

PRINGSHEIM: I should like very much to know what you consider to be primitive, it all depends on your definition of primitiveness, and I do not know that we have any primitive organisms left.

10

Photoreceptors: Comparative Studies*

Jerome J. Wolken

*Biophysical Research Laboratory, Eye and Ear Hospital, and the
University of Pittsburgh Medical School, Pittsburgh, Pennsylvania*

The stimulation by light of living organisms is mediated through photosensitive pigments within photoreceptors. These responses are known primarily as phototropisms, photosynthesis, and vision. The photoreceptors for photosynthesis are the chromatophores and chloroplasts; for vision there are the retinal rods and cones in the vertebrates, and sensory cells, ocelli, ommatidia, rhabdomeres, and other structures in the invertebrates.

A comparative study of these plant and animal photoreceptors has been in progress in our laboratory (Wolken, 1956b, 1957, 1958b; Wolken et al., 1957a,b). The structure of the photoreceptors and their pigments in relation to their function are discussed in several recent symposia (Whitelock, 1958; Brookhaven Symposia in Biology, 1958; Reviews of Modern Physics, 1959). It is the purpose here then to review briefly our microscopic studies, particularly electron microscopy of the photoreceptors' structure, to describe the usefulness of a microspectrophotometer developed for the study of the pigments within the photoreceptors *in vivo*, and then to indicate some physico-chemical properties of the photoreceptor pigment complex.

The techniques of microscopy (i.e., polarization, fluorescence, phase and interference) together with microspectrophotometry provide the most direct approach to obtaining information on the photoreceptors in their natural state. These data, together with newer information obtained from electron microscopy of fixed structures *in situ*, are helping us elucidate the structure of the photoreceptors at a molecular level.

Electron Microscopy

The methods of electron microscopy have been discussed in numerous publications and are only briefly noted here. For electron microscopy

---

* Aided in part by grants from U. S. P. H. S. Institute Neurological Diseases and Blindness (B-397 C-5), National Council to Combat Blindness (G-199 C7), and the McClintic Endowment.

the structures are usually fixed with agents that increase the electron density; 1% osmium tetroxide buffered with acetate-veronal, pH 7.0–8.0, has proved most successful. Other fixatives such as potassium permanganate, potassium chromate and dichromate, uranyl nitrate, and combinations of these have been used. In our studies the osmotic changes in the tissue cells were partially controlled by adding sucrose (0.15 M) to the fixative. The fixed cells were embedded in a resin (n-butyl methacrylate, methyl methacrylate, or mixtures of these) that when polymerized possessed the right properties of hardness and ductility for sectioning. Thin sections, less than 0.05 µ in thickness, were cut with a glass knife. These thin sections permit electron penetration and allow us to see through the fixed photoreceptor itself rather than a shadow or replica of it. The necessity for fixation, dehydration, and embedding suggests that new techniques of freeze-drying, specific staining, and the use of water-soluble resins should be investigated. However, these methods are greatly aiding our studies of the molecular structure of photoreceptors.

Microspectrophotometry

The use of sensitive high-speed spectrophotometry for the investigation of pigments within living plant and animal cells has only recently begun. A microspectrophotometer using solid state electronic components has been constructed by Dr. G. K. Strother in our laboratory and is being applied to the investigation of pigments in living cells (Strother and Wolken, 1959a). The light-sensitive element is a cadmium selenide photoconductive cell. For some time, photoconductive cells have been used for spectral measurements in the infrared region. More recently they have been used for x-ray spectroscopy (Henry and Cole, 1959). However, the present application of these cells to visible and ultraviolet microspectrophotometry is a novel use. The spectral response of the cadmium selenide cells extends all the way from the far ultraviolet to the near infrared, with a rather sharp cut-off at about 1 µ. The peak response is in the visible region of the spectrum. At high light intensity, the photocell time constant is on the order of milliseconds, but increases to seconds at low light levels. Sensitivity of the photocells per unit surface area is equivalent to photomultiplier tube performance. The microspectrophotometer uses either a xenon arc or a tungsten ribbon filament light source, a Bausch & Lomb 250 mm. grating monochromator and reflecting microscope optics (50 ×, with Numerical Aperture of 0.5).

In Fig. 1 is shown a diagram of the instrument. Light from source L enters a grating monochromator G as shown by the dotted line. The exit slit of the monochromator is focused by a quartz lens on the condenser of the microscope M. After passing through the specimen, the light beam hits the photosensitive cell C, accurately positioned relative to the specimen image and located 25 cm. above the microscope eyepiece. The electrical signal from the photocell is amplified by a d.c. amplifier, A, and displayed on a Dumont Model 403 oscilloscope, D. The oscilloscope is used here because of its high d.c. amplification capabilities; a high impedance voltmeter would be just as satisfactory.

The instrument is usable over the wavelength range 200 to 990 mμ in a single sweep for specimen areas of 16 μ². For specimen areas of the order of 2 μ², the useful range is 270 to 990 mμ. The time required to

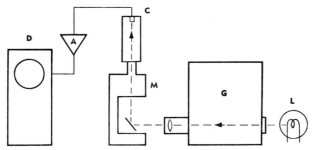

FIG. 1. Diagram of microspectrophotometer. L, light source; G, monochromator; M, microscope with reflecting optics; C, photoconductive cell; A, amplifier; D, oscilloscope.

determine a spectral curve depends upon the number of wavelength settings desired. The maximum magnification employed to date is 500 ×. At this magnification, spectra from areas of 2 μ² are easily obtained. Optical alignment is not critical. No optical beam chopping is used because the spectra are obtained manually. The photocell amplifier has a gain of 300 and is battery powered. At present, the noise level is determined by the light source and amplifier fluctuations rather than by the photocell.

Plant Photoreceptors

CHLOROPLASTS

The photoreceptors of plant cells, from the photosynthetic bacteria to the higher plants, exist in various shapes and sizes. These structures have been given a variety of names, i.e., chromatophores, grana, megaplasts, plastids, and chloroplasts, depending on their phylogenetic position and internal organization. All of these structures, except where the photosynthetic pigments are located in small granules called *chromatophores*, are referred to as *chloroplasts*.

It is experimentally difficult to isolate chloroplasts free from other cellular components, and precise data on their chemical composition are not easily obtainable. A summary of available data on isolated chloroplasts from a variety of plants is given by Rabinowitch (1945). A range of values on a dry weight basis from 35% to 55% for the protein, 18% to 37% for the lipids, and 5% to 8% for the inorganic materials was found. Studies to determine kinds of proteins and their amino acid composition from a variety of chloroplasts are in progress. Two cytochromes, cytochrome f and cytochrome b_3, together can make up as much as 20% of the chloroplasts' total protein content. Nucleic acids (both RNA and DNA) are also reported to be in the chloroplasts; experimental values range from 0.3% to 3.5% on a dry weight basis.

The pigments make up about 8% of the chloroplast; the chlorophylls averaging about 6% (concentrations as high as 20% have been reported) and the carotenoids about 2%. Chlorophyll a occurs in all plant chloroplasts but other isomers are also found; in the higher plants there are chlorophyll a and b, in a ratio of about 3:1. The number of chlorophyll molecules per chloroplast is $\sim 1 \times 10^9$. The carotenoids, including β-carotene and the xanthophylls, particularly lutein and zeaxanthin, are intimately arranged with chlorophyll in the chloroplast. Other photosynthetic pigments, phycoerythrin and phycocyanin, are present in the red and blue-green algae. In all of these analytical studies it has been found that the state of the plant cells, age, nutrition, temperature, and light conditions can bring about changes in the pigments and over-all chemistry of the cells.

CHLOROPLAST STRUCTURE

Previously, it was proposed, from studies of negative birefringence, that the chloroplast is a lamellar structure of 20 to 30 lipid layers each 50 Å in thickness and separated from layers 250 Å in thickness by monomolecular films of chlorophyll molecules (Frey-Wyssling, 1957). There is now excellent proof of this lamellar structure from electron microscopic studies of a great variety of plants, and x-ray diffraction techniques have corroborated the electron microscopic observations, showing a repeating unit of $\sim$ 250 Å. Observations with the electron microscope show a basic internal organization. The chloroplasts can be described as lamellar chloroplasts and, in the higher plants, as grana lamellar chloroplasts. Although we have investigated the structure of the chloroplasts in a variety of algae and higher plants, the algal flagellate *Euglena* was found to be especially adaptable for experimental studies on chloroplast structure. *Euglena* is a photosynthetic plant cell in the light and behaves as an animal cell in the dark; when light-grown, it contains many green elongated cylindrical chloroplasts that are from 5 to 10 μ in length and from 0.5 to 2 μ in diameter. Changes in the environmental conditions (light, darkness, temperature, and drugs) are reflected in its chemistry and chloroplast structure. Here, we were able to study the chloroplast by various microscopic techniques and to correlate these structural observations as far as possible with the pigment synthesis in the light ↔ dark adaptation (Wolken, 1956a; Wolken et al., 1955; Wolken and Mellon, 1956; Wolken and Palade, 1953; Wolken and Schwertz, 1953).

In order to determine more precisely the lamellar structure and to test the pigment monolayer hypothesis, the diameter, length, number, and thickness of the lamellae of the chloroplasts from numerous electron micrographs were measured and statistically evaluated (Wolken and Schwertz, 1953). For example, the chloroplasts of *Euglena gracilis* consist of 21 dense layers ∼ 250 Å in thickness with less dense interspaces of 300 to 500 Å in thickness. Each dense layer appears to be covered on both sides by thinner and denser layers (lamellae) 50 to 100 Å in thickness. The average thickness measured for a variety of plant chloroplasts ranges from 20 to 100 Å for the electron dense lamellae (Leyon, 1956; Sager, 1958; Sager and Palade, 1957). Frey-Wyssling and Steinman had previously noted (Frey-Wyssling, 1957) that the thin lamellae consist of spherical particles 65 Å in diameter, indicated to be the protein or lipoprotein macromolecules. In *Euglena granulata*, the chloroplast lamellae appear as small granules ∼ 100 Å in diameter (Wolken, 1960). There is experimental evidence to indicate that the chlorophyll molecules are preferentially oriented parallel to the lamellae and are within them (Thomas, 1955). The areas of least density are considered to be aqueous proteins, enzymes, and dissolved salts. From the geometry of the chloroplast, the number of dense layers, and the chlorophyll concentration per chloroplast, we have calculated the cross-sectional area occupied by each chlorophyll molecule to be 222 $Å^2$ for the *Euglena* chloroplast and 246 $Å^2$ for the *Poteriochromonas* chloroplast (Table 1). Elbers and associates (1957) have since collected data on chlorophyll concentrations and geometry of chloroplasts in a number of plant species, and by similar analogy calculated the mean area available for the chlorophyll molecule in the monolayer to be ∼ 200 $Å^2$. Studies of the dichroism, birefringence, and polarization of fluorescence in *Mougeotia* chloroplasts also indicate that chlorophyll resides as a monolayer on the lamellar surface and the area available per chlorophyll molecule was calculated to be ∼ 250 $Å^2$ (Goedheer, 1955). Since the cross-sectional area of the porphin head of the chlorophyll molecule is known from x-ray studies to be about 225 to

TABLE I
DIMENSIONS OF PIGMENT MACROMOLECULES CALCULATED FROM THE PHOTORECEPTOR GEOMETRY

Photoreceptor	Cross-section area, A, of macromolecules (Å^2)	Diameter, d_m, of macromolecules (Å)	Length, $T/2$, of macromolecules (Å)
Chloroplast (*Euglena gracilis*)	222	17	121
Chloroplast (*Poteriochromonas stipitata*)	246	18	195
Retinal rod (*Rana pipiens*)	2620	51	100
Retinal rod (Cattle)	2500	50	100

242 Å^2, these results indicate that all the available chlorophyll molecules could be packed into the interfacial area and cover all of the dense surface of the lamellae as a monolayer.

On the basis of these calculations, a simplified schematic molecular model was proposed (Fig. 2). The suggestion that four chlorophyll molecules are united to form tetrads in which the reactive isocyclic rings turn toward each other was employed (Wolken and Schwertz, 1953). Interaction between the phytol tails was eliminated in the model by arranging the tetrads in such a way that one, and only one, of the phytol tails is located at each intersection in the rectangular network. If the chlorophyll was packed as a monolayer, as illustrated in the schematic molecular network in Fig. 2, there would still be space available for the carotenoid molecules at the interstitial positions between the chlorophyll molecules. If these spaces are occupied as shown, there will be one carotenoid molecule for at least every three chlorophyll molecules in the network. This kind of close packing of the chlorophyll and carotenoid molecules in the pigment monolayers would also permit energetic interaction between them. Other models with slight modifications from that presented here indicate that the chlorophyll molecules are also turned inward as well as being oriented on the surface (Hodge *et al.*, 1955). Calvin has recently suggested a similar model in which the porphin heads lie at an angle of 45° and in which one protein layer contains CO_2-reducing enzymes and another protein layer O_2-evolving enzymes (Calvin, 1959).

There are several possible ways in which the chlorophyll molecules could be oriented in the lamellae. If the porphin heads of the chlorophyll

10. PHOTORECEPTORS: COMPARATIVE STUDIES 151

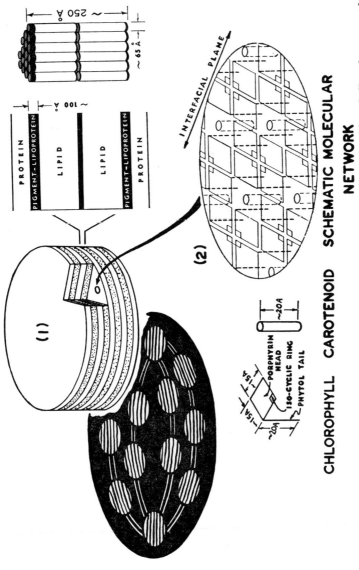

FIG. 2. Model of chloroplast, showing a schematic molecular network of the chlorophyll and carotenoid molecules in the chloroplast layers.

molecules lie at 0° as flat plates as indicated in Fig. 2, their greatest cross section would be available and it would be most efficient. However, if they were oriented within the lamellae at increasing angles up to 90°, the cross-sectional area available would be decreasing and hence less efficient. Studies of chlorophyll monolayers on various liquid surfaces suggested that the chlorophyll molecules probably lie at an angle of 35° to 55° within the chloroplast, hence reducing the above calculation for the cross section of the chlorophyll molecule to 100 Å^2 (Trurnit and Colmano, 1958). It is very likely that the absorption oscillators of these pigment molecules are arranged with an orderly orientation in such a way that maximum absorption will be observed for incident light polarized in a certain direction.

Pigments

Closely connected with the problem of chloroplast structure is that of the synthesis of chlorophyll. In order to follow more directly the pigments and their synthesis within a single chloroplast, we have applied the microspectrophotometer to the problem (Strother and Wolken, 1959). In Fig. 3 are shown the absorption spectra obtained with this instrument for the chloroplasts in two different *Euglena* cells. The magnification was 250 ×, representing an area of 8 µ^2. The absorption spectra obtained by this method for the chloroplasts begin to show *fine structure*. The major absorption peaks are near 675 and 435 mµ with other peaks at 430, 590, and 625 mµ. Since these are *in vivo* spectra, some shifts in peaks are expected from those of the pigments dissolved in organic solvents. These spectra closely approximate previous absorption spectra for the *Euglena in vivo* and for digitonin extracts of its chloroplasts (Wolken, 1960). The concentration of chlorophyll calculated from the absorption data corroborates the previous analysis that there are $\sim 1 \times 10^9$ chlorophyll molecules per chloroplast. By scanning across the chloroplast at 0.5 µ intervals, at 675 mµ (the absorption maximum), and at 550 mµ (the absorption minimum), it was found that chlorophyll is evenly distributed throughout the whole chloroplast (Fig. 4). Because of the limitation in the resolving power of the instrument it is not possible to determine precisely whether chlorophyll in the chloroplast is in the lamellae or not.

In the alga *Chlamydomonas*, whether grown in light or in darkness, a complex chloroplast organization of lamellae, pyrenoid, and eyespot occurs within a limiting membrane (Sager, 1958; Sager and Palade, 1957). In the absence of chlorophyll the lamellae are not formed, although the eyespot, starch grains, and the pyrenoid are still found within the intact chloroplast membrane. Similarly in *Euglena*, we have found that the lamellae are also dependent on chlorophyll, for as soon as we detect chlorophyll spectroscopically the lamellae of the chloroplast are observed in the electron micrographs (Wolken and Palade, 1953; Wolken, 1956a). The action spectrum obtained for chlorophyll formation in *Euglena gra-*

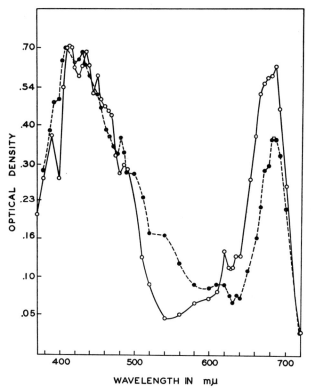

Fig. 3. *In vivo* spectra from two different *Euglena* chloroplasts taken with the microspectrophotometer. (8 μ^2 area and 4 mμ entering half-band width.)

cilis (Wolken and Mellon, 1956) corresponds with the action spectrum for "greening" of etiolated *Avena* seedlings (Frank, 1946) and with the absorption spectrum of protochlorophyll (Koski and Smith, 1948; Noack and Kiessling, 1929). In addition, mixed etioporphyrins have been extracted from the dark-adapted organisms as well as a substance fluorescing in the range between 520 and 580 mμ.

Strain has suggested that the carotenoids are involved in chlorophyll formation, based on the reports that the spectral bands that are most effective in chlorophyll formation are relatively ineffective in carotenoid formation (Strain, 1938). Granick reported a significant decrease in light absorption by carotenoid as Mg-vinyl pheoporphyrin is transformed to chlorophyll in a mutant of *Chlorella* (Granick, 1948). A similar decrease in carotenoid absorption was observed in oat seedlings at high light intensities, and it was suggested that the substance responsible for the carotenoid decrease is a porphyrin type compound concerned with

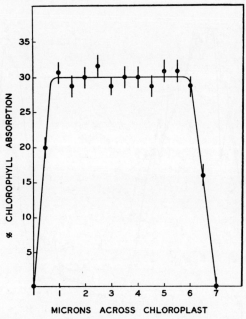

FIG. 4. Absorption across a single chloroplast at 0.5 μ intervals, measured with the microspectrophotometer at 675 mμ relative to 550 mμ, indicating the distribution of chlorophyll in the chloroplast.

"esterifying" the porphyrin molecule and converting Mg-vinyl pheoporphyrin to protochlorophyll, its phytol ester (Frank, 1951).

Frank observed that carotenoids decrease at high intensities of both red and blue light, suggesting that the synthesis of carotenoids in light is probably not autocatalytic, but is influenced by a pigment which absorbs light in both the red and blue portions of the spectrum (Frank, 1946). The "action spectra" obtained for carotenoid synthesis seem to substantiate this finding (Wolken and Mellon, 1956). The curve, representing the action spectrum for net carotenoid removal, when extrapolated in both the blue and red regions of the spectra, indicated that there might be two other absorption bands, one in the red and the other in the near ultraviolet. It was suggested that there may be a porphyrin-like system involved in removing carotenoids and subsequently phytolizing the precursor before its conversion to chlorophyll (Frank, 1946). We have found no evidence of the "phytolization." However, the great decrease in carotenoid concentration in a "chlorophyll-free" mutant, accompanied by a rise in porphyrin concentration, would seem to indicate that the two processes of chlorophyll and carotenoid synthesis are closely related and that perhaps a similar porphyrin-like molecule influences the synthesis of both pigments (Wolken and Mellon, 1956). It should now be possible to corroborate these studies of pigment synthesis within single living cells by microspectrophotometric analyses.

Animal Photoreceptors

INVERTEBRATES

The invertebrates possess the greatest variety of photoreceptors and their structure and chemistry is least understood. Eyespots, sensory cells, ocelli, and compound eyes have arisen among annelids, molluscs, and arthropods, in each instance with differences in organization (Fig. 5).

Eyespots. The eyespot is a photoreceptor for light perception that has been described and studied in *Chlamydomonas, Volvox, Euglena,* and other microorganisms. The eyespot in *Euglena* is an orange-red area at the anterior part of the organism which appears to be intimately linked to the flagellum at its base. In the electron microscope, the eyespot area for *E. gracilis* is about 2×3 μ and consists of 40–50 packed granules (rods) 100–300 mμ in diameter embedded in a matrix (Wolken and Palade, 1953; Wolken, 1956a). It has been suggested that there may be at least two pigments in the eyespot. Red variants of *Euglena* contain astaxanthin, an animal pigment, and some spectra suggest that it may be the phototactic pigment. Other experiments indicate that β-carotene is present in the eyespot. Microspectrophotometric analysis of the eyespot of *E. gracilis in vivo* shows a general absorption from 460–480 mμ, with the major peak near 480 mμ, a secondary peak at 425 mμ, small minor peaks near 500, 530, and 590 mμ, and a broad band at 630 mμ (Fig. 6). These peaks would seem to indicate β-carotene, but the spectra also could indicate the presence of astaxanthin, porphyrins, or other pigments. It is interesting to compare the spectrum obtained with the microspectrophotometer to that of the action spectrum for swimming and phototaxis of *E. gracilis* (Wolken and Shin, 1958). Here the absorption peaks are at 420, 465, 490 mμ with a broad band near 630 mμ (Fig. 7). However, no one so far has isolated astaxanthin or other photosensitive pigments from eyespots of green euglenas.

Sensory Cells. In the flatworm, *Planaria,* the two eyes consist of pigment granules and sensory cells. The ends of the sensory cells continue as nerves which enter the brain. The pigment granules shade the sensory cells from light in all directions but one, and so enable the animal to respond in a negative way to the direction of light. Behind the dense pigment granules are sensory cells that are structurally similar to retinal

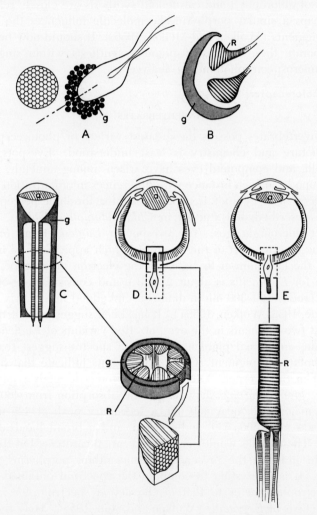

Fig. 5. Variety of photoreceptor structures. A, eyespot; B, flatworm; C, insect; D, mollusc; E, vertebrate retinal rod; *a*, lens; *g*, pigment granules; *R*, retinal rods. It is to be noted that the visual photoreceptors of animals are structurally similar to the photosynthetic chloroplast structure of plants. (From Wolken, 1958b.)

rods of the vertebrates. These retinal rods are about 5 µ in diameter with a variable length averaging approximately 35 µ. These rods consist of layers (lamellae). The total thickness of the dense layers is ~ 400 Å, and each lamella is about 100 Å in thickness. There are 8–10 dense layers, providing 16–20 surfaces per micron. Little is known of the

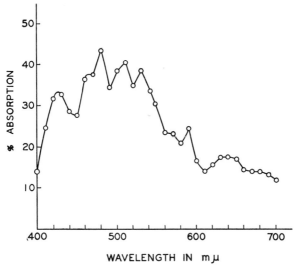

Fig. 6. Average spectrum of *Euglena* eyespot obtained with the microspectrophotometer. (2 µ² area and 4 mµ entering band width.)

photosensitive pigment within these receptors. Pirenne and Marriot, using the aquatic planarian *Dendrocoelem lacteum*, found that its action spectrum has two main absorption peaks, one at 510 mµ, suggesting a rhodopsin, and another in the ultraviolet at 370 mµ (Pirenne and Marriott, 1955).

Ommatidia. The compound eye of the insect consists of ommatidia, each ommatidium made up of retinula cells; the differentiated part of the retinula cell is the rhabdomere. The rhabdomeres taken together form a rhabdome. The rhabdome has been considered the "light trapping" area where the visual process is initiated. The eye of the *Drosophila* is composed of approximately 700 ommatidia, each ommatidium consisting of seven retinula cells radially arranged, forming a cylinder. Electron microscope studies indicate that the rhabdomeres are structurally packed rods or tubes; the thickness of the edges of these tubes averages 120 Å and the interspaces vary from 200–400 Å (Wolken *et al.*, 1957a, b). Similar rod or tube structures for the rhabdomeres have been shown in other insect rhabdomeres (i.e., housefly, dragonfly, honeybee, spider) and in the crustacean, *Limulus*.

From the eyes of the lobster, a crustacean, Wald and Hubbard (1957) isolated a rhodopsin which has the prosthetic group, retinene$_1$. Recently a photosensitive pigment has been isolated from honeybee and housefly

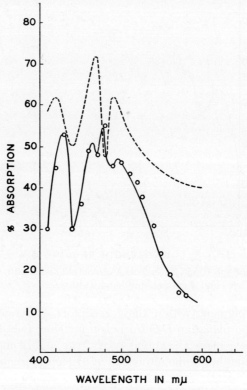

FIG. 7. Comparison of *Euglena* eyespot absorption spectrum (—·—·—) to that of the combined action spectrum for *Euglena* photokinesis and phototaxis (Wolken and Shin, 1958) – – – –.

eyes. In the honeybee, Goldsmith (1958) found that he could extract a photosensitive pigment with an absorption maximum of 440 mµ; this peak was associated with the retinene$_1$ complex. Similarly, we have isolated from the housefly a photosensitive pigment with an absorption maximum at 437 mµ, which can not be identified with any known photosensitive pigment (Bowness and Wolken, 1959). However, by techniques of acetone extraction and chromatography it was found that the housefly head contains retinene$_1$, the aldehyde of vitamin A$_1$ (Wolken *et al.*,

1960). In three eye-color mutants of *Drosophila melanogaster* we found that the action spectrum was indicative of a pigment absorbing at 508 mµ, suggesting a rhodopsin, but no such pigment complex has yet been isolated (Wolken et al., 1957b, 1960).

Retina. The cephalopod molluscs *Octopus* and *Sepia* have single lens eyes, resembling the vertebrate eye in physical organization. However, the retina is not inverted as in the vertebrate eye and the photoreceptors are directly exposed to the incident light. Electron microscopy of the retina shows that the arrangement of the retinal cells is very much like that of the arthropod ommatidia with retinula cells and rhabdomeres. Four rhabdomeres make up each rhabdome. There is a striking similarity in *fine structure* to that of the insect and the *Limulus* visual organ (Wolken et al., 1957a; Goldsmith and Philpott, 1957; Miller, 1957). In both *Octopus* and *Sepia* each retinal rod (rhabdomere) averages ~ 1 µ in diameter. Its microstructure is that of densely packed rods or tubes (there are about 20 rods per micron), each ~ 200 Å in diameter (Wolken, 1958a,b). The visual pigment here also has been shown to be a rhodopsin (Brown and Brown, 1958).

Vertebrates

Retinal Rods and Cones. All the vertebrate retinal rods that have been recently studied by electron microscopy (the retinal rods of frog, chicken, guinea pig, rabbit, perch, whale, cattle, and monkey) appear as lamellae, 100–200 Å in thickness with less dense interspaces 200–500 Å (Wolken, 1958b; Review of Modern Physics, 1959). A schematic molecular model for the retinal rods, not unlike that of the chloroplast, is illustrated in Fig. 8. Fewer data are available for the cone structure, but it too is a lamellar structure of dense and less dense layers ~ 200 Å.

The visual pigment complex, rhodopsin, is contained only in the outer segments of the retinal rods of the eye. In all the animal photoreceptors (retinal rods) studied there have been found from 1×10^6 to 1×10^9 pigment molecules per receptor. The photoreceptor geometry (length, diameter, thickness, and number of dense layers) was used similarly to that of the chloroplast for calculating the cross-sectional area of the pigment macromolecule (Table I). For example, for cattle and frog rhodopsin, where the cross-sectional area is ~ 2500 Å^2, the diameter of the molecule would be ~ 50 Å. This seems to be about the right order of magnitude for the rhodopsin molecule (Wald, 1954).

Retinal Colored Globules. Within the retina of birds there are brightly colored globules; photomicrographs indicate that these globules are located near or between the inner and the outer cone segments (Det-

160   JEROME J. WOLKEN

wiler, 1953). The relative ease with which the colored globules may be identified *in situ* in the retina is another example of the application of the microspectrophotometer (Strother and Wolken, 1960). In the chicken retina these globules range from 3–6 μ in diameter. The predominant colors observed in a freshly excised chicken retina are red, green, and yellow, with other mixed colors. What function the globules may have in the visual process and in the retina is still unknown, but they have long been suspected of acting as color filters for the retinal cones (Wald and Zussman, 1938). Since a red globule, for instance, absorbs light somewhere in the blue-green region, the location of the globule between the inner cone segment and the light source makes a

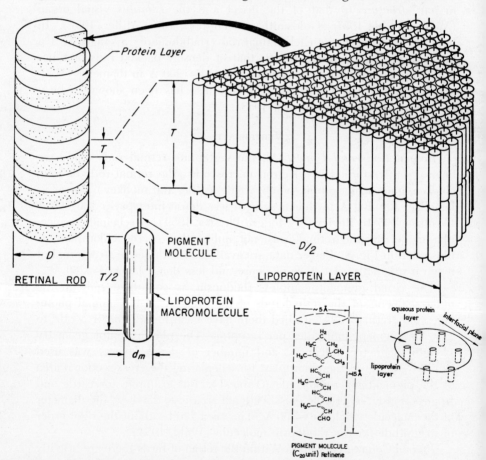

Fig. 8. Schematic model of retinal rod. (From Wolken, 1956b.) D, diameter of outer segment; T, thickness of dense lamellae; d^m, diameter of the rhodopsin molecule.

color filter theory reasonable. However, there is no direct correlation between the color of the globule and the color discrimination of animals possessing them. Wald and Zussman (1938) isolated from the chicken retina three carotenoids, lutein, zeaxanthin, astaxanthin, and a new carotenoid, galloxanthin (Wald, 1947).

The absorption spectra for individual colored globules were measured over the range 340 to 700 mμ. For mixtures of the globules, absorption spectra from 280 to 700 mμ were obtained. The red globules have a broad general absorption maximum near 500 mμ. This is indicative of the carotenoid astaxanthin, which is believed to be the main pigment of the red colored globule (Wald and Zussman, 1938). The green globules have a general absorption in the region 370–460 mμ; their major absorption peak is near 420 mμ, with another strong single peak at 480 mμ, indicating that the major component is the carotenoid galloxanthin (Wald, 1947). The yellow globules are more difficult to identify because of the large number of carotenoids absorbing in the 400–500 mμ region. However, the two carotenoids lutein and zeaxanthin are believed to be in the yellow globules (Fig. 9).

Some observations of bleaching indicate that the green globules are the most unstable of the three major color types. When all the pigmented globules are mixed together, a reddish-orange color results. Upon exposure to white light this mixture bleaches first to orange, then to yellow, then becomes colorless. The time required is on the order of 15–30 minutes, depending upon the intensity of the light.

It is interesting to compare the spectra of these pigment globules (Fig. 9) with those of the eyespot granules of *Euglena* (Fig. 6), since they could be related (from the evolutionary standpoint) in the development of the visual pigment system.

Pigment Complex

The photosynthetic and visual pigments, chlorophyll, carotenoids, and retinene, are in their natural state bound to proteins or lipoproteins. With 1.0–2.0% digitonin a photosynthetic pigment complex, chloroplastin (chlorophyll, carotenoid plus protein) can be extracted from chloroplasts and a visual pigment complex, rhodopsin (retinene plus opsin) can be extracted from retinal rods. The absorption spectra of these complexes are comparable to those of the chloroplast and retinal rod pigments *in vivo*. Sedimentation in the analytical ultracentrifuge and the electrophoretic patterns of these complexes indicate that they are homogeneous (Wolken, 1956b, 1960).

The physicochemical properties of *Euglena* chloroplastin — such as

(1) the kinetics of bleaching, (2) the rate of photoreduction of the dye 2,6-dichlorobenzenoneindeophenol, (3) the evolution of oxygen, and (4) the conversion of inorganic phosphate to organic phosphate — were investigated (Eversole and Wolken, 1958; Wolken and Mellon, 1957). The results indicated that chloroplastin has physiological activity. In the

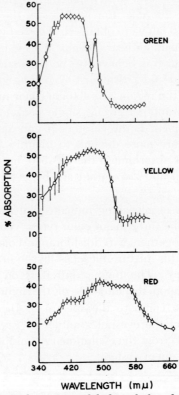

Fig. 9. Spectra of retinal pigment globules of the chicken obtained with the microspectrophotometer.

visual pigment complex, rhodopsin, a light-dependent reaction analogous to that occurring in the intact retina can be measured spectroscopically. This photochemical reaction "bleaches" rhodopsin and results in a shift of the absorption peak (from around 500 mμ to around 365 mμ). When chloroplastin was bleached by light no similar shift in spectrum was observed. Chloroplastin is "bleached" by wavelengths lower than 560 mμ; above 560 mμ it requires heat energy in addition to light energy. The bleaching is a steady decrease in optical density at 675 mμ, the chlorophyll absorption maximum, with the eventual disappearance of absorption at this wavelength. The total activation energy of bleaching was calculated to be 48.3 kcal./mole (Wolken and Mellon, 1957). (Table II). During the Hill reaction with the dye, it was found that an absorption peak at 488 mµ (one of the *Euglena* carotenoid absorption peaks) increases in optical density in the light and then decreases in the dark to its original density. This reaction could be repeated by placing the reactants alternately in light and darkness without further addition of dye, indicating that there is an increase in the amount of carotenoid absorption at 488 mµ simultaneously with the dye reduction. This reaction is analogous in some ways to the bleaching of rhodopsin.

TABLE II
ACTIVATION ENERGY OF BLEACHING

	Comparative rate of bleaching at all temperatures in white light	Calculated experimental activation energy for thermal bleaching in darkness (k cal./mole)	Experimental wave length at which temperature dependence of bleaching begins (mµ)	Calculated experimental activation energy for light bleaching (k cal./mole)
Chloroplastin (*Euglena*)	1	48.2	560	48.3
Rhodopsin[a] (frog, cattle)	1	44.0[b]	590	48.5

[a] St. George (1952).
[b] Lythgoe and Quilliam (1938).

The fact that physiological activity is retained by rhodopsin and chloroplastin may be due to the orientation of the pigment complex within the digitonin micelles. Digitonin exhibits paracrystalline properties and when evaporated from solution will form periodic rings. A dye or chlorophyll added to a digitonin solution will become oriented within the digitonin rings. Similarly chloroplastin, when evaporated from solution, forms periodic rings within which the chlorophyll complex becomes oriented (Fig. 10). The structural relationship of the pigment complex is therefore analogous to the *fine structure* of chloroplasts and retinal rods (Wolken, 1960).

The molecular weights of chloroplastin and rhodopsin have also been calculated from their sedimentation constants as measured in the analytical ultracentrifuge and from their nitrogen content (Table III). Miller and Anderson (1942) and Smith and Pickels (1940) have shown that digitonin forms micelles of minimal molecular weight of 75,000.

Hubbard (1954) has demonstrated that three such micellar units of digitonin are probably associated together with an average molecular weight of 225,000. If this molecular weight is subtracted from the average molecular weight calculated for the pigment complex, a molec-

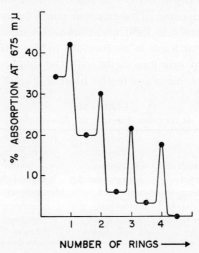

Fig. 10. Chloroplastin scanned at 675 mμ with the microspectrophotometer showing that chlorophyll is located within the ring structures.

ular weight of about 40,000 appears to be of the right order of magnitude for the pigment protein. The average molecular weight of chloroplastin calculated (from the geometry of the chloroplast and from the analytical data on the pigment concentration, nitrogen, and dry weight of

TABLE III
ANALYTICAL ULTRACENTRIFUGE DATA[a]

Pigment complex in digitonin	Average sedimentation ($S_{20} \times 10^{13}$)	Average complex micelle weight, M'	Average molecular weight pigment complex, M
Digitonin	7.1	155,000	155,000
Euglena chloroplastin	13.5	290,000	40,000
Cattle rhodopsin	9.77	275,000	40,000
Frog rhodopsin	12.1	295,000	67,000

[a] M is calculated from M' using the dry weights and per cent nitrogen as used by Hubbard (1954). Smith and Pickels (1940) have shown that digitonin forms micelles of minimal molecular weight 75,000.

the complex) was of the order of 21,000 to 40,000. The molecular weight for frog and cattle rhodopsin has been calculated to be 67,000 and 40,000 respectively (Wolken, 1956b).

Molecular weights have been determined for other chlorophyll-protein complexes, e.g., a chlorophyll-lipoprotein complex was crystallized out from leaf extracts and from diffusion studies a molecular weight of 19,200 was calculated (Takashima, 1952). A protochlorophyll "holochrome" with a molecular weight of about 400,000* was isolated by Smith and collaborators (1958) from bean seedlings (*Phaseolus vulgaris*) in glycerine — KOH, pH 9.6. Data obtained with the interference microscope on the chloroplast of *Euglena in vivo* were used to calculate a molecular weight of ~ 16,000. Frey-Wyssling predicts that there would be 16 molecules of chlorophyll to one protein or lipoprotein macromolecule on the basis that the macromolecule is 65 Å in diameter; therefore the chlorophyll complex would have a molecular weight of 68,000 (Frey-Wyssling, 1957). However, the nature of the protein or lipoprotein is yet to be determined for the pigment complex of the chloroplast.

REFERENCES

Bowness, J. M., and Wolken, J. J. (1959). *J. Gen. Physiol.* **42**, 779-792.
Brookhaven Symposia in Biology No. **11** (1958).
Brown, P. K., and Brown, P. S. (1958). *Nature* **182**, 1288-1290.
Calvin, M. (1959). *Rev. Modern Phys.* **31**, 147-156, 157-161; (1958). *Brookhaven Symposia in Biol. No.* **11**, 160-180.
Detwiler, S. R. (1953). "Vertebrate Photoreceptors." Macmillan, New York.
Elbers, P. F., Minnaert, K., and Thomas, J. B. (1957). *Acta Botan. Neerl.* **6**, 345-350.
Eversole, R. A., and Wolken, J. J. (1958). *Science* **127**, 1287-1288.
Frank, S. R. (1946). *J. Gen. Physiol.* **29**, 157-179.
Frank, S. R. (1951). *Arch. Biochem.* **30**, 52.
Frey-Wyssling, A. (1957). "Macromolecules in Cell Structure," pp. 49-69. Harvard Univ. Press, Cambridge, Massachusetts.
Goedheer, J. C. (1955). *Biochim. et Biophys. Acta.* **16**, 471-476; (1957). Thesis, State University of Utrecht, Netherlands.
Goldsmith, T. H. (1958). *Proc. Natl. Acad. Sci. U. S.* **44**, 123-126.
Goldsmith, T. H., and Philpott, D. E. (1957). *J. Biophys. Biochem. Cytol.* **3**, 429-438.
Granick, S. (1948). *J. Biol. Chem.* **175**, 333-342; (1950) **183**, 713-730.
Henry, B., and Cole, H. (1959). *Rev. Sci. Instr.* **30**, 90-92.
Hodge, A. J., McLean, J. D., and Mercer, F. V. (1955). *J. Biophys. Biochem. Cytol.* **1**, 606-614; Hodge, A. J. (1959). *Revs. Modern Phys.* **31**, 331-341.
Hubbard, R. (1954). *J. Gen. Physiol.* **37**, 381-399.
Koski, V. M., and Smith, J. H. C. (1948). *J. Am. Chem. Soc.* **70**, 3558-3562.

---

* The molecular weight of this complex is now reported to be 1×10^6 (Smith, 1959).

Leyon, H. (1956). *Svensk Kem. Tidskr.* **68**, 70-89.
Lythgoe, R. S., and Quilliam, J. P. (1938). *J. Physiol. (London)* **93**, 24.
Miller, G. L., and Anderson, K. J. I. (1942). *J. Biol. Chem.* **144**, 475-486.
Miller, W. H. (1957). *J. Biophys. Biochem. Cytol.* **3**, 421-427.
Noack, K., and Kiessling, W. (1929). *Z. physiol. Chem.* **182**, 13-49; (1930). **193**, 97-137.
Pirenne, M. H., and Marriott, F. H. C. (1955). *Nature* **175**, 642.
Rabinowitch, E. I. "Photosynthesis and Related Processes," Vol. I; (1945), Vol. II, Pt. 1; (1951) Vol. II, Pt. 2 (1956). Interscience, New York.
Reviews of Modern Physics. (1959). **31**, Nos. 1 and 2 (Wiley, New York, 1960).
Sager, R. (1958). *Brookhaven Symposia in Biol. No.* **11**, 101-117.
Sager, R., and Palade, G. E. (1957). *J. Biophys. Biochem. Cytol.* **3**, 463-448.
St. George, R. C. (1952). *J. Gen. Physiol.* **35**, 495.
Smith, E. L., and Pickels, E. G. (1940). *Proc. Natl. Acad. Sci. U. S.* **26**, 272-277.
Smith, J. H. C. (1958). *Brookhaven Symposia in Biol. No.* **11**, 296-302.
Smith, J. H. C. (1959). *Proc. 9th Intern. Botan. Congr. Montreal.*
Strain, H. (1938). *Carnegie Inst. Wash. Publ. No.* **490**.
Strother, G. K., and Wolken, J. J. (1959). *Science* **130**, 1084-1088.
Strother, G. K., and Wolken, J. J. (1960). *Exptl. Cell Research* (in press).
Takashima, S. (1952). *Nature* **169**, 182-183.
Thomas, J. B. (1955). *Progr. in Biophys. and Biophys. Chem.* **5**, 109-139; (1958). *Endeavor* **17**, 156-161.
Trurnit, H. J., and Colmano, G. (1958). *Biochim. et Biophys. Acta* **30**, 435-447.
Wald, G. (1947). *J. Gen. Physiol.* **31**, 377-383.
Wald, G. (1954). *Science* **119**, 887-892.
Wald, G., and Hubbard, R. (1957). *Nature* **180**, 278-280.
Wald, G., and Zussman, H. (1938). *J. Biol. Chem.* **122**, 449-460.
Whitelock, O. V., ed. (1958). *Ann. N. Y. Acad. Sci.* **74**, 161-406.
Wolken, J. J. (1956a). *J. Protozool.* **3**, 211-221.
Wolken, J. J. (1956b). *J. Cellular Comp. Physiol.* **48**, 349-370.
Wolken, J. J. (1957). *Trans. N. Y. Acad. Sci. Ser. II* **19**, 315-327.
Wolken, J. J. (1958a). *J. Biophys. Biochem. Cytol.* **4**, 835-838.
Wolken, J. J. (1958b). *Ann. N. Y. Acad. Sci.* **74**, 164-181.
Wolken, J. J. (1960). In "Origin and Role of Complex Macromolecular Aggregates in Development" (M. V. Edds, Jr., ed.). Ronald Press, New York (in press).
Wolken, J. J., and Mellon, A. D. (1956). *J. Gen. Physiol.* **39**, 675-685.
Wolken, J. J., and Mellon, A. D. (1957). *Biochim. et Biophys. Acta* **25**, 267-274.
Wolken, J. J., and Palade, G. E. (1953). *Ann. N. Y. Acad. Sci.* **56**, 873-881.
Wolken, J. J., and Schwertz, F. A. (1953). *J. Gen. Physiol.* **37**, 111-120.
Wolken, J. J., and Shin, E. (1958). *J. Protozool.* **5**, 39-46.
Wolken, J. J., Mellon, A. D., and Greenblatt, C. L. (1955). *J. Protozool.* **2**, 89-96.
Wolken, J. J., Capenos, J., and Turano, A. M. (1957a). *J. Biophys. Biochem. Cytol.* **3**, 441-448.
Wolken, J. J., Mellon, A. D., and Contis, G. (1957b). *J. Exptl. Zool.* **134**, 383-410.
Wolken, J. J., Bowness, J. M., and Scheer, I. J. (1960). *Biochim. et Biophys. Acta* (in press).

Discussion

MILLOTT: The central fact still remains that the connections between light absorption and all the biochemical edifice that has been erected around it, and nerve excitation, are theoretical. I wonder what you think of this work of Noel's and if there has been any further addition to it. He found that if you feed rats on a certain diet or give them certain drugs, the outer segment — the pigment bearing segment of the rods — degenerates but the electroretinogram still remains.

WOLKEN: It is very difficult to interpret electroretinograms. The interesting thing is that, as in the chloroplast, if you do something to destroy the pigment, you destroy the photoreceptor structure itself.

FRENCH: While you were speaking rather rapidly about visual pigments, you said something about getting Hill activity in digitonin extracts of chlorophyll. We have not been able to do this.

WEBER: It is a moot point whether the protein is there merely to support pigment in a certain position, a certain orientation, or whether the protein is performing an active role in the photochemical process. There is no molecular property that one knows of protein that would help in this respect.

11

Recent Studies of Chlorophyll Chemistry*

A. S. HOLT AND H. V. MORLEY†

Division of Applied Biology, National Research Council, Ottawa, Canada

The finding of new chlorophylls by Manning and Strain (1943) and Strain *et al.* (1943), and the presence in photosynthetic green bacteria of chlorophylls unlike any previously characterized (Goodwin, 1955; Larsen, 1953) indicate the need for further studies of chlorophyll chemistry. This paper will describe the results of the oxidation of chlorophyll *a* by permanganate which have led to the elucidation of the structure of chlorophyll *d*, and will summarize some results obtained in a study of a new chlorophyll isolated from *Chlorobium thiosulfatophilum* (Strain VN).

Oxidation of Chlorophyll A

Fischer and Walter (1941) first showed that the 2-vinyl group of the *a* series of pigments can be oxidized by potassium permanganate (see Fig. 1). Magnesium-free derivatives which lacked the C-10 carbomethoxy group, e.g., chlorin-e_6-trimethyl ester and methyl pyropheophorbide *a* were chosen as starting materials because permanganate was believed to attack Ring V preferentially leaving the vinyl group intact. Introduction of magnesium into the oxidation products was accompanied by a side reaction because subsequent removal of magnesium did not yield the parent oxidation product. The 2-(1,2-dihydroxyethyl), -formyl,- and -carboxy derivatives were isolated.

Recent work (Holt and Morley, 1959) has shown that oxidation of chlorophyll *a* does not necessarily cause allomerization, i.e., oxidation of Ring V. Four products were found, all of which gave a positive Molisch phase test (see below). Because one of these could be extracted from ether by aqueous alkali, it was assumed to be the 2-carboxy derivative and remains to be purified. The other three products were separated by adsorption chromatography on a sucrose column.

---

* Contribution from the Division of Applied Biology, National Research Council, Ottawa, Canada. Issued as N.R.C. No. 5703.

† National Research Council Postdoctorate Fellow, 1957–1959. *Present address:* Department of Chemistry, McMaster University, Hamilton, Ontario, Canada.

The lowest zone contained a pigment which had a visible absorption spectrum identical with that of chlorophyll d, shown in Fig. 2. The ready interconversion of pheophytin d (Fig. 3) and a product, "isopheophytin d," by alternate treatment with methanolic hydrogen chloride and

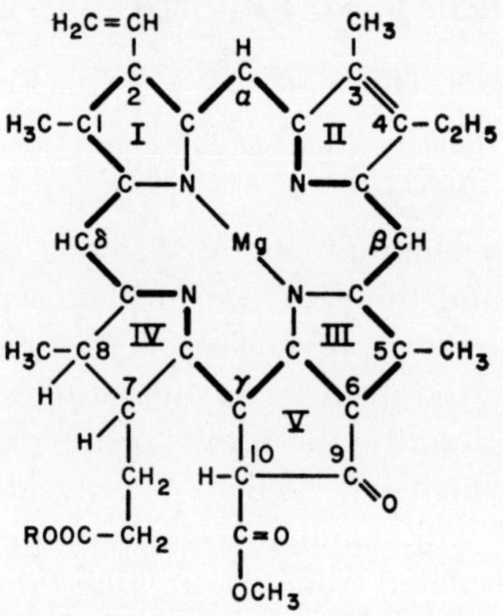

COMPOUND	R	Mg
CHLOROPHYLLIDE	H	+
PHEOPHORBIDE	H	−
CHLOROPHYLL	$C_{20}H_{39}$	+
PHEOPHYTIN	$C_{20}H_{39}$	−
METHYL CHLOROPHYLLIDE	CH_3	+
METHYL PHEOPHORBIDE	CH_3	−
b COMPOUNDS: $3-CH_3$ IS REPLACED BY CHO "PYRO" COMPOUNDS: $10-CO_2CH_3$ IS REPLACED BY H "MESO" COMPOUNDS: $2-CH=CH_2$ IS REPLACED BY C_2H_5 CHLORIN E_6- TRIMETHYL ESTER: $R=CH_3$, RING V IS REPLACED BY $\gamma - CH_2CO_2CH_3$ AND $6-CO_2CH_3$		

FIG. 1. The structure of chlorophyll a and its derivatives.

aqueous hydrochloric acid (Manning and Strain, 1943) was also observed with the pheophytin of the above oxidation product. Treatment of the phyllin with potassium borohydride (Holt, 1959b) gave a "phase test-positive" product which had a spectrum almost identical with that of mesochlorophyll a (Fischer and Spielberger, 1935). These results suggested a 2-formyl group, a conclusion which was supported by the presence of a C=O absorption band at 1675 cm.$^{-1}$ in the infrared spectrum of the magnesium-free derivative [Fig. 4 (3)]. This structure was confirmed by elementary analysis and by comparison of the properties of its chlorin-trimethyl ester derivative with those previously described

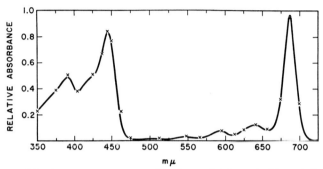

FIG. 2. Visible absorption spectrum of 2-desvinyl-2-formyl chlorophyll *a* in ether (Holt and Morley, 1959).

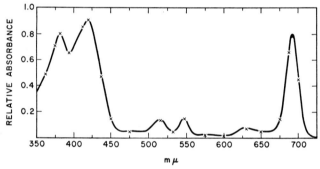

FIG. 3. Visible absorption spectrum of 2-desvinyl-2-formyl pheophytin *a* in ether (Holt and Morley, 1959).

for 2-desvinyl-2-formyl-chlorin-e_6-trimethyl ester (Fischer and Walter, 1941). The product equivalent to "isopheophytin *d*" was thus the dimethyl acetal derivative. Subsequent work has shown that 2-desvinyl-2-formyl chlorophyll *a* obtained as above and chlorophyll *d* isolated from red algae are apparently identical (Holt, 1959a).

The top zone contained pigment with a visible absorption spectrum almost identical to that of the borohydride reduction product of chlorophyll *d*, mentioned above. The chromatographic behavior plus the fact that the acid number of the methyl pheophorbide derivative was 9 suggested the presence of a hydroxyl group(s). [The acid number of methyl pheophorbide *a* is 15 (Willstätter and Stoll, 1913).] The properties of its chlorin-trimethyl ester derivative corresponded to those of 2-desvinyl-2-(1,2-dihydroxy-ethyl)chlorin-e_6-trimethyl ester (Fischer and Walter, 1941). Treatment of the pheophytin with periodic acid at room temperature readily yielded a product with the visible spectrum of pheophytin *d*, indicating the presence of the dihydroxyethyl group. The infrared spectrum of the pheophytin [Fig. 4(*1*)] showed only the C=O

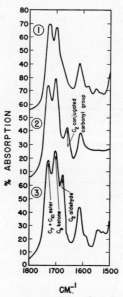

Fig. 4. Infrared absorption spectra of products of permanganate oxidation of chlorophyll *a*. Solvent: CCl_4. (*1*) 2-Desvinyl-2-(1,2 dihydroxyethyl)pheophytin *a*. (*2*) See text. (*3*) 2-Desvinyl-2-formyl pheophytin *a*.

bands characteristic of the C-7 and C-10 ester groups and the C-9 keto group (Holt and Jacobs, 1955), thus eliminating any possibility of an hydroxyaldehyde structure. Analysis of the monodinitrobenzoate derivative also confirmed the dihydroxyethyl structure.

The pigment in the middle zone has yet to be identified. The visible absorption spectrum resembles that of chlorophyll *d*, and also has its "red" absorption maximum at 686 mμ. It is converted by borohydride into a "phase-test-positive" product whose visible and infrared spectra are identical with those of the dihydroxyethyl derivative described above. Its chromatographic behavior and the fact that the acid number of the methyl pheophorbide is 12, compared to 9 and 16 for the methyl pheophorbides of the dihydroxyethyl and formyl derivatives, respectively, suggests the presence of a hydroxyl group and indicates a keto-alcohol structure. The infrared absorption spectrum of the pheophytin [Fig. 4 (2)] shows the absorption band of the 2-conjugated carbonyl group at 1663 cm.$^{-1}$. Treatment of the pheophytin with periodate yields a product which can be extracted from ether into aqueous alkali, and which has a visible spectrum corresponding to that of 2-desvinyl-2-carboxy-methyl-pyropheophorbide a (Fischer and Walter, 1941).

Chlorobium Chlorophyll

In an earlier paper of this volume, Stanier* demonstrated that two different chlorophylls may be isolated from two different strains of a green-sulfur bacterium, *Chlorobium thiosulfatophilum*. Examination of

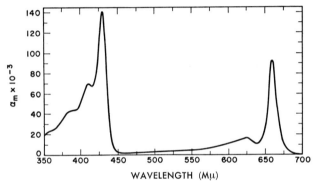

FIG. 5. Molar extinction curve of *Chlorobium* chlorophyll (660) in ether.

the literature indicates that, with the exception of Kaplan and Silberman, (1959), previous studies were made using organisms which contained the pigment to be described below (Goodwin, 1955; Katz and Wassink, 1939; Larsen, 1953; Metzner, 1922). This pigment will be called *Chlorobium* chlorophyll (660), abbreviated hereafter to CbCh. The figure in the brackets designates the wavelength in millimicrons of the "red" absorption maximum of an ether solution. The metal-free derivative will be referred to as Cb"Ph"; the quotation marks are used to show that the molecule may not contain phytol.

The visible absorption spectra of CbCh and Cb"Ph" are given in Figs. 5 and 6. They differ from those of Larsen (1953) and Goodwin (1955) chiefly by being free of absorption between 700 and 800 mµ. They are similar to those of the a, b, and d series of pigments insofar as they are characteristic of dihydroporphyrins (Stern and Wenderlein, 1935). The wavelengths of the "red" and "blue" maxima are identical,

---

* See Paper No. 5 in this volume.

or almost so, with those of chlorophyll *a* and pheophytin *a*, respectively (see Figs. 7 and 8). However, the band at 547 mµ of Cb"Ph" is unusual in that it absorbs more intensely than those at 610, 515, and 483 mµ, and indicates that CbCh must have quite a different structure from that of chlorophyll *a*. Below is a brief summary of some of the results which substantiate this conclusion.

I. Tests for the Presence of Ring V with Its C-10 Hydrogen Atom and Carbomethoxy Group

A. *Molisch Phase Test*

The addition of alkali to a solution of chlorophyll *a*, *b*, or *d* generates an intermediate which is stable in the absence of molecular oxygen or

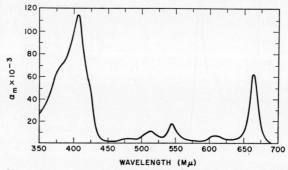

Fig. 6. Molar extinction curve of magnesium-free derivative of *Chlorobium* chlorophyll (660) in ether.

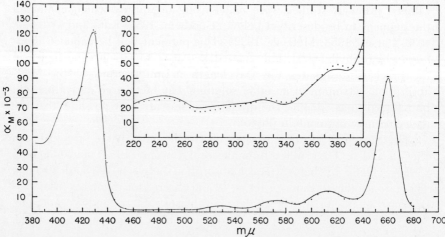

Fig. 7. Molar extinction curve of ethyl chlorophyllide *a* in ether (Holt and Jacobs, 1954).

other oxidants (Holt, 1958; Molisch, 1896; Weller, 1954). Relative to the spectrum of untreated pigment that of the intermediate is characterized by enhanced absorption in the "green" and diminished absorption in the "blue" and "red" regions of the spectrum. Its color can vary from yellow-green to red depending on the compound and upon the solvent and base used.

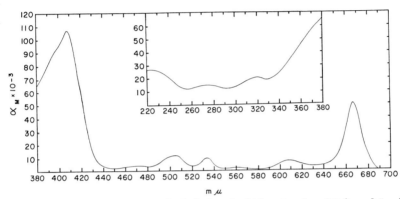

Fig. 8. Molar extinction curve of ethyl pheophorbide *a* in ether (Holt and Jacobs, 1954).

The addition of a methanolic solution of magnesium methoxide to a pyridine solution of CbCh caused no spectral change during 72 hours. Under these conditions CbCh differs from chlorophyll *a* since the latter is oxidized immediately to purpurin derivatives (Holt, 1958). The addition of methanolic KOH (30% w/v) to an ether solution of CbCh also gave no immediate color change. However, after several hours in the alkaline layer, it was converted to a product with a "red" maximum at ca. 645 mµ. The addition of 1–2 drops of the same reagent to a pyridine solution accelerated the formation of the "645 mµ" product but again caused no immediate color change. Preliminary observations have shown that an uptake of oxygen is probably involved in the reaction and suggests a similarity to allomerization. However, a spectrum characteristic of the phase test intermediate has not been observed in any of the above experiments.

B. *Methanolysis with Diazomethane*

Compounds which possess Ring V intact are converted to their chlorin-triester derivatives by diazomethane in methanol (Fischer and Riedmair, 1933). This reaction is readily detectable spectroscopically. Conditions which effected conversion of pheophytin *a* in ca. 100% yield in less than 24 hours did not change the visible spectrum of Cb"Ph" even after 72 hours.

C. Analysis of the Zinc Salt of Cb"Ph"

This showed the absence of a methoxyl group.

From these Experiments (IA, B, and C) it must be concluded that Ring V, if present at all, is greatly modified insofar as it lacks the carbomethoxy group and/or the C-10 hydrogen atom.

II. Tests for a β-Vinyl Substituent
A. Oxidation

Under conditions used for the oxidation of the 2-vinyl group of chlorophyll *a*, no spectral shift was observed with CbCh. Extending the reaction time or heating resulted only in gradual decomposition to product(s) having no visible absorption maxima.

B. Reduction

Additional support for the absence of a vinyl group is the fact that the visible spectrum of CbCh remained unchanged when catalytic hydrogenation (Fischer and Spielberger, 1935) was attempted under conditions which converted chlorophyll *a* entirely to meso-chlorophyll *a*.

III. Tests for Other Components
A. Metal

Magnesium was determined quantitatively by a procedure slightly modified from that of Smith and Benitez (1955). The molar extinction coefficients in Fig. 5 were calculated on the basis of a 1:1 ratio of magnesium to CbCh; the coefficients of the "blue" and "red" maxima of CbCh are slightly lower than those of chlorophyll *a* (Fig. 7).

B. Conjugated Carbonyl Group

Upon reduction of Cb"Ph" with potassium borohydride, the C=O band at 1693 cm.$^{-1}$ disappeared (Fig. 9), and the "red" maximum shifted from 666 mμ to ca. 648 mμ. Chromic acid in acetic acid–benzene reoxidized the reduced product to one which had a visible spectrum identical with that of Cb"Ph". The product could then be reduced again by borohydride. Such a cycle is analogous to that between pheophorbide *a* and 9-hydroxydesoxopheophorbide *a*.

The assumption that only one conjugated group is present seems valid because during borohydride reduction only a direct conversion of Cb"Ph" was observed spectroscopically. Stepwise reduction of the carbonyl groups of pheophytins *b* and *d* is known to occur; the formyl group is far more rapidly reduced than is the ring ketone (Holt, 1959b). Conclusive evidence awaits the results of further work now in progress.

C. Ester Group

The presence of an ester group was indicated by the fact that Cb"Ph" was not extractable from ether solution by aqueous alkali. Acid hydrolysis

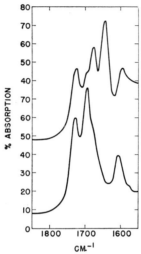

FIG. 9. Infrared absorption spectra of *Chlorobium* chlorophyll (660), *upper curve*, and its magnesium-free derivative, *lower curve*, in CCl_4.

of Cb"Ph" yielded a product extractable from ether by dilute alkali; its visible spectrum was identical with that of Cb"Ph".

A small amount of the alcohol was obtained by alkaline hydrolysis of Cb"Ph". Its infrared spectrum was compared to that of phytol obtained from pheophytin *a* and was found to be different. Gas-liquid partition chromatography showed also that it had a different retention volume (relative to that of quinoline) than that of phytol. Whether this difference will explain why the acid number of Cb"Ph" is 19 while that of pheophytin *a* is 29 (Willstätter and Stoll, 1913) remains to be determined.

IV. Products of Chromic Acid Oxidation

The products isolated from chromic acid oxidation (Fischer and Wenderoth, 1939; Muir and Neuberger, 1949) of the "free" acid obtained from Cb"Ph" by acid hydrolysis, confirmed the fact that CbCh must be different from any previously characterized chlorophyll. Dihydrohematinic acid imide was obtained from the "acid" fraction of the oxidation mixture. Two main products were obtained, in approximately 2:1 ratio, from the "neutral" fraction. The major constituent has been identified as methyl ethyl maleimide. The minor constituent has been identified as methyl *n*-propyl maleimide (Fischer et al., 1931) by melting point, elementary analysis, and nuclear magnetic resonance spectrum. This imide has not been obtained previously from a naturally occurring porphyrin (Holt and Morley, 1960; Morley, Cooper and Holt, 1959).

The above results permit the partial formula (A) for Cb"Ph" (where R is probably a long branched chain other than the phytyl chain).

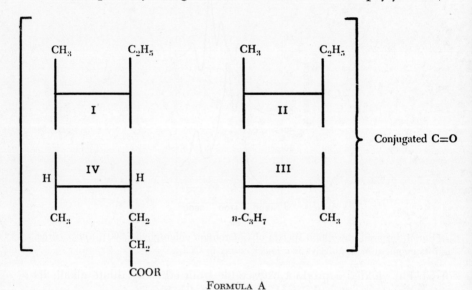

FORMULA A

It is expected that further studies of *Chlorobium* chlorophylls will reveal other similarities and differences between them and other known chlorophylls. The results may well be significant in the eventual understanding of how chlorophyll functions in photosynthesis.

REFERENCES

Fischer, H., and Riedmair, J. (1933). *Ann.* **506**, 107-123.
Fischer, H., and Spielberger, G. (1935). *Ann.* **515**, 130-148.
Fischer, H., and Walter, H. (1941). *Ann.* **549**, 44-79.
Fischer, H., and Wenderoth, H. (1939). *Ann.* **537**, 170-177.
Fischer, H., Goldschmidt, M., and Nüssler, W. (1931). *Ann.* **486**, 1-54.
Goodwin, T. W. (1955). *Biochim. et Biophys. Acta* **18**, 309-310.
Holt, A. S. (1958). *Can. J. Biochem. Physiol.* **36**, 439-456.
Holt, A. S. (1959a). Unpublished results.
Holt, A. S. (1959b). *Plant Physiol.* **34**, 310-314.
Holt, A. S., and Jacobs, E. E. (1954). *Am. J. Botany.* **41**, 710-717.

Holt, A. S., and Jacobs, E. E. (1955). *Plant Physiol.* **30**, 553-559.
Holt, A. S., and Morley, H. V. (1959). *Can. J. Chem.* **37**, 507-514.
Holt, A. S., and Morley, H. V. (1960). *J. Am. Chem. Soc.* **82**, 500.
Kaplan, I. R., and Silberman, H. (1959). *Arch. Biochem. Biophys.* **80**, 114-124.
Katz, E., and Wassink, E. C. (1939). *Enzymologia* **7**, 97-112.
Larsen, H. (1953). *Kgl. Norske Videnskab. Selskab. Skr.* pp. 1-205.
Manning, W. M., and Strain, H. H. (1943). *J. Biol. Chem.* **151**, 1-19.
Metzner, P. (1922). *Ber. deut. botan. Ges.* **40**, 125-129.
Muir, H. M., and Neuberger, A. (1949). *Biochem. J.* **45**, 163-170.
Molisch, H. (1896). *Ber. deut. botan. Ges.* **14**, 16-18.
Morley, H. V., Cooper, F. P., and Holt, A. S. (1959). *Chem. & Ind. (London)* p. 1018.
Smith, J. H. C., and Benitez, A. (1955). "Modern Methods of Plant Analysis," pp. 142-196. Springer, Berlin.
Stern, A., and Wenderlein, H. (1935). *Z. physik. Chem.* **A174**, 321-334.
Strain, H. H., Manning, W. M., and Hardin, G. (1943). *J. Biol. Chem.* **148**, 655-668.
Weller, A. (1954). *J. Am. Chem. Soc.* **76**, 5819-5821.
Willstätter, R., and Stoll, A. (1913). "Untersuchungen über Chlorophyll," p. 269. Springer, Berlin.

12

Chemical Studies of Phycoerythrins and Phycocyanins

COLM Ó HEOCHA

Chemistry Department, University College, Galway, Ireland

The work of Engelmann (1883), performed over 75 years ago, indicated that the biliproteins, phycoerythrins and phycocyanins, are photoreceptive pigments in photosynthesis in blue-green and red algae. Modern studies, to be reviewed later in this Symposium, have fully borne out Engelmann's conclusions.*

Several terms have been suggested to designate phycoerythrins and phycocyanins in general. These include *phycochromoproteids* (Kylin, 1910), *bilichromoproteins* (Haxo et al., 1955), *tetrapyrrl* proteins (Haurowitz, 1958) and *biliproteins* (Ó hEocha, 1958). The latter term has the advantage of being brief and specific and is used in this discussion. The generic term *phycobilins* is reserved exclusively for the biliprotein chromophoric groups. (Lemberg, 1928; Lemberg and Bader, 1933).

Occurrence and Preparation

In addition to the blue-green and red algae, the cryptomonads, a group of widely distributed marine microflagellates, are known to contain biliproteins (Allen et al., 1959; Haxo and Fork, 1959; Ó hEocha and Raftery, 1959). It is most likely that the biliproteins are located, with the other photosynthetically active pigments, within the chromoplasts (Rhodophyta and Cryptophyceae) or chromatophores (Cyanophyta).

In attached marine algae, the phycoerythrin content increases while the phycocyanin content decreases with increasing depth of growth (see, for example, Jones and Blinks, 1957); indeed, some deep-growing Rhodophyta are reported to contain phycoerythrin only (Kylin, 1931; Svedberg and Eriksson, 1932). Halldal (1958) is the most recent worker to produce spectral evidence showing that the qualitative biliprotein composition of some blue-green algae depends on the light intensity at which they are grown. At low intensity (75 foot-candles) he found that *Anabaena* sp., but not *Anacystis nidulans*, formed a considerable amount of phycoerythrin. In addition, phycocyanin was present at all intensities.

---

* See Paper No. 21 by F. T. Haxo.

The attached marine alga *Ceramium rubrum* was reported to contain a little less than 2% biliproteins, mostly phycoerythrin, in early spring (Kylin, 1910; Lemberg, 1928). The biliprotein content of cultured algae is variable; under optimum conditions, involving low light intensity, phycocyanin accounted for 24% of the dry weight and 40% of the total cell protein of *Anacystis nidulans* (Myers and Kratz, 1955).

Fresh undried algae form the best source of biliproteins. In many cases, these pigments leach into distilled water, while other more resilient plants require such treatment as grinding or repeated freezing and thawing before they release their biliproteins. Pigments may be purified from the aqueous extract by fractional precipitation with ammonium sulfate (Kylin, 1910) and/or by chromatography on tricalcium phosphate gel (Swingle and Tiselius, 1951; Krasnovskii et al., 1952; Haxo et al., 1955). The biliproteins may be crystallized from ammonium sulfate solution.

Physical Properties

The absorption spectra of phycoerythrins and phycocyanins show maxima in the ultraviolet at about 275 and 365 mµ, the former maximum being attributable to aromatic amino acids in the protein (Svedberg and Katsurai, 1929; Haxo et al., 1955; Bannister, 1954). The phycoerythrin spectra of Svedberg and Katsurai (1929) display ultra-violet absorption maxima at these wavelengths only, whereas the phycoerythrins we have examined (including that of *Ceramium rubrum*, studied by Svedberg) have a third maximum in the ultraviolet at 305–310 mµ (Fig. 1). Shibata and associates (1954) found that the spectrum of intact *Porphyridium cruentum* cells displays a slight shoulder at 310 mµ. This was not found in the case of a phycocyanin-containing blue-green alga examined and it seems certain that it is attributable to phycoerythrin absorption. It corresponds to an absorption maximum at 312 mµ in the phycoerythrin chromophore spectrum (see below).

Visible absorption spectra are among the most characteristic properties of individual biliproteins and are used to differentiate between them (Haxo and Ó hEocha, 1960). For example, R-phycocyanin (λ_{max} ca. 550 and 615 mµ) C-phycocyanin (λ_{max} ca. 615 mµ) and allophycocyanin (λ_{max} ca. 650 mµ) may be easily characterized by the positions of their absorption maxima. Biliprotein extinction coefficients have been published by Lemberg (1928) and Svedberg and Katsurai (1929). The exact positions of the absorption maxima are somewhat variable, depending on the source, and even the method of preparation (Fig. 2) (Haxo et al., 1955; Airth and Blinks, 1956; Halldal, 1958; Ó hEocha and Haxo, 1960). The greatest variability in spectral shape is found among the R-phycoerythrins, which may be characterized by three peaks (λ_{max} ca. 495, 540, and 565 mμ). The extinctions at these wavelengths vary greatly and in some proteins the maximum at 540 mμ is missing or is replaced by a shoulder (Svedberg and Eriksson, 1932). There is evidence that the extinction at this wavelength varies with season in the case of R-phycoerythrin from *Rhodymenia palmata* (Ó hEocha and Ó Reachtaire, 1958) (Fig. 1). However, it cannot be stated with certainty that the phyco-

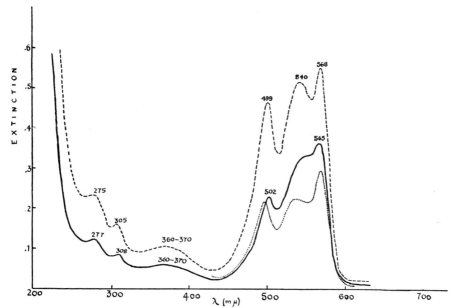

Fig. 1. Absorption spectra of chromatographically purified R-phycoerythrin in phosphate buffer, pH 6.5. Key: ————, from *Ceramium rubrum;* - - - - -, from *Rhodymenia palmata* (collected June); ......, from *R. palmata* (collected January).

erythrin spectrum undergoes changes *in vivo* because other constituents of the alga, seasonal in their occurrence, may effect the spectrum of the released phycoerythrin.

If the 495 mμ peak is lacking, the pigment is called B-phycoerythrin (Airth and Blinks, 1956) (Fig. 3). This biliprotein was first isolated by Airth and Blinks from *Porphyra naiadum* var. *naiadum* collected at Pacific Grove, California, and the spectral difference between it and R-phycoerythrin led to an investigation of the life-cycle of the alga, which has resulted in its being placed in a new genus, *Smithora* (Hollenberg, 1959). *S. naiadum* var. *australis,* which grows off Southern California, contains the same phycoerythrin as the northern variety; they both contain allophycocyanin also, but, while chromatography of var. *naiadum* yields C-phycocyanin, var. *australis* gives a labile pigment which is more reminiscent of R-phycocyanin, but distinctive from it (Ó hEocha and Haxo, 1960) (Fig. 3). Another instance of the possible significance of

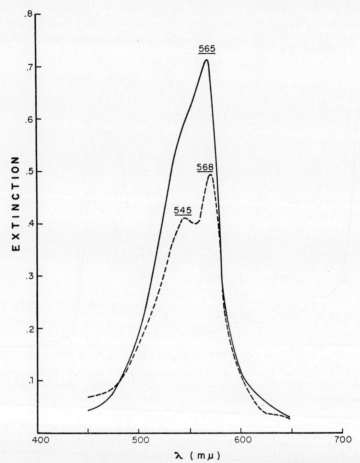

FIG. 2. Absorption spectra of aqueous extracts of *Phormidium ectocarpi*. KEY: ————, alga extracted in the cold for 2–3 hours; - - - - - -, alga extracted in the cold overnight. [The spectral change is probably enzymatically induced.]

biliprotein spectral characteristics in algal classification may be cited: the phycocyanin of the cryptomonad *Hemiselmis virescens* Droop (Millport No. 64) has three absorption maxima (580, 620–625, 645 mμ), as reported by Allen et al. 1959, while the biliprotein obtained from the Plymouth strain of this alga absorbs maximally at 588 and 615 mμ (Ó hEocha and

Raftery, 1959) (Fig. 4). The phycocyanins from both strains differ in their spectra from those of the blue-green and red algae.

The brilliant colors of the algal biliproteins attracted the attention of many physical chemists interested in studying the properties of pro-

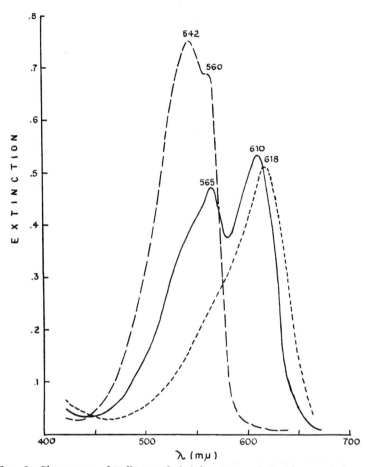

FIG. 3. Chromatographically purified biliproteins of *Smithora naiadum*. KEY: — — —, B-phycoerythrin from *S. naiadum* var. *australis* (in PO_4 buffer); ———, purple fraction from *S. naiadum* var. *australis* (in PO_4 buffer); - - - - - -, C-phycocyanin from *S. naiadum* var. *naiadum* (in saline). (From Ó hEocha and Haxo, 1960.)

teins. Tiselius (1930) showed by means of his electrophoresis apparatus that their isoelectric points lie in the pH region 4.25–4.85. His results and others placed phycoerythrin near the lower figure and phycocyanins near the higher value.

Svedberg and co-workers (Svedberg and Lewis, 1928; Svedberg and Katsurai, 1929; Svedberg and Eriksson, 1932; see also Eriksson-Quensel, 1938) applied their ultracentrifugation technique to a study of the molecular weights of the biliproteins under various conditions. They reported greatest stability near the pI of the proteins. R-Phycoerythrin

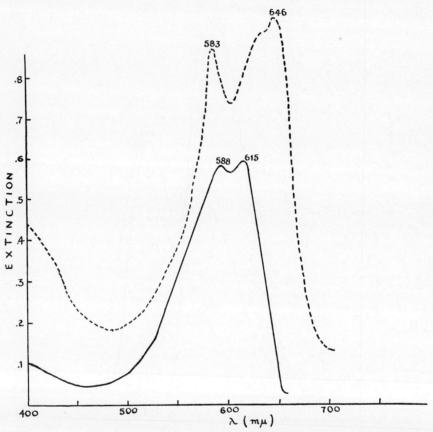

FIG. 4. Chromatographically purified phycocyanin from (- - - - - -) *Hemiselmis virescens* Droop (Millport no. 64); (———) *Hemiselmis virescens* (?) (Plymouth no. 157).

was stable at molecular weight 291,000 in the pH range 3–10 and R-phycocyanin at molecular weight 273,000 in the range 2.5–6.0. The phycoerythrin molecule was reported to be spherical, of density 1.33. The molecular weight of phycoerythrin from other sources was found to be very close to the Svedberg value by Krasnovskii *et al.* (1952) and Airth and Blinks (1956). Haxo *et al.* (1955) and Tiselius *et al.* (1956) found that fresh phycoerythrin extracts, which had not been exposed to extreme pH values, formed a number of zones on column chromatography. The spectral properties of the various phycoerythrin fractions were identical. The latter authors felt that, in view of Svedberg's stability studies, mentioned above, the multiple chromatographic zones were not due to dissociation of R-phycoerythrin, which is ultracentrifugally homogeneous under the experimental conditions used. They may be caused by heterogeneity of protein binding due to there being a number of sites in the molecule capable of attachment to the adsorbent, each with its own binding energy (Sober and Peterson, 1957).

Airth and Blinks (1957) found that fresh B-phycoerythrin preparations formed two zones on chromatography. The pigments of the two fractions differed markedly in electric charge, since they could be readily separated also by electrophoresis. The charge difference disappeared on storage at pH 7.0.

Protein Structure

The bulk of the weight of phycoerythrins and phycocyanins can be accounted for in terms of amino acid residues. Many biliproteins have been subjected to amino acid analyses in recent years (Wassink and Ragetli, 1952; Sisakyan et al., 1954; Fujiwara, 1956; Jones and Blinks, 1957; Kimmel and Smith, 1958; Raftery and Ó hEocha, 1958). The usual amino acids of plant proteins were reported to be present in nearly all the pigments, and Kimmel and Smith, having compared the products of hydrolysis after 20, 70, and 140 hours, suggested that at least some of the unidentified compounds reported by other workers were peptides resulting from incomplete hydrolysis of the biliproteins.

R-Phycoerythrin from *Porphyra tenera* (Nori) has been examined in three different laboratories (Fujiwara, 1956; Kimmel and Smith, 1958; Raftery and Ó hEocha, 1958) and the results differ considerably, the one from the other. The discrepancies may be due to differences in the preparation of the commercial plant source (Nori), as well as the methods of hydrolysis and estimation of the amino acids, but there may also be a seasonal and environmental influence on algal biliprotein composition. There is evidence that mineral nutrition, for example, exerts some control over the quantitative composition of other plant proteins (Schütte and Schendel, 1958; Pleshkov and Fowdon, 1959).

Some of the amino acid analyses of hydrolyzed biliproteins obtained by Raftery and Ó hEocha are given in Table I. The two-dimensional paper chromatographic method of Levy and Chung (1953) was used to separate the amino acids which were estimated by reaction with nin-

TABLE I
AMINO-ACID COMPOSITION OF BILIPROTEINS[a,b]

Amino acid	R-PE Rhodymenia palmata	C-PE Phormidium persicinum	R-PE Porphyra tenera	B-PE Porphyridium cruentum	R-PC Rhodymenia palmata	C-PC Arthrospira maxima	Allo-PC Arthrospira maxima
Aspartic acid	10.7	12.5	10.4	11.1	11.2	12.7	9.2
Glutamic acid	9.4	10.3	4.6	8.9	13.5	12.0	14.0
Serine	7.1	8.0	6.6	7.9	9.0	4.0	6.1
Threonine	5.1	3.5	2.6	5.8	6.1	5.6	9.3
Glycine	5.5	8.6	6.8	10.0	4.8	5.8	6.0
Alanine	11.3	15.3	16.3	11.9	16.6	12.5	9.6
Valine	9.9	9.0	11.8	7.1	5.5	8.3	7.9
Isoleucine	4.1	5.0	4.8	4.9	3.9	5.3	3.7
Leucine	11.1	8.5	9.0	7.2	10.4	10.0	9.7
Phenylalanine	4.0	3.2	2.6	4.0	1.6	5.1	2.3
Tyrosine	4.1	2.0	2.0	3.7	3.6	2.6	5.4
Proline	5.4	2.3	4.5	4.3	1.0	5.9	4.9
Histidine	1.7	1.8	3.9	2.0	1.5	2.3	1.6
Lysine	3.9	4.5	2.9	6.2	6.0	2.9	2.6
Arginine	3.7	4.6	7.8	7.0	3.5	3.1	4.0
Methionine	2.2	1.0	2.6	1.0	1.3	2.3	2.2
Cystine	0.8	0.5	0.8	0.5	0.5	0.3	1.5

[a] Percentage by weight of recovered amino acids. (PE = phycoerythrin; PC = phycocyanin.)
[b] Raftery and Ó hEocha (1958).

hydrin (Naftalin, 1948). Tryptophan was not estimated because of complications due to the presence of the tetrapyrrolic chromophore, which interferes with the determination (Fujiwara, 1956). These results are mainly of comparative interest now, since the ion-exchange resin methods give more accurate absolute values for amino acid composition.

We have found, in common with other workers, that the acidic amino acids are present in relatively large amounts in all the biliproteins examined. The R-phycoerythrin and R-phycocyanin of *Rhodymenia palmata* (Table I) contain an excess of dicarboxylic over basic amino acids whereas the reverse is true of the bulk proteins of this alga (Coulson, 1955; Smith and Young, 1955). Another feature of biliproteins from all sources is their high alanine and leucine contents.

Kimmel and Smith (1958) failed to account for the elementary sulfur content of either R-phycoerythrin or C-phycocyanin (from *P. tenera*) in terms of half-cystine and methionine. In this they agree with Fujiwara (1956), who suggested on the basis of this evidence that the chromophores may be attached to half-cystine units through thioether linkages, as in the case of cytochrome *c*. These linkages would not cleave during hydrolysis and consequently the cystine yield would be low. However, as pointed out by Kimmel and Smith (1958) the possibility of contamination by bound sulfate (from ammonium sulfate used during isolation of the protein) cannot be overlooked. We have prepared R-phycoerythrin (from *Ceramium rubrum*) without recourse to ammonium sulfate precipitation (Raftery and Ó hEocha, 1958). The aqueous algal extract was purified by chromatography on tricalcium phosphate (Haxo et al., 1955), followed by the ion-exchanger Amberlite IRC- 50, and then dialyzed against distilled water until it was free of phosphate and sodium. The protein was finally precipitated with a large excess of acetone and washed with acetone and ether before drying. In this experiment cystine was determined as cysteic acid (Schram *et al.*, 1954), which method—although not quantitative—gives higher values than those reported in Table I. Assuming a recovery value of 88% for cystine (Huisman, 1959), only 65% of the total sulfur content of the phycoerythrin (1.66%) could be accounted for as S-containing amino acids. Fujiwara (1955) reported the presence of carbohydrate in her purest phycoerythrin, and if the carbohydrate were a sulfate ester, characteristic of algae, it could account for the high sulfur content, as well as the fact that the total weight of phycoerythrin cannot be accounted for in terms of amino acid residues. It remains to be seen whether or not phycoerythrin is a true glycoprotein.

Kimmel and Smith (1958) calculated the isolectric point of the two *P. tenera* chromoproteins they analyzed from their content of acidic and basic amino acids and their amide nitrogen. The calculated figure for C-phycocyanin agreed remarkably well with the experimentally determined one of 4.76 (Lemberg, 1930b) but the analysis of R-phycoerythrin would indicate an excess of basic groups with resultant basic pI, whereas the experimental value is 4.25 (Lemberg, 1930b). This discrepancy could be explained, according to these authors, if some of the basic groups were bound or if the phycoerythrin contained a non-amino-acid acidic component.

The terminal groups of some biliproteins have been determined by Raftery and Ó hEocha (1958). The N-terminal groups were determined by the DNP-method of Sanger and the phenylisothiocyanate method of Edman (Fraenkel-Conrat et al., 1955). The C-terminal sequence was determined by the carboxypeptidase method (Fraenkel-Conrat et al., 1955) and in the case of R-phycoerythrin by the hydrazinolysis method (Bradbury, 1958). Carboxypeptidase hydrolysis indicated that alanine was C-terminal in R- and B-phycoerythrin and R-phycocyanin. This amino acid was also obtained from R-phycoerythrin (*C. rubrum*) by hydrazinolysis; this treatment yielded a derivative of mesobilirubin as well. On the basis of her hydrazinolysis experiments, Fujiwara (1957) deduced that the chromophore was C-terminal in R-phycoerythrin (*Porphyra tenera*).

We find approximately 20 C-terminal alanine residues per molecular weight unit while the N-terminal amino acids determined in the case of R-phycoerythrin were leucine (10) and serine (9). The minimal molecular weight calculated for R-phycoerythrin from these data agrees well with the value of 14,600 which was calculated from its histidine content by Kimmel and Smith (1958).

Phycoerythrin is partially decomposed into six fragments at pH 11.0 (Svedberg and Katsurai, 1929). In view of their dissociation at this pH, these fragments are probably attached through tyrosine–carboxyl linkages in the intact chromoprotein and may consist in turn of 3-4 subunits, joined through disulfide bridges.

Phycobilins

The bile pigments are open-chain tetrapyrroles which may be classified according to the number of doubly-bonded methine groups (=CH—) joining their pyrrole rings (Lemberg and Legge, 1949; Gray, 1953). Bilitrienes contain the maximum number, three, bilidienes, two, bilenes, one, while bilans contain three joining methylene groups (—CH_2—). The convention employed in the numbering of substituted β-carbon atoms is indicated in the bilitriene formula of Table II. Nat-

TABLE II
THE BILE PIGMENTS[a]

Name	Formula	Number of conjugated double bonds	Color [and $\lambda_{max.}$ in mμ]	Fluorescence of zinc-complex salt
Bilitriene Mesobiliverdin		10	Green to blue [640; 392 (methanol)]	None
Bilidiene Mesobiliviolin		8	Violet [570–575 (chloroform)]	Red
Mesobilirhodin		7	Red [ca. 575 (chloroform)]	Yellow
Mesobilirubin		5	Yellow to orange [425 (chloroform)]	None
Bilene i-Urobilin		5	Yellow [452 (dioxan)]	Green
Bilan Mesobilirubinogen		2	Colorless	None

[a] Based on Lemberg and Legge (1949) and Gray (1953).

urally occurring bile pigments have methyl groups in positions 1,3,6,7 and propionic acid chains in positions 4 and 5. Carbon atoms 2 and 8 may have vinyl or ethyl substituents. This arrangement is denoted by the suffix IXα. Where there are vinyl groups on C-2 and C-8, the prefix proto may be used, although it usually is not; when there are ethyl groups present the prefix meso is applied. Thus biliviolin has vinyl groups on C-2 and C-8, while mesobiliviolin has ethyl radicals attached to these carbon atoms.

In Table II the bile pigment structures are presented in the usual lactim form but the bislactam structure (I) accounts better for some of the properties of the bile pigments (Gray and Nicholson, 1958).

(I)

MESOBILIVIOLIN
(bislactam form)

The bilan, mesobilirubinogen, is very unstable and forms i-urobilin on dehydrogenation. Urobilin gives a mixture of mesobiliviolin and mesobilirhodin on mild dehydrogenation, while stronger ferric chloride treatment yields mesobiliverdin. The bilidienes were separated chromatographically by Siedel, who also synthesized them from dipyrrylmethenes (Siedel, 1935; Siedel and Möller, 1940). Siedel's work indicated that the isomeric mesobiliviolin and mesobilirhodin were derived from urobilin by dehydrogenation of one or other of the side methylene groups; however, the mesobilirhodin structure is still open to question (Lemberg and Legge, 1949; Rabinowitch, 1956). Recent work shows that biliviolinoid pigments are also obtained from urobilin by isomerization under nonoxidizing conditions (Gray and Nicholson, 1958).

Strong oxidizing agents attack the methine group which is oxidized, through a number of intermediate stages, to a carbonyl group. Thus mesobiliverdin yields mesobilipurpurin (II), while mesobiliviolin forms "oxo"-urobilin (III). The spectral properties and fluorescent derivatives of these carbonyl compounds resemble those of naturally occurring bile pigments with similar conjugated double-bond systems, but many of them have been incompletely characterized and the literature concerning them is difficult to correlate (Lemberg and Legge, 1949).

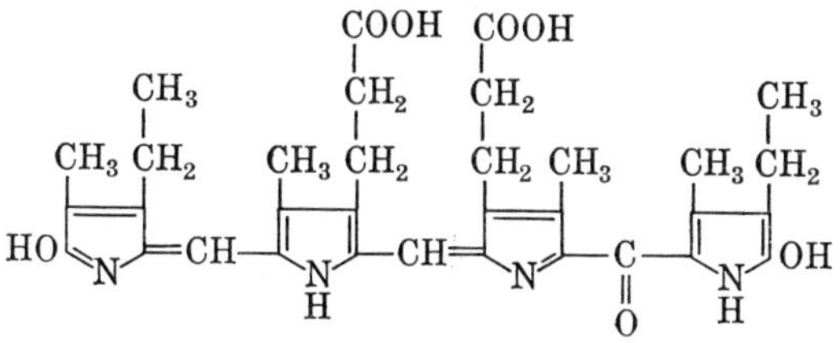

(II)

MESOBILIPURPURIN

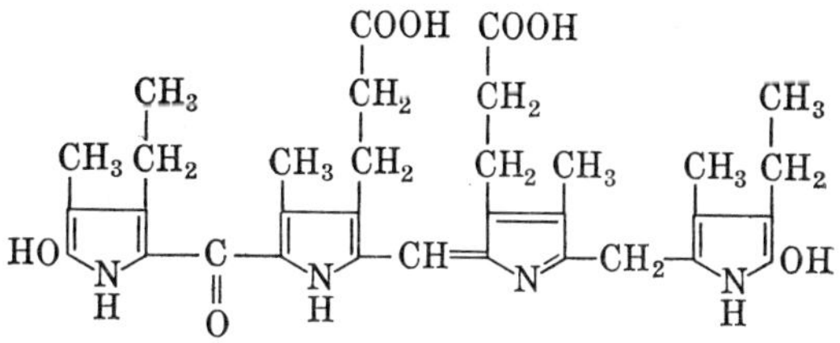

(III)

"OXO"-UROBILIN

A. PHYCOCYANOBILIN

The identification of the chromophores of phycoerythrins and phycocyanins as bile pigments derives from the work of Lemberg (1928, 1930a; Lemberg and Bader 1933). He coined the term *phycobilins* to indicate the algal origin of these pigments and differentiated between *phycocyanobilin* (chromophore of C-phycocyanin) and *phycoerythrobilin* (chromophore of R-phycoerythrin). Lemberg concluded that phycocyanobilin was identical with mesobiliviolin. His latest estimate is that there are 16 residues of mesobiliviolin per molecule of chromoprotein (Lemberg and Legge, 1949). It was found by Clendenming, Ke, and Curry (quoted by Ó hEocha, 1958) that Lemberg's original determination (1930a) was too low and their value of about 4% chromophore by weight of phycocyanin agrees with his present estimate, as does the value deduced from a study of polarization of C-phycocyanin fluorescence (Goedheer, 1957). These results, when taken in conjunction with the terminal group determinations of Raftery and Ó hEocha (1958) indicate that each minimal molecular weight unit of C-phycocyanin contains a chromophoric group.

The phycobilins are strongly bound to the proteins, and Lemberg used concentrated hydrochloric acid at 80°C to obtain mesobiliviolin from C-phycocyanin. Methanolic sodium hydroxide treatment yielded mesobiliverdin, which was also obtained on ferric chloride oxidation of the pigment from the acid hydrolyzate, thereby confirming its violin structure (rather than a purpurin structure) (Lemberg and Legge, 1949). The meso arrangement of mesobiliverdin from phycocyanin, and

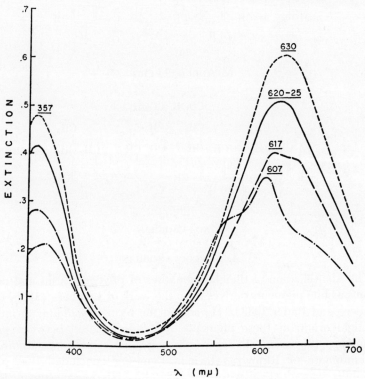

FIG. 5. Absorption spectra in acid chloroform of phycocyanobilin (top curve) and its breakdown products formed in concentrated hydrochloric acid at room temperature. C-phycocyanin (*Arthrospira maxima*) hydrolyzed for: - - -, 30 minutes; ———, 60 minutes; ― ― ―, 2 hours; —·—, 7 hours (mesobiliviolin). (From Ó hEocha, 1958.)

therefore of phycocyanobilin, was confirmed by Siedel and Melachrinos (quoted by Siedel, 1944).

There is recent spectral evidence to indicate that mesobiliviolin is not the true chromophore of C-phycocyanin but an artifact formed from phycocyanobilin in concentrated hydrochloric acid solution (Fig. 5) (Ó hEocha, 1958). The chloroform-soluble phycocyanobilin was obtained on short-term hydrolysis of C-phycocyanin from *Arthrospira maxima* in concentrated hydrochloric acid. Its structural relationship to mesobiliviolin is not yet established, but since it forms a red-fluorescing Zn-complex, it is assumed to possess a violin structure. Allophycocyanin yielded a somewhat similar pigment (λ_{max} at 640 mμ), but the R-variety differs appreciably in its chromophore system from the other phycocyanins.

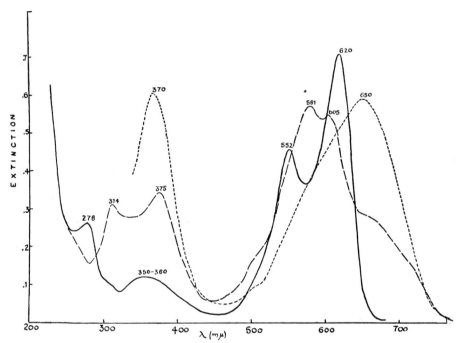

Fig. 6. R-phycocyanin and its phycobilins. Key: ———, R-phycocyanin (phosphate buffer pH 6.5; ———, chlorofrom-soluble phycobilins; - - - - -, acetone-soluble phycobilin. [Phycobilins obtained from 20 minute R. T. hydrolyzate.]

R-Phycocyanin from *Ceramium rubrum* was hydrolyzed under the conditions described elsewhere, i.e., concentrated hydrochloric acid at room temperature for 20–30 minutes, followed by dilution and extraction with chloroform (Ó hEocha, 1958). The blue chloroform extract had a complex absorption spectrum (Fig. 6), and was apparently a mixture. On the addition of zinc acetate, orange fluorescence appeared despite the red fluorescence of the parent chromoprotein. It seems probable that phycoerythrobilin (see below) forms a major component of this extract. A precipitate was formed when the acid hydrolyzate of R-phycocyanin was diluted, and when this precipitate was treated with acetone, a solution of a blue pigment was obtained (Fig. 6) which did not form a fluorescent zinc-complex. Some of its properties resemble those of mesobiliverdin, which is only slightly soluble in chloroform. It seems likely that R-phycocyanin has two different chromophoric groups: phycoerythrobilin and a blue pigment possibly related to mesobiliverdin. An acetone-soluble blue pigment (λ_{max} 650 mμ) was also obtained from C-phycocyanin from *Nostoc muscorum* and *Anabaena* spp. on acid hydrolysis. Phycocyanin from these algae also gave a low yield of chloroform-soluble phycobilin which forms a red-fluorescing zinc complex but is distinguished by the position of its absorption maximum (660 mμ) from *Arthrospira* phycocyanobilin (Fig. 5).

B. Phycoerythrobilin

Lemberg's preparation of phycoerythrin chromophore was probably still attached to a small peptide chain, as only its methyl ester was chloroform-soluble (Lemberg and Legge, 1949). Using milder hydrolyzing conditions than those of Lemberg, Ó hEocha (1958) succeeded in obtaining a chloroform-soluble phycoerythrobilin from all the phycoerythrin types characteristic of the red and blue-green algae. This has also been obtained from cryptomonad phycoerythrin (Ó hEocha and Raftery, 1959). Phycoerythrobilin was characterized by absorption maxima at 312 and 576 mμ in acid chloroform (Fig. 7) and 593 mμ in

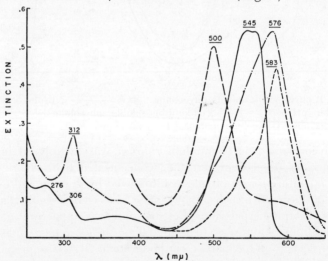

Fig. 7. *Porphyridium* phycoerythrin and its phycobilins. Key: ———, Phycoerythrin (in phosphate buffer pH. 6.5); — · — · — ·, phycoerythrobilin (in acid chloroform); - - - - - -, phycoerythrobilin zinc-complex (in chloroform); — — —, oxidized phycoerythrobilin. (From Ó hEocha, 1958.)

neutral chloroform. Its zinc-complex salt was orange-fluorescent, and from its behavior on paper chromatography it was assumed to contain two free carboxylic acid groups among its side chains.

The phycoerythrin chromophore is very labile; even bubbling nitrogen through its chloroform solution caused a pronounced shift in

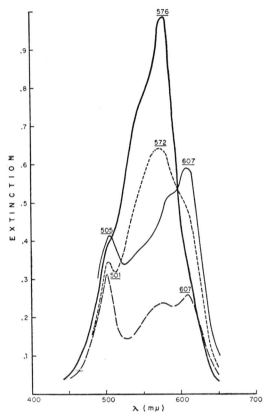

Fig. 8. Absorption spectra in acid chloroform of *Porphyridium* phycoerythrobilin (top curve) and its breakdown products formed in concentrated hydrochloric acid. Phycoerythrin hydrolyzed for: ————, 20 minutes; - - - - -, 2 hours; ————, 5½ hours; — — —, 18 hours. (From Ó hEocha, 1958.)

λ_{max}; heating or long standing in concentrated hydrochloric acid resulted in the spectral changes shown in Fig. 8. The final curve represents a mixture—separable into three zones when chromatographed on talc; and partially separated into two zones, characterized by greenish-yellow and red fluorescence, when paper chromatographed in water-saturated 2,4-lutidine (Ó Carra and Ó hEocha, 1959). The substance absorbing maximally at 501 mμ could be removed from ethereal solution by washing with distilled water. Urobilin (λ_{max} at 499 mμ in acid chloroform) can be separated from biliviolinoid pigments in this way, (Gray and Nicholson, 1958), and it seems likely that phycoerythrobilin is converted, at least in part, to a urobilinoid pigment on standing in acid. A drop of concentrated nitric acid converts phycoerythrobilin immediately to urobilinoid pigment (λ_{max} 500 mμ) (Fig. 7), while both phycoerythrobilin and phycocyanobilin zinc-complex salts are converted by iodine to a green-fluorescing substance with λ_{max} at 510 mμ [somewhat reminiscent of "oxo"-urobilin (III) which was obtained from mesobiliviolin by a similar treatment (Siedel and Möller, 1940)]. The principal absorption maximum and the green fluorescence of Lemberg's "phycoerythrobilin"–Zn complex, may have been due to urobilinoid and/or "oxo"-urobilinoid pigments which are readily formed from phycoerythrobilin.

As is true of R-phycocyanin, R-phycoerythrin produces an acetone as well as a chloroform-soluble pigment on acid treatment. The acetone solution is characterized by absorption maxima at 565 and 500 mμ (slight) and the pigment bears some resemblance to mesobilirhodin. It turns blue on standing, when the main peak shifts to 600 mμ, with a shoulder at 650 mμ.

One phycoerythrin [that from *Rhodymenia palmata* (Fig. 1)] has been encountered which differs from the rest in that it releases only a very small amount of pigment when subjected to acid hydrolysis. The chloroform-soluble material thus obtained absorbs much more strongly in the red than the usual phycoerythrobilin, while its zinc-complex is red-fluorescing (Fig. 9). It would appear that this R-phycoerythrin releases a biliviolinoid pigment, in addition to a small amount of phycoerythrobilin on acid hydrolysis. However, the bulk of the biliprotein chromophore is not released by this treatment (Ó Carra and Ó hEocha, 1959).

While the pigment yield from *Rhodymenia palmata* phycoerythrin is particularly low, our hydrolysis treatment does not release the total phycobilin content of any biliprotein we have examined. An attempt was made to increase the phycoerythrobilin yield from *C. rubrum* phycoerythrin by extracting the hydrolyzate with chloroform every 20 minutes. The unhydrolyzed protein precipitated out when the hydrolyzate was diluted with water and the precipitate was centrifuged off and rehydrolyzed. Four such hydrolyses yielded chloroform-soluble phycoerythrobilin, but additional acid treatment yielded no further pigment. The concentrated acid solution of the residual unhydrolyzed protein was reddish-brown in color and, on dilution, showed maximum absorption at 500 mμ with a slight shoulder at 550–560 mμ. This solution turned violet on standing at room temperature for a day. Stercobilin, a tetrahydro derivative of urobilin, undergoes such a change (Watson, 1948); but since our material is not extractable into organic solvents, it may be a urobilinoid pigment which is still attached to protein fragments.

Part of this urobilinoid pigment is undoubtedly formed from phyco-

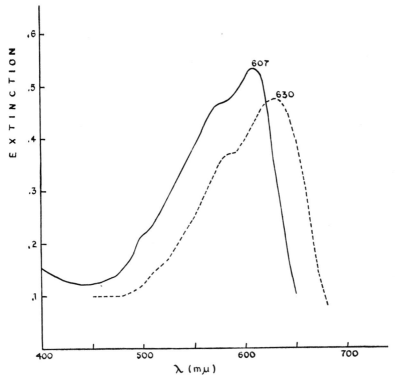

Fig. 9. Absorption spectra of acid-hydrolyzed, chloroform-soluble pigments from R-phycoerythrin (*Rhodymenia palmata*). Key: ———, pigment in acid chloroform; - - - - -, zinc-complex in chloroform. (Ó Carra and Ó hEocha, 1959.)

erythrobilin, but a urobilinoid phycobilin may form one component of the native R-phycoerythrin chromophore system accounting for the λ_{max} at 495 mµ. If so, urobilinoid pigment would be expected to be absent from B- and C-phycoerythrins in their natural states.

The protein-urobilinoid bonds (natural or artifact) must be much stronger than those linking the protein with the other phycobilins, since no urobilinoid pigment is released as such as a result of our acid hydrolysis. The yield of bile pigment from choleglobin, a biliverdin-containing product of hemoglobin oxidation, is very low. The explanation appears to be that as the choleglobin is denatured the prosthetic group becomes more firmly attached through the formation of sulfur bridges with cysteine of the denatured protein (Lemberg, 1956). It may be that secondary linkages, formed as a result of acid denaturation of the protein, are also responsible for the difficulty encountered in increasing the phycobilin yield from the algal biliproteins. In this connection, it is of interest that a 4-year-old dried sample of C-phycocyanin, which had been a good source of phycocyanobilin when first dried, yielded no phycobilin when treated with concentrated HCl.

Protein-Chromophore Attachment

The comparatively drastic treatment necessary to cleave the prosthetic groups from the biliproteins led Lemberg (1930a) to postulate a peptide linkage between the propionic acid side chains of the phycobilins and amino groups in the protein. Lysine, with its ϵ-amino group, is the likely amino acid to be involved in this linkage. However, in the case of R-phycoerythrin from *C. rubrum* we have failed to obtain any evidence for participation of the lysine ϵ-amino groups in peptide bonds (Ó Carra and Ó hEocha, 1959). The phycoerythrin was treated with dinitrofluorobenzene, which reacts with the ϵ-amino group of lysine in addition to the terminal amino groups of the protein (Fraenkel-Conrat *et al.*, 1955). The DNP-protein was then hydrolyzed, but its yield of lysine, as ϵ-DNP derivative, did not differ significantly from the lysine content of the untreated protein. It appears therefore that lysine ϵ-amino groups are not involved in chromophore bonding in R-phycoerythrin.

We have attempted to get confirmation for Fujiwara's (1956) suggestion that the phycoerythrin chromophore may be attached to the protein through thioether bonds. R-Phycoerythrin (*Rhodymenia palmata*) was treated with the mild oxidizing reagent silver sulfate, used by Paul (1950) to release the cytochrome *c* prosthetic group, but we failed to release any pigment from the biliprotein (Ó Carra and Ó hEocha, 1959).

The biliproteins are strongly fluorescent substances. All the phycoerythrins examined by French displayed fluorescence maxima at 578 mμ, while maxima have been reported at 637 and 663 mμ for C-phycocyanin and allophycocyanin respectively (French *et al.*, 1956; French and Young, 1956). Dilute acid quenches biliprotein fluorescence with a concomitant shift in absorption maxima; if the acid treatment is mild (e.g., standing at pH 3.4 for a few hours), fluorescence may be restored by raising the pH again to 6–7. At extreme pH values there are remarkable changes in fluorescence color; for example, on the addition of a few drops of alkali to R-phycocyanin its fluorescence is decreased but changed in color from red to brilliant orange, while alkali changes the fluorescence color of R-phycoerythrin from orange to brilliant red. Lemberg (1930a) found that C-phycocyanin, the fluorescence of which was quenched by $N/10$ HCl, reacted with ammoniacal zinc acetate to form a fluorescent zinc-complex. It was concluded by Lemberg that in quenching fluorescence, the acid unmasked the pyrrole nitrogens of the chromophore which then reacted with zinc to form another fluorescent complex. This did not involve replacement of another metal by zinc, because it appears from the analytical data that the chromophore is metal-free in the native state, at least in R-phycoerythrin (Lemberg, 1928). Lemberg interpreted his study to mean that fluorescence depends on a second, more labile linkage between chromophore and protein.

We find that 8 M urea quenches the red fluorescence of phycocyanins instantly, while phycoerythrin fluorescence is quenched after about a day at 0°C. This hydrogen-bond-cleaving reagent also unmasks metal-binding sites in protein-bound phycoerythrobilin, and it appears that the fluorescence of native biliprotein solutions depends on the existence of such bonding between chromophores and proteins.

Acknowledgment

This work was supported in part by the U. S. Department of the Army, through its European Research Office.

References

Airth, R. L., and Blinks, L. R. (1956). *Biol. Bull.* **111**, 321-327.
Airth, R. L., and Blinks, L. R. (1957). *J. Gen. Physiol.* **41**, 77-90.
Allen, M. B., Dougherty, E. C., and McLaughlin, J. J. A. (1959). *Nature* **184**, 1047-1049.
Bannister, T. T. (1954). *Arch. Biochem. Biophys.* **49**, 222-233.
Bradbury, J. H. (1958). *Biochem. J.* **68**, 482-486.
Coulson, C. B. (1955). *J. Sci. Food Agr.* **6**, 674-682.
Engelmann, T. W. (1883). *Botan. Z.* **41**, 1-13, 17-29; (1884). **42**, 81-93, 97-105.
Eriksson-Quensel, I-B. (1938). *Biochem. J.* **32**, 585-589.
Fraenkel-Conrat, H., Harris, J. I., and Levy, A. L. (1955). *Methods of Biochem. Anal.* **2**, 359-425.
French, C. S., and Young, V. M. K. (1956). *Radiation Biol.* **3**, 343-392.
French, C. S., Smith, J. H. C., Virgin, H. I., and Airth, R. L. (1956). *Plant Physiol.* **31**, 369-374.
Fujiwara, T. (1955). *J. Biochem. (Tokyo)* **42**, 411-417.
Fujiwara, T. (1956). *J. Biochem. (Tokyo)* **43**, 195-203.
Fujiwara, T. (1957). *J. Biochem. (Tokyo)* **44**, 723-733.
Goedheer, J. C. (1957). Doctoral Thesis, State University, Utrecht, pp. 37-38.
Gray, C. H. (1953). "The Bile Pigments." Methuen, London.
Gray, C. H., and Nicholson, D. C. (1958). *J. Chem. Soc.* pp. 3085-3099.
Halldal, P. (1958). *Physiol. Plantarum* **11**, 401-420.

Haurowitz, F. (1958). *In* "Handbuch der Pflanzenphysiologie" (W. Ruhland, ed.), Vol. VIII, pp. 338-340. Springer, Berlin.
Haxo, F. T., and Fork, D. C. (1959). *Nature* **184**, 1051-1052.
Haxo, F. T., and Ó hEocha, C. (1960). *In* "Handbuch der Pflanzenphysiologie" (W. Ruhland, ed.), Vol. V, pp. 497-510. Springer, Berlin.
Haxo, F., Ó hEocha, C., and Norris, P. S. (1955). *Arch. Biochem. Biophys.* **54**, 162-173.
Hollenberg, G. J. (1959). *Pacific Naturalist* **1**, 1.
Huisman, T. H. J. (1959). *In* "Sulfur in Proteins" (R. Benesch *et al.*, eds.), p. 161. Academic Press, New York.
Jones, R. F., and Blinks, L. R. (1957). *Biol. Bull.* **112**, 363-370.
Kimmel, J. R., and Smith, E. L. (1958). *Bull. soc. chim. Biol.* **40**, 2049-2065.
Krasnovskii, A. A., Evstigneev, V. B., Brin, G. P., and Gavrilova, V. A. (1952). *Doklady Akad. Nauk. S. S. S. R.* **82**, 947-950.
Kylin, H. (1910). *Z. physiol. Chem.* **69**, 169-239.
Kylin, H. (1931). *Z. physiol. Chem.* **197**, 1-6.
Lemberg, R. (1928). *Ann. Chem. Liebigs* **461**, 46-89.
Lemberg, R. (1930a). *Ann. Chem. Liebigs* **477**, 195-245.
Lemberg, R. (1930b). *Biochem. Z.* **219**, 255-257.
Lemberg, R. (1956). *Revs. Pure and Appl. Chem. (Australia)* **6**, 1-23.
Lemberg, R., and Bader, G. (1933). *Ann. Chem. Liebigs* **505**, 151-177.
Lemberg, R., and Legge, J. W. (1949). "Hematin Compounds and Bile Pigments." Interscience, New York.
Levy, A. L., and Chung, D. (1953). *Anal. Chem.* **25**, 396-399.
Myers, J., and Kratz, W. A. (1955). *J. Gen. Physiol.* **39**, 11-22.
Naftalin, L. (1948). *Nature* **161**, 763.
Ó Carra, P., and Ó hEocha, C. (1959). Unpublished observations.
Ó hEocha, C. (1958). *Arch. Biochem. Biophys.* **73**, 207-219.
Ó hEocha, C., and Haxo, F. T. (1960). *Biochim. et Biophys. Acta* (in press).
Ó hEocha, C., and Ó Reachtaire, M. (1958). *Abstr. 4th Intern. Congr. Biochem.* Sec. **11-55** p. 147.
Ó hEocha, C., and Raftery, M. (1959). *Nature* **184**, 1049-1051.
Paul, K. G. (1950). *Acta Chem. Scand.* **4**, 239-244.
Pleshkov, B. P., and Fowdon, L. (1959). *Nature* **183**, 1445-1446.
Rabinowitch, E. (1956). "Photosynthesis," Vol. II, Pt. II. Interscience, New York.
Raftery, M., and Ó hEocha, C. (1958). Unpublished observations.
Schram, E., Moore, S., and Bigwood, E. J. (1954). *Biochem. J.* **57**, 33-37.
Schütte, K. H., and Schendel, H. E. (1958). *Nature* **182**, 958-959.
Shibata, K., Benson, A. A., and Calvin, M. (1954). *Biochim. et Biophys. Acta* **15**, 461-470.
Siedel, W. (1935). *Z. physiol. Chem.* **237**, 8-34.
Siedel, W. (1944). *Ber. deut. chem. Ges.* **77A**, 21-42.
Siedel, W., and Möller, H. (1940). *Z. physiol. Chem.* **264**, 64-90.
Sisakyan, N. M., Bezinger, E. N., and Kivkutsan, F. R. (1954). *Doklady Akad. Nauk S. S. S. R.* **98**, 111-114.
Smith, D. G., and Young, E. C. (1955). *J. Biol. Chem.* **217**, 845-853.
Sober, H. A., and Peterson, E. A. (1957). *In* "Ion Exchangers in Organic and Biochemistry" (C. Calmon and T. R. E. Kressman, eds.), p. 340. Interscience, New York.

Swingle, S. M., and Tiselius, A. (1951). *Biochem. J.* **48**, 171-174.
Svedberg, T., and Eriksson, I-B. (1932). *J. Am. Chem. Soc.* **54**, 3998-4010.
Svedberg, T., and Katsurai, T. (1929). *J. Am. Chem. Soc.* **51**, 3573-3583.
Svedberg, T., and Lewis, N. B. (1928). *J. Am. Chem. Soc.* **50**, 525-536.
Tiselius, A. (1930). *Nova Acta Regiae Soc. Sci. Upsaliensis* **7**, 1-107.
Tiselius, A., Hjertén, S., and Levin, Ö. (1956). *Arch. Biochem. Biophys.* **65**, 132-155.
Wassink, E. C., and Ragetli, H. W. J. (1952). *Koninkl. Ned. Akad. Wetenschap. Proc.* **55C**, 462-470.
Watson, C. J. (1948). *Harvey Lectures Ser.* **44**, 41-83.

Discussion

NEILANDS: If the prosthetic group in these pigments is so firmly bound, can you hydrolyze with a proteolytic enzyme and get out a peptide analogous to that from cytochrome *c*?

Ó HEOCHA: It should be possible. It has not been done. The peptides have been obtained by proteolytic enzymes, but most people have been trying to break the link between chromophore and protein rather than to get peptides. Peptides have been obtained on a number of occasions, but nobody has analyzed the amino acids.

13

Biosynthesis of Carotenoids

G. MACKINNEY AND C. O. CHICHESTER

*Department of Food Science and Technology, University of California,
Berkeley and Davis, California*

Introduction

Because of our interest in this Symposium in the carotenoid pigments as photoreceptors, it will be possible to omit much that can be classified here as controversial detail, important though it may be in other contexts. Thus we can deal cursorily with such problems as whether alicyclic carotenoids arise from aliphatic ones, and with even narrower problems such as uncertainty as the correct formula for one compound or the precise location of the hydroxyl group in another, because they are not really germane to the present purpose and are discussed elsewhere. Thus Goodwin (1959) has recently reviewed problems in biosynthesis and function and Mackinney (1960) prepared a chapter entitled "Carotenoids and Vitamin A" in *Metabolic Pathways*.

Nevertheless, the evidence on which some conclusions have been drawn is not as adequate or decisive as one might wish, and there is currently much work under way in different laboratories dealing particularly with the synthesis of lycopene or of β-carotene and, until these findings are published, comment must necessarily be tentative.

A comparative biochemist must begin by presenting a picture of carotenoid distribution in the plant and animal worlds which involves consideration of an imposing array of compounds, many of them species-specific; this tends to confuse the issue for the nonspecialist, particularly when dealing with nonphotosynthetic organisms or tissues. The question then arises whether we can state the problem more simply.

Carotenoids are an essential part of the photosynthetic apparatus in plants, and are needed as sources of provitamin A in animals. Otherwise, carotenoid production in plants or its storage (whether unchanged or modified) in animals is nowhere a biochemical necessity for survival of the species, and numerous instances may be recalled: e.g., the white-rooted carrot, or the loss of carotenoid in captivity of the brilliantly colored marine goldfish. This is not to imply that because of a given genetic make-up, or the presence of some nutrient not normally utilized or needed, a particular strain or variety may not be under the necessity of synthesizing carotenoids, and conditions under which this may happen will be reviewed in subsequent sections.

Insofar as a central thesis can be established, differing from previous reviews prepared for other purposes, we shall consider the biosynthesis of the carotenoids in relation to their function.

The Carotenoid Structure

It is necessary first to examine the carotenoid structure to determine its potential usefulness to the cell. A property common to all conjugated polyenes is their ability to form carbonium ions in the presence of a proton donor or an electron acceptor. This property undoubtedly increases the ease with which are formed what Mulliken (1952) has termed *charge-transfer complexes*.

The characteristic feature of complexes of this type, between two ions, molecules or compounds A and B, is that an intense electronic spectrum exists for AB, nonexistent for either partner alone. These complexes vary greatly in stability and we may mention two examples: (*a*) the carotene-protein chromatophore found in spinach leaves by Nishimura and Takamatsu (1957), maximum for the complex ca. 530 mμ, cf. maximum for carotene ca. 480 mμ; (*b*) retinene, ca. 378 mμ, and the protein opsin, combined as rhodopsin, maximum ca. 500.

Mulliken's concept has been further developed by Platt (1958), who proposed a trimolecular charge complex, the macrostructure of which requires the carotenoid to serve as a link between electron donor and acceptor in rigidly oriented, highly organized structures such as the chloroplasts. It is possibly not surprising that experimental difficulties have hitherto precluded successful attempts to demonstrate conversion of compounds such as acetate or mevalonate into carotenoid in these structures. As shown by Goodwin (1958), CO_2 is by far the most readily incorporated into plastid carotene, and he suggests that failure with the other compounds may merely mean that they are spatially removed from the synthetic sites.

If the rigid spatial restrictions for photosynthetic activity are no longer needed, as in a green tomato fruit during ripening, it is by no means certain that the steps in carotenoid synthesis will necessarily be the same, in whole or in part, as in the normal green leaf. This applies also to nonphotosynthetic tissues and organisms.

In the photosynthetic tissue, the complex chromatophores perform highly specialized functions. Any major tampering with these structures will affect, quantitatively and possibly qualitatively, the performance of these functions. Thus a mutation affecting the carotenoid also affects the chlorophyll, and Bergeron (1959) has observed that loss of the ability of the carotenoid to transfer energy to the bacteriochlorophyll in a *Chromatium* mutant is correlated with the increase in absorption at 800 mµ, and the decreases at 850 and 880 mµ.

Although one may assume that some of the associated enzymes remain functional as the chloroplast structure disintegrates, we must recognize a serious limitation in presupposing that the biochemical steps in nonphotosynthetic tissues will necessarily parallel those in the green leaf. With this in mind, we may consider the effects of genetic and cultural factors.

Gene Control of Carotenoid Differences

Mutations involving changes in carotenoid pigmentation in *Rhodopseudomonas, Neurospora, Chlorella,* and the fruits of the tomato have received the most detailed attention. With respect to the last mentioned, the ripe fruit of several wild species is devoid of carotenoid, although in two (both native to the Galapagos Islands), β-carotene is found.

Lycopene predominates in the red-fruited species. Here two independent mutations occurred, prior to the conquest of the New World, giving rise to a weakly colored mutant, yellow, with only traces of the colorless polyenes, and tangerine, containing predominantly ζ-carotene, prolycopene, and phytoene. Even here, where only a two-gene difference exists, there is an anomaly. The genotypes may be represented respectively by alleles rt^+ and r^+t. It is generally accepted that r^+ controls polyene production, and t^+ the lycopene configuration. We should not then expect the yellow-tangerine, doubly recessive, of genotype rt, to contain any more pigment than the yellow rt^+, yet such is the case.

An even more serious anomaly involves the ghost gene, gh, also recessive, described by Mackinney *et al.* (1956). The effect of this gene is to eliminate virtually all the colored polyenes, replaced in varying degree in the different combinations (e.g., red ghost, r^+gh; yellow ghost, $r\ gh$; tangerine ghost, $t\ gh$) by phytoene.

Thus the only conclusion we can draw is that there is a considerable amount of gene interaction. This leaves us with the conviction that the gene action appears complicated, not because it really is, but because we are analyzing an end-result. We infer that what the gene controls must lie a considerable way back in the sequence of biochemical events, possibly an event prior to the formation of the C_{40} chain. In any event, there is no reason to suspect that genetic evidence in the other cases will be capable of a radically different or simpler explanation. The effects described in the tomato are all single gene differences, verified by traditional genetic procedures, laborious and slow though they be in practice.

Effects of Culture Conditions

NUTRIENTS

The effect of thiamine concentration on carotenoid pigmentation in *Corynebacteria* was studied by Braun (1949) and Starr and Saperstein (1952). This determines whether aliphatic or alicyclic carotenoids predominate.

Of interest to Goodwin and ourselves (see Goodwin and Lijinsky, 1951; Yokoyama *et al.*, 1957) has been the effect of leucine on β-carotene production in *Phycomyces*. The level of β-carotene in this organism may vary from less than 100 to 3500 µg. per gram dry mycelium, depending upon the culture conditions. Studies with labeled leucine showed a twofold preference for the C-4 as contrasted with the C-3 in the resulting labeling of the carotene. Labeling in the 2, 3, 4, or 5 positions leads to labeled β-hydroxy-β-methylglutaric acid (HMG). The quantity of HMG synthesized in the absence of leucine is scarcely to be detected. Consequently it is possible to explain the preferential use of the C-4 in terms of how the mold reacts to the potential threat of acetone bodies, as a detoxification mechanism. If we accept such an explanation, we need not be surprised at the variety of carotenoids of a species-specific character in nonphotosynthetic tissues and organisms. The end-result is merely the visible expression of the ability of the organism to handle the fragments involved.

As shown by Yokoyama *et al.* (1960), mevalonate is convertible to HMG by cell-free preparations of *Phycomyces*. A multiplicity of effects can be explained in terms of whether a given metabolite contributes, or is inhibited from contributing, to a general pool from which the carotenoid precursors must be drawn. The pool undoubtedly includes acetoacetate. Thus we can fit another fact to our inferences, namely, the effect of streptomycin in reducing the level of carotenoid (see, for example, Goodwin and Griffiths, 1952) and the suggestion of two pathways becomes unnecessary.

OXYGEN

Wong (1953) found that the carotenoid synthesis in *Phycomyces* was unaffected by lowered oxygen tensions, down to levels at which the mold would not grow, between 0.3 and 0.7% oxygen. Where some other cause for growth retardation existed, e.g., addition of β-ionone, higher levels of oxygen were required, above 3%.

Light

It is well established that carotene is formed in *Phycomyces* cultured completely in the dark. A light effect is observable, but the magnitude is small, of the order of 20 to 30% in experiments reported by Chichester *et al.* (1954). No significant difference was observed between continuous exposure to light during the day, and exposure for 4 hours, 16 hours after inoculation. Since at this stage, there is negligible carotene synthesis, the effect must be on some other photoreceptor, presumably a flavin or a pteridine. In any event, the effect is small compared with that of ionone or methylheptenone.

Ionone and Methylheptenone

The presence of traces of ionone in the culture medium causes the most remarkable increase in β-carotene production in the mold, and we have had yields as high as 0.35% on a dry-weight basis, under the most favorable conditions. Effects of ionone and light are additive, but whereas ratios for lighted to dark were ca 1.25 to 1, those for ionone-treated to control were ca. 4.6 to 1.

Methylheptenone causes a marked increase in the more saturated polyenes. Mackinney *et al.* (1954) therefore applied heptenone-ionone mixtures to determine to what extent they acted independently. The results obtained, summarized as follows, unfortunately did not include phytoene, which probably accounts for the low total polyene reported for heptenone alone. In μg. per gram, values for the control were: 163.5 for β-carotene; 172.4 for the total polyene. Corresponding figures for ionone alone were 522.4 and 529.6; for an ionone-heptenone mixture (2:1), 536.3 and 587.4; for a mixture (1:1), 440.2 and 487.7; for a mixture (1:9), 362.2 and 520.3; for heptenone alone, 67.7 and 217.2. The effects of the two compounds are clearly independent. This led us to a tentative schematic representation of the syntheses as follows:

Step I: $A_1 \longrightarrow A_n$
Step IIa: $A_n \longrightarrow X; X_n \longrightarrow$ phytofluene, ζ-carotene
Step IIb: $A_n \longrightarrow Y; Y_n \longrightarrow$ β-carotene

Step I is merely a formal representation of all steps in common up to a branch point. Ionone-heptenone mixtures stimulated production of both the more saturated aliphatic hydrocarbons and the alicyclic β-carotene to a much greater extent than did the control, yet neither component of the mixture could exert its full effect; competition for precursor A_n in Steps IIa and IIb would appear to be the simplest explanation for this phenomenon.

Diphenylamine

As explained by Turian and Haxo (1952), diphenylamine directs metabolism toward the formation of less oxidized representatives of the C_{40} series. Thus Step IIa, above, would be favored over Step IIb. However the effect of starvation (Varma et al., 1959) has caused some modification of our views. Carotene disappears from a starved mycelial mat without concomitant rise in any other polyene. When the starved mat is then placed in an energy-rich medium, i.e., glucose, carotene levels are slowly raised. If diphenylamine is present, the level is restored to normal much more rapidly, within 12 to 24 hours. If the diphenylamine is exerting its effect as an antioxidant, it can only be by preventing oxidative destruction of preformed endogenous precursors. In the case of unstarved mats, it may be effective in controlling the metabolism in diverting intermediate A_n to X rather than to Y in the schematic diagram, following the suggestion made by Turian and Haxo.

The Formation of the C_{40} Chain

It is not possible at this juncture to give an unequivocal answer to the question as to whether there is one C_{40} precursor common to all natural polyenes, or even whether there is more than one C_{20} unit, assuming 2 C_{20} units condense tail-to-tail, as required by Lynen et al. (1958) for farnesol C_{15}, to squalene C_{30}. Our own unease with the evidence so far accumulated and published on mevalonate stems in part from the leucine data for carbons 3 and 4. The evidence indicates recycling of HMG with the acetoacetate pool (Chichester et al., 1959), and greater assurance is needed that mevalonate is the take-off point in all cases. Furthermore, to accept the hypothesis of a single pathway would require an involved explanation to account for the data of Purcell et al. (1959). These workers injected labeled mevalonate into ripening tomatoes and measured the specific activities of the various carotenoid components. We are left with no clue as to the identity of the common precursor for alicyclic and aliphatic forms.

Our approach to the problem is by use of C^{14}-labeled tritiated substrates. Whether this will prove profitable is necessarily conjectural at this stage, but it would at the least be reassuring to ascertain the T/C^{14} ratios found in lycopene and β-carotene in the ripening tomato for HMG, mevalonate, and possibly also for acetate.

Transformations in the C_{40} Molecule

IN PURPLE BACTERIA

The most thorough rate study yet made, by Jensen et al. (1958), establishes a sequence from phytofluene, through ζ-carotene to neurosporene, lycopene, and finally spirilloxanthin in the purple bacterium, *Rhodospirillum*. When diphenylamine was added to photosynthetically growing cultures there was a rapid accumulation of the more saturated carotenoids. The diphenylamine was then removed and the cells resuspended in buffer and incubated anaerobically in the light. An endogenous synthesis of normally occurring carotenoids was then demonstrated to occur at the expense of the more saturated carotenoids which had accumulated.

Of more restricted nature is the yellow-red pigment conversion in *Rhodopseudomonas* noted by van Niel (1947), though this may be typical of many more cases where light or dark and aerobic or anaerobic conditions prevail. These pigments Y and R are monomethoxy dihydro derivatives of lycopene and lycopenone respectively. The red pigment differs from the yellow by one oxygen atom, the respective formulas being $C_{41}H_{58}O_2$ and $C_{41}H_{60}O$.

On the basis of their studies on *Rhodopseudomonas* mutants (Griffiths and Stanier, 1956), it was possible for them to combine the two sets of data, to show divergence at neurosporene thus:

Common precursors $\longrightarrow$ neurosporene $\longrightarrow$ (a) or (b)
(a) = pigments Y and R (*Rhodopseudomonas*)
(b) = lycopene $\longrightarrow$ P_{481} $\longrightarrow$ Spirilloxanthin (*Rhodospirillum*)

XANTHOPHYLLS AND EPOXY-DERIVATIVES

There is as yet no unequivocal evidence as to the mode of formation of the 3-monohydroxy and 3,3'-dihydroxy derivatives present in green leaves. Whether the oxygen originates in water or molecular oxygen has not yet been shown. It will be recalled that the oxygen in lanosterol, which in turn derives from the hydrocarbon squalene, comes from molecular oxygen, not water. *Chlorella* mutants which favor xanthophyll formation would probably be the most suitable test organisms. The hydroxy derivatives of the purple bacterial carotenoids probably have their substituent hydroxyls located at C-1, in which respect they differ from their alicyclic counterparts.

The epoxy derivatives found in green leaves are formed by the addition of oxygen across the 5,6 double bond.

Light Effects

Claes and Nakayama (1959) have recently shown that conversion of *cis* to *trans* forms occurs in the presence of chlorophyll when the system is irradiated with red light. Still more recently (Claes and Nakayama, 1960) they have shown that the protective effect of carotene on photo-oxidation of chlorophyll observed by Aronoff and Mackinney (1943) can be exactly duplicated for neurosporene, lycopene, dehydrosqualene, lutein, and β-carotene. Rate losses were identical. However, ζ-carotene, phytofluene, phytoene, and squalene had no effect and gave the same curve as the chlorophyll alone. It would seem from this that a molecule with at least 9 bonds in conjugation is required to secure this protective effect.

In view of the foregoing, we recently exposed solutions of pure chlorophyll *a* and β-carotene in petroleum ether containing 10% acetone to red light of wavelengths longer than 600 mμ until the control, chlorophyll *a* alone, was virtually colorless. The chlorophyll *a*–carotene mixture was still colored, and from this solution was isolated a small quantity of yellow pigment, absorption maxima at 425 and 400 mμ, with a shallow intervening minimum. The pigment reacted strongly with 20% hydrochloric acid to give a deep blue coloration. Approximately 50% of the β-carotene was recovered unchanged, and we have here a possible clue as to the mechanism of the protective effect.

Summary and Conclusions

A review on the biosynthesis of carotenoids cannot be definitive at this stage, although many promising approaches are being explored. The most attractive, analogous to the farnesol condensation to yield squalene, involves geranyl-geraniol, C_{20}, which would condense to yield the desired C_{40} precursor. Whether all carotenoid syntheses can be explained in terms of a single C_{40} precursor or whether modification of the C_{20} molecule will yield different C_{40} precursors are questions as yet unanswerable. If we accept data indicating radically different activities for the various tomato carotenoids derived from labeled mevalonate, a single pathway will not explain the data. Neither does a critical evaluation of the genetic results with tomatoes permit a conclusive answer. It will be noted however that in *Rhodopseudomonas* mutants, in going from neurosporene to pigments Y and R, we go from a hydrocarbon to a methoxylated derivative, and finally to a ketone, and primary control of these transformations would lie right at these points, and not, as postulated for the tomato, far back in the biochemical sequence.

The effect of cultural conditions on carotenoid accumulation in nonphotosynthetic tissue is the easiest to observe experimentally. If carotenoids play no essential role in plants apart from their function in photosynthesis, then their accumulation in nonphotosynthetic organisms can be regarded as a detoxification mechanism, particularly when one considers how the mold *Phycomyces* handles the terminal iso-C_3 fragment of leucine.

With regard to biochemical sequences in various transformations, there is now convincing evidence for the sequence beginning with phytofluene, a branching at neurosporene, proceeding in *Rhodospirillum* to lycopene, and finally spirilloxanthin; in *Rhodopseudomonas* proceeding to pigments termed for convenience Y and R. The fact that some *Chlorella* mutants also contain the phytofluene-to-lycopene sequence constitutes a plausible argument for a general carotenoid pattern following a single pathway, but a more conservative view requires only that we regard it as one among several possibilities.

References

Aronoff, S., and Mackinney, G. (1943). *J. Am. Chem. Soc.* **65**, 956.
Bergeron, J. A. (1959). Division of Biol. Chem., Am. Chem. Soc. Abstr. of Papers 61c, September Meeting, Atlantic City, New Jersey.
Braun, A. C. (1949). *Phytopathology* **39**, 171.
Chichester, C. O., Wong, P. S., and Mackinney, G. (1954). *Plant Physiol.* **29**, 238.
Chichester, C. O., Yokoyama, H., Nakayama, T. O. M., Lukton, A., and Mackinney, G. (1959). *J. Biol. Chem.* **234**, 3.
Claes, H., and Nakayama, T. O. M. (1959). *Nature* **183**, 1053.
Claes, H., and Nakayama, T. O. M. (1960). *Z. Naturforsch.* **14b**, 746.
Goodwin, T. W. (1958). *Biochem. J.* **70**, 612.
Goodwin, T. W. (1959). *Advances in Enzymol.* **21**, 295.
Goodwin, T. W., and Griffiths, L. A. (1952). *Biochem. J.* **51**, xxxiii.
Goodwin, T. W., and Lijinsky, W. (1951). *Biochem. J.* **50**, 268.
Griffiths, M., and Stanier, R. Y. (1956). *J. Gen. Microbiol.* **14**, 698.
Jensen, S. L., Cohen-Bazire, G., Nakayama, T. O. M., and Stanier, R. Y. (1958). *Biochim. et Biophys. Acta* **29**, 477.
Lynen, F., Eggerer, H., Henning, U., and Kessel, I. (1958). *Angew. Chem.* **70**, 738.
Mackinney, G. (1960). *In* "Metabolic Pathways" (D. M. Greenberg, ed.), p. 481. Academic Press, New York.
Mackinney, G., Chichester, C. O., and Wong, P. S. (1954). *Arch. Biochem. Biophys.* **53**, 479.
Mackinney, G., Rick, C. M., and Jenkins, J. A. (1956). *Proc. Natl. Acad. Sci. U. S.* **42**, 404.
Mulliken, R. A. (1952). *J. Am. Chem. Soc.* **74**, 811.
Nishimura, M., and Takamatsu, K. (1957). *Nature* **180**, 699.
Platt, J. R. (1958). *Science* **129**, 372.
Purcell, A. E., Thompson, G. A., and Bonner, J. (1959). *J. Biol. Chem.* **234**, 1081.
Starr, M. P., and Saperstein, S. (1952). *Arch. Biochem. Biophys.* **43**, 157.

Turian, G., and Haxo, F. T. (1952). *J. Bacteriol.* **63**, 690.
van Niel, C. B. (1947). *Antonie van Leeuwenhoek J. Microbiol. Serol.* **12**, 156.
Varma, T. N. R., Chichester, C. O., and Mackinney, G. (1959). *Nature* **183**, 188.
Wong, P. S. (1953). M.S. Thesis, Univ. of Calif., Berkeley, California.
Yokoyama, H., Chichester, C. O., Nakayama, T. O. M., Lukton, A., and Mackinney, G. (1957). *J. Am. Chem. Soc.* **79**, 2029.
Yokoyama, H., Chichester, C. O., and Mackinney, G. (1960). *Nature* **185**, 687.

14

Biosynthesis and Possible Relations among the Carotenoids and between Chlorophyll a and b[*]

J. M. ANDERSON, U. BLASS,[†] AND M. CALVIN

Lawrence Radiation Laboratory, University of California, Berkeley, California

It is generally conceded that the carotenoids must have some indispensable function in photosynthesis; this has been variously formulated in terms of physical or chemical roles. We are concerned with one of these ideas, namely whether or not the carotenoids are involved in photosynthetic oxygen transport. This is quite an old idea; some years ago Dorough and Calvin (1951) looked at the isotopic oxygen content of the carotenoids from algae photosynthesized in H_2O^{18}, but the techniques necessary to answer this question were not refined enough at that time. We decided to look at the carbon-14 labeling of the carotenoids instead.

During the course of this work Sapozhnikov et al. (1957) reported a reciprocal relationship between lutein and the diepoxide, violaxanthin, which was influenced by light; in the dark the lutein concentration decreased and that of the violaxanthin increased, while the reverse reaction was induced by light. This would have as a consequence a requirement that the specific radioactivity of C^{14}-labeled violaxanthin and lutein be nearly equal.

As we proceeded it appeared that there was a marked difference between the activities of chlorophyll *a* and *b* so it seemed worthwhile to look at this also. It was the purpose of this research to try to find rapid and accurate techniques for the determination of the specific radioactivities of the algal pigments (Blass et al., 1959).

The algae used in these experiments were *Chlorella pyrenoidosa* Chick and *Scenedesmus obliquus* (Turpin) Kuetzing, grown in the continuous, constant-density culture apparatus by the methods currently used in our laboratory (Bassham and Calvin, 1957). The algal cells (1.2 ml wet packed cells/80 ml nutrient solution) were allowed to photosynthesize in the "lollipop" of the "steady state" apparatus which was

---

[*] The work described in this paper was sponsored in part by the United States Atomic Energy Commission and in part by the Department of Chemistry, University of California, Berkeley, California.

[†] *Present address:* Sandoz A. G., Basel, Switzerland.

illuminated with about 4,000 foot-candles of light and contained circulating 2% CO_2-in-air (Anderson, 1959); the carbon-14 was supplied as either $NaHC^{14}O_3$, 1 ml of 0.036 M solution (5.5 μcuries/μmole), or as $C^{14}O_2$ (11.1 μcuries/μmole) inserted in a spiral included in the circulating gas system of the "steady state" apparatus. Samples of algae were withdrawn at various time intervals, centrifuged, and the algal cells extracted twice with boiling methanol (1.5 ml methanol/0.1 ml wet packed cells); diethyl ether (ether-methanol, 1:1) was added to the combined cooled, methanolic extract and sodium chloride solution (5%) cautiously poured into the tubes until two layers were obtained. After centrifugation the ethereal layer containing the pigments was removed, washed with water several times by centrifugation, dried with a little anhydrous Na_2SO_4, and kept in the dark at 0° until ready for use.

A preliminary attempt to separate the algal pigments by paper chromatography was unsuccessful even with prior partial separation by partition methods. Despite the excellent resolution of the pigments on one- and two-dimensional chromatograms, the corresponding radioautograms were disappointing. Figure 1 shows a typical result obtained from the separation of the chlorophylls; the complete streaking of the radioactivity from the chlorophyll a to b is illustrated, and indicates the danger of measuring the radioactivity on the paper. The longer time required for better chromatographic separation of the pigments from the colorless contaminants was sufficient for marked decomposition of the pigments to occur, and yet the whole success of the method depended on adequate chromatographic resolution; thus an impasse had been reached.

In order to have greater amounts of material to work with, the extracts were first separated by column chromatography. For this purpose two adsorbents were used concurrently; namely, polyethylene and cellulose powder. One half of the methanolic algal extract was placed on a polyethylene column (15 gm powder) which had been previously packed as a slurry in 80% aqueous methanol under slight pressure. [In the case of the carotenoid extracts it was found to be advantageous to prewash the column with a solution of EDTA-Na_4* in methanol-water (1:1).] The polyethylene column was then developed with 80% aqueous methanol; after the elution of the chlorophylls, the concentration of methanol was gradually increased to 95% aqueous methanol. The other half of the methanolic extract was extracted with redistilled petroleum ether (b.p. 70°) and, after washing the petroleum ether extract several times with sodium chloride solution (5%) and drying over anhydrous

---

* EDTA-Na_4 is ethylenediaminetetraacetic acid – tetrasodium salt.

Na₂SO₄, the extract was applied to a cellulose column which had been previously packed under slight pressure as a slurry with petroleum ether (b.p. 70°). The column was developed with petroleum ether, which could be reinforced with 1% isopropanol after the chlorophyll *a* and *b* bands were beginning to travel down the column.

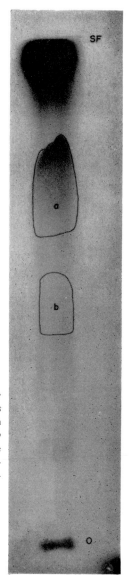

Fig. 1. Radioautograph of the chlorophylls from *Scenedesmus* after 2 hours photosynthesis with $C^{14}O_2$. The lines around *a* and *b* represent the green areas of the corresponding paper chromatogram. Solvent: toluene-acetone-isopropanol (100:2.5:2.5).

The separation pattern of the pigments on these two columns is shown in Fig. 2; it can be seen that the development of the pigment zones on the polyethylene is the reverse of that found with cellulose or sucrose adsorbents. This is an advantage because the colorless contaminants may run in different positions on these columns and a check on the purity and radioactivity measurements may be made for each pigment which has been separated on two different adsorbents.

The radioactivity of the successive fractions from the columns was determined. Very little correspondence between the chromatographic distribution of the radioactivity and the light-absorbing pigments was

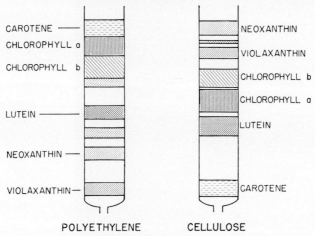

FIG. 2. Separation of the pigments of *Chlorella* on a polyethylene column (left) with 80% aqueous methanol and on a cellulose column (right) with petroleum ether.

found with either polyethylene or cellulose columns. It is obvious that although the pigments may be obtained in a spectroscopically pure state by the column separations, no validity can be attached to radioactive measurements at this point. It was, therefore, necessary to purify the pigments from the colorless contaminants which accompanied them; this was achieved by two-dimensional paper chromatography.

Aliquots of the pigment solutions were applied to Whatman No. 3 MM paper and descending chromatograms were made in the dark, using toluene as the first-dimension solvent and petroleum ether-isopropanol (100:2.5) as the second-dimension solvent. The radioactive areas on the paper chromatograms were detected by exposure to DuPont x-ray film type 507. The pigment spots were eluted from the paper in the following manner: with methanol-benzene (4:1) for the carotenoids, and isopropyl ether-isopropanol (9:1) for the chlorophylls. In the case of the carotenoids it was necessary to use calcium hydroxide chromatography (petroleum ether) in order to separate the carotenes into the α- and β-isomers.

From the resulting solutions the pigment concentrations were determined in the Cary spectrophotometer, Model 14, and the radioactivity determined by planchet counting or by liquid scintillation techniques. The spectroscopic data and radioactive techniques are fully reported elsewhere (Anderson, 1959).

The molar radioactivities of chlorophyll a and b from *Scenedesmus* are shown in Table I after 7.5 and 8.5 hours of photosynthesis with

TABLE I

Specific Radioactivity of the Pigments from *Scenedesmus* after Photosynthesis with $NaHC^{14}O_3$ (11.1 μcuries/μmole)

Chromatographic separation	Chlorophyll a	Chlorophyll b
7.5 hours photosynthesis		
Cellulose	1.98	1.49
	1.67	1.71
	1.83	1.89
Polyethylene	1.81	1.28
	1.72	1.42
Paper	1.31	0.89
(Petroleum ether-n-butanol)	1.22	0.62
100:3	1.17	0.44
8.5 hours photosynthesis		
Polyethylene	2.12	1.71
	1.79	(0.29)
	2.38	1.35
Paper	1.48	0.41
(Petroleum ether-n-butanol)	0.92	0.67
100:3	1.47	0.59

$C^{14}O_2$. This table illustrates the lowering of the specific activities of the chlorophylls after the second separation by two-dimensional paper chromatography. It may be noted that the specific activity of chlorophyll a is about twice that of chlorophyll b, even after this relatively long time of photosynthesis.

Table II illustrates the molar radioactivities of the pigments from *Scenedesmus* after one hour of photosynthesis with $C^{14}O_2$. Here the specific radioactivity of chlorophyll a is about three times greater than that of chlorophyll b. In the carotenoid series, three groups with high, medium, and low radioactivity can be distinguished; the first group contains the α- and β-carotenes, the second contains lutein and the diepoxide, violaxanthin, while the neoxanthin has a very low activity.

TABLE II
Molar Radioactivity of *Scenedesmus* Pigments after 1 Hour of Photosynthesis with $NaHC^{14}O_3$ (11.1 μcuries/μmole)

Pigment	μcuries/μmole
Violaxanthin	0.074
Neoxanthin	0.006
Lutein	0.078
α-Carotene	0.814
β-Carotene	0.700
Chlorophyll *a*	0.147
Chlorophyll *b*	0.056

The inherent weakness in the above method lies in the time lag incurred in the paper chromatographic separation of the spectroscopically pure pigments, which was necessary to separate the pigments from the colorless contaminants which are associated with them. After the elution of the pigment spot from the paper chromatogram at least 30% of the pigment had been lost, either in the development of the paper or in the elution.

In the hopes of eliminating the above-mentioned time lag centrifugally accelerated paper chromatography was investigated. This method was developed by McDonald *et al.* (1957) and used for the separation of amino acids with aqueous solvents (McDonald *et al.*, 1958). Excellent results were obtained here with nonpolar solvents; any solvent system which resolves the pigments by conventional paper chromatographic methods may be used in the chromatofuge.* The greatest advantage was the development time which is ten minutes instead of three hours; moreover, the pigment zones obtained were uniform and narrow. The pigments could be easily eluted from the paper, and only 5 to 10% of the pigments were lost in the development and elution, which is some 20% better than with conventional paper chromatography. Details of the solvents used, the R_F† values of the pigments, and so on, will be published elsewhere (Anderson, 1960).

A typical radioautograph of the total pigment extract from *Chlorella* is shown in Fig. 3; two unknown colorless substances are present near the solvent front just above the carotenes. The considerable activity at

---

* Chromatofuge No. 5060, Labline Inc., Chicago, Illinois.
† R_F, distance the compound travels relative to the solvent front.

the origin could be decreased by adequate washing of the ethereal extracts before application to the paper. Some typical results for the molar radioactivities of some of the pigments from *Chlorella* are listed in Table III. It may be seen that the specific radioactivity of chlorophyll *a* is greater than that of chlorophyll *b*, while violaxanthin and lutein

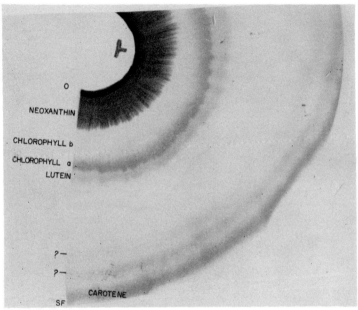

Fig. 3. Radioautograph of the total pigment extract from *Chlorella* after 2 hours photosynthesis with $C^{14}O_2$. The corresponding centrifugally accelerated paper chromatogram was developed with petroleum ether-isopropanol (100:2.5) for 10 minutes; flow rate of solvent 2.5 ml per minute; origin to solvent front, 16.5 cm; Whatman No. 3 MM paper.

show a parallel increase in activity with time. Some values show wide deviation, which indicate errors over which we have no control as yet; however, if the data are compared from the many experiments done, it can be definitely said that the specific activities of lutein and violaxanthin are very nearly equal, while the reverse is true for those of chlorophyll *a* and *b*.

The large differences between the specific activities of chlorophyll *a* and *b* found in algae confirm the findings of Shlyk et al. (1957), who reported chlorophyll *a* to be more radioactive than chlorophyll *b* in some higher plants, even after 24 hours of exposure to $C^{14}O_2$. Firstly it may be said that the possibility of conversion of chlorophyll *b* to chlorophyll *a* is definitely excluded. Also, it may be unequivocally stated that chlorophyll *a* and *b* are not rapidly interconverted in algae. Two possibilities remain: either chlorophyll *a* is a precursor of chlorophyll *b*, or the chlorophylls are derived from a common precursor (either at the

TABLE III

Molar Radioactivity of the Pigments from *Chlorella* after Photosynthesis with $C^{14}O_2$ (11.1 μcuries/μmole) and Separation by Centrifugally Accelerated Paper Chromatography

	Carotenoid extract: Toluene at 700 rpm		
Photosynthesis time in hours	Chlorophyll *a*	Chlorophyll *b*	Lutein
1.25	0.76	0.36	0.77
2.25	1.16	0.15	0.98
3.25	0.96	—	0.98
4.25	1.43	0.71	2.12
	1.62		2.61
	1.45		
5.25	1.85	0.46	1.78
	2.03		1.50

	Total extract: Petroleum ether-isopropanol (100:2.5) at 400 rpm		
Photosynthesis time in hours	Neoxanthin	Violaxanthin	Lutein
1.0	0.11	0.44	0.27
	0.19	0.87	0.59
2.0		1.12	0.94
3.0	0.63	1.96	1.7
		1.24	
		0.73	
6.0	1.71	1.36	0.55
	1.34	0.69	1.06
		1.65	1.62

porphobilinogen or porphin ring stage) and are formed independently of one another in the latter stages of the biosynthetic chain. These two possibilities are further complicated because the turnover rate of chlorophyll *a* may be greater than that of chlorophyll *b*. An answer to this problem should be provided by following the increase with time of specific activity of chlorophyll *a* as compared to that of *b*. It is believed that such measurements could be obtained by the methods of column chromatography followed by centrifugally accelerated chromatography.

It is also interesting to speculate on the possible relations which might exist among the carotenoids. From the simultaneous determinations of the pigment concentrations of algae, one half of which was kept in the light and the other half in the dark, the concentrations of the carotenoids were determined. No significant changes between the light-treated and dark-treated algae were found, except for the diepoxide, violaxanthin, which showed a marked increase in concentration in the dark (see Table IV). The depression of the violaxanthin concentration

TABLE IV
VIOLAXANTHIN CONCENTRATION AFTER 30-MINUTE PERIODS IN THE LIGHT AND DARK[a]

Chlorella		Scenedesmus	
Dark	Light	Dark	Light
42.6	22.0	56.2	48.4
52.0	24.0	52.8	43.6

[a] Concentrations of violaxanthin in 10^{-8} mole per milliliter wet packed algal cells.

in the light confirms the data of Sapozhnikov et al. (1957, 1958), though we were unable to see the corresponding smaller percentage rise in the concentrations of lutein in our algae experiments. The proposed interrelationship between these two compounds is strongly supported by the fact that their specific activities are very nearly equal. Such a light-dark interconversion of lutein and violaxanthin is indicated in Fig. 4. The expulsion of molecular oxygen from violaxanthin, giving back lutein, would be a light-dependent transformation, while the hydration of lutein

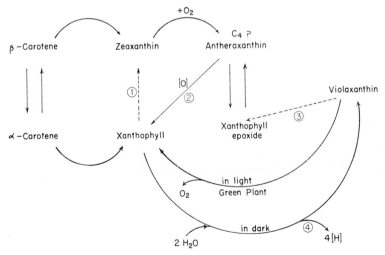

FIG. 4. Possible dynamic relations between the carotenoids: (1) rate-limiting in green plants; (2) [O] available as oxidant for other systems; (3) possible slow step; (4) ultimately available for [H] reduction.

and its dehydrogenation to violaxanthin would be dark, enzymatic reactions. A similar conclusion, however, could be obtained if we supposed that there is a dark, direct oxidation of lutein by molecular oxygen and a photoreduction of violaxanthin. An unequivocal distinction between these two alternatives would be provided by the demonstration of isotopic oxygen content in violaxanthin isolated from plants which had been photosynthesizing in the presence of H_2O^{18}.

If the oxygen transport system proposed by Cholnoky et al. (1956, 1958) is functioning as shown in Figure 4, whereby zeaxanthin is oxygenated to the monoepoxide, antheraxanthin, which may then give up its oxygen to some other substrate and return to lutein, we might have a weak coupling between this system and the lutein-violaxanthin system in plants which contain both mono- and diepoxides. Finally, the higher specific activities of the carotenes might implicate these compounds as precursors to the more oxygenated forms just discussed.

REFERENCES

Anderson, J. M. (1959). Ph.D. Thesis, University of California, Berkeley, California.

Anderson, J. M. (1960). *J. Chromatog.* (in press).

Bassham, J. A., and Calvin, M. (1957). "The Path of Carbon in Photosynthesis," pp. 29-32. Prentice-Hall, Englewood Cliffs, New Jersey.

Blass, U., Anderson, J. M., and Calvin, M. (1959). *Plant Physiol.* **34**, 329-333.

Cholnoky, L., Gyorgyfy, C., Nagy, E., and Pánczél, M. (1956). *Nature* **178**, 410-411.

Cholnoky, L., Szablcs, J., and Nagy, E. (1958). *Ann. Chem. Liebigs* **616**, 207-218.

Dorough, G. D., and Calvin, M. (1951). *J. Am. Chem. Soc.* **73**, 2362-2365.

McDonald, H. J., Bermes, E. W., and Shepherd, H. G. (1957). *Chromatog. Methods* **1**, 1-5.

McDonald, H. J., McKendell, L. V., and Bermes, E. W. (1958). *J. Chromatog.* **1**, 259-265.

Sapozhnikov, D. I., and Bazhanova, N. V. (1958). *Doklady Akad. Nauk. S.S.S.R. Botan. Sci. Sect. (English Translation)* **120**, 162-164.

Sapozhnikov, D. I., Krasovskaya, T. A., and Maevskaya, A. N. (1957). *Doklady Akad. Nauk S.S.S.R., Botan. Sci. Sect. (English Translation)* **113**, 74-76.

Shlyk, A. A., Godnev, T. N., Rotfarb, R. M., and Lyakhnovich, Y. P. (1957). *Proc. Acad. Sci. U.S.S.R. Biochem. Sect. (English Translation)* **113**, 103-106.

DISCUSSION

STANIER: How much biosynthesis of the photosynthetic pigments was occurring during the exposure to $C^{14}O_2$?

ANDERSON: The amount of incorporation into these pigments is of the order of 1% to 5%.

STANIER: It seems to me you could perfectly well be dealing here largely with exchange reactions which might be completely irrelevant to the problem of biosynthesis; that there is no appreciable synthesis of pigment under the conditions of the experiment.

REDFEARN: I am just a bit surprised that you have so much more radioactivity in the carotene than you do in the xanthophylls. Some years ago I did some experiments in which we found that the activity of the β-carotene was very small. We also found that we had comparable activity in the xanthophylls, while in your experiment you have 10 times as much radioactivity in the β-carotene. Another thing is, we found that in individual detached leaves left for 24 hours in the dark the β-carotene level dropped to about 10% of the normal in light. At the same time the carotene epioxides went up. So it seemed to us that β-carotene was by far the most labile of the carotenoids and the first stage in its destruction was the formation of the epioxides. When we placed the leaves in light, for a light period of about 24 hours, β-carotene came up to the original level again and the epioxides then disappeared. I think we looked upon this as an irreversible process, that the β-carotene which appeared in the light was formed by a new synthesis rather than by reversal of the dark reactions.

15

The Biosynthesis of Protochlorophyll[*]

LAWRENCE BOGORAD

Department of Botany, University of Chicago, Chicago, Illinois

Introduction

Before starting on the principal subject of this paper, the author would like to add a small section on porphyrins to the preceding discussions which dealt with the distribution of other classes of photoreceptive pigments. This exercise in natural history will be abridged but will serve, at least, as a brief review of the structure of some biologically important porphyrins.

Comfort (1950) reports that the shell of the pearl oyster, *Pinctada vulgaris*, contains uroporphyrin (Uro) isomer I (Fig. 1), an octacar-

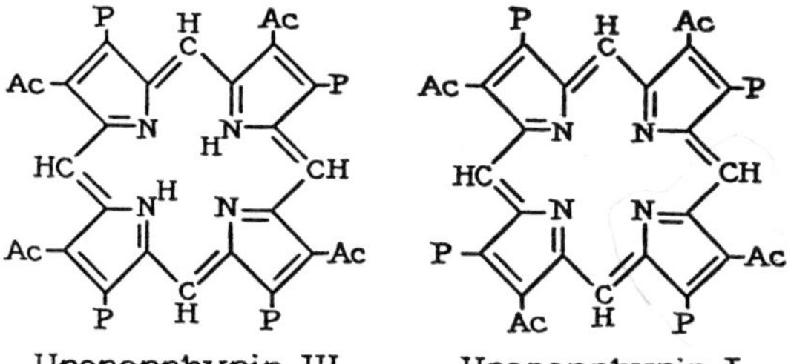

Uroporphyrin III   Uroporphyrin I

Ac = -CH$_2$-COOH
P = -CH$_2$-CH$_2$-COOH

FIG. 1.

boxylic porphyrin with one acetic and one propionic acid side chain on each pyrrole residue. The planarian, *Dugesia dorotocephala*, contains the III as well as the I isomer of Uro (E. K. MacRae, unpublished). Of the four possible isomers of Uro only I and III (Fig. 1) have been found in biological material. Isomer III differs from I in the arrangement of the

---

[*] The work reported from the author's laboratory was supported by grants from the National Science Foundation and the National Institute of Arthritis and Metabolic Diseases, United States Public Health Service (A-1010). It was also supported in part by the Dr. Wallace C. and Clara A. Abbott Memorial Fund of the University of Chicago.

side chains. The I isomer is completely symmetrical; the side chains on three of the pyrroles of Uro III are arranged like those on Uro I but on pyrrole D (see Fig. 2 for designation of rings) the positions of the acetic and propionic acid side chains are reversed. Uroporphyrins have been found in many other organisms but in most cases the isomer numbers are not reported.

Coproporphyrin (Copro) III has been found in the annelids *Nereis diversicolor, Chaetopterus variopedatus,* and *Myxicola infundibulum* (Kennedy and Vevers, 1954). In Copro III (Fig. 2) methyl groups are

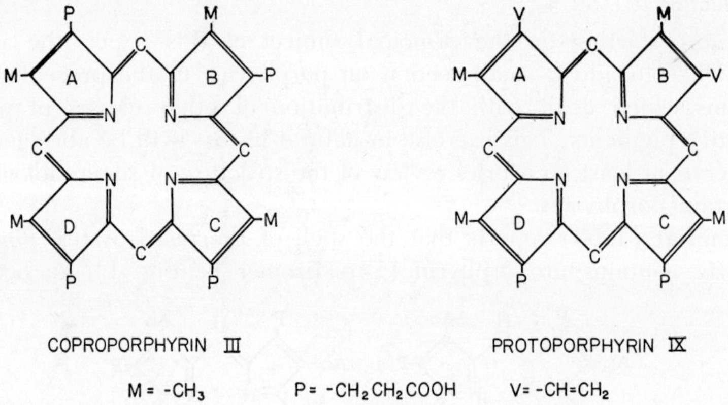

COPROPORPHYRIN III   PROTOPORPHYRIN IX

M = -CH$_3$ P = -CH$_2$CH$_2$COOH   V = -CH=CH$_2$

FIG. 2.

found in place of the acetic acid substituents of Uro III; these two porphyrins are identical in all other respects. Copro I of biological origin is also known.

Protoporphyrin (Proto) IX (Fig. 2) is found in *Lumbricus* and in *Eisenia*. This porphyrin differs from Copro III in that vinyl substituents are present on rings A and B in the positions occupied in Copro III by propionic acid residues. It can be seen that the propionic acid residues on rings C and D of Proto IX, Copro III, and Uro III are arranged in the same manner with respect to one another while in Uro I they are placed differently. In the chlorophylls the propionic acid residue on ring C is contorted into part of the cyclopentanone ring and esterified with methanol. Thus, fundamentally the arrangement of the propionic acid residues of Proto IX and of the chlorophylls are the same and these two porphyrins bear a closer resemblance to Uro III than to any other of the four possible Uro isomers.

Many aspects of research in porphyrin biosynthesis have been closely tied to studies of hereditary porphyria diseases and porphyrin excretion in some types of metal and barbiturate poisoning in humans and other animals (Lemberg and Legge, 1949). Uro, as its name indicates, was isolated first from urine (Fischer, 1915a); Copro, from feces (Fischer, 1915b). Higher plants, mammalian marrow and red cells, duck and chicken erythrocytes, algae, bacteria, and pulverized earthworms have all been used as research material. The diversity of organisms and tissues studied is almost matched by the number of different techniques which have been employed in investigations of this biosynthetic chain.

The two most abundant porphyrins in nature are heme (iron Proto IX) and chlorophyll a (Fig. 3). The transformation of protochlorophyll a

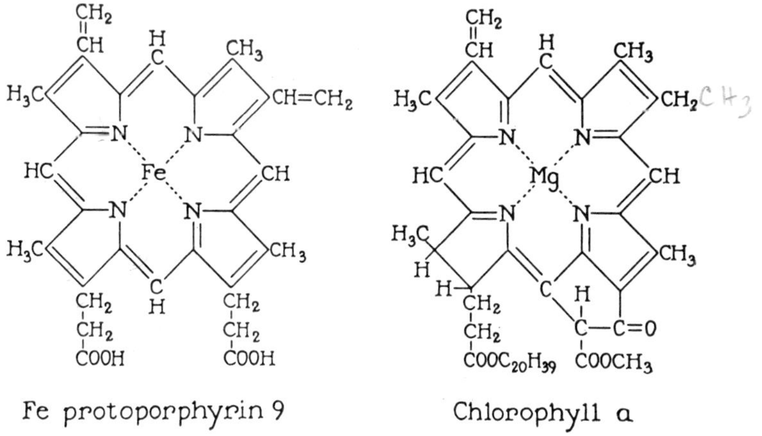

Fe protoporphyrin 9   Chlorophyll a

FIG. 3.

into chlorophyll a is discussed by Dr. Smith in the next paper. The present paper includes a brief review of current knowledge, plus a few speculations, on the formation of heme and of protochlorophyll. The discussion will begin with an over-all view of this field and continue with a more detailed examination of some current investigations and problems.

The Biosynthesis of Protoporphyrin IX: Earlier Information and Views

After Fischer had established the structure of heme and it was clear that Proto contained two kinds of pyrroles (rings A and B with methyl and vinyl side chains, rings C and D with methyl and propionic substituents), the earliest proposals regarding the mechanism of Proto IX biosynthesis were based on the reasonable supposition that first the two different kinds of pyrroles are made and then, somehow, these pyrroles are assembled in the correct proportions and in the proper order to produce the porphyrin (Lemberg and Legge, 1949). Data from subsequent biochemical investigations contradicted such proposals.

Shemin and Rittenberg (1946) tested a number of amino acids as sources of nitrogen for the formation of heme by suspensions of avian erythrocytes and found glycine to be the most effective. Ultimately Wittenberg and Shemin (1949) and Muir and Neuberger (1949) demonstrated that glycine nitrogen was used equally well for the formation of all four pyrrole rings of this porphyrin. Meanwhile, a number of investigations of the source of the carbon atoms of heme showed that the carboxyl carbon atom of glycine was not incorporated (Grinstein *et al.*, 1949; Radin *et al.*, 1950), but that one of the four carbon atoms of each pyrrole ring as well as the carbon atoms of the bridges between pyrroles were derived from the methylene carbon atom of glycine (Wittenberg and Shemin, 1950; Muir and Neuberger, 1950). Data from the elegant experiments of Shemin and Wittenberg (1951) were pertinent to the problem of whether the four pyrrole residues of Proto IX arise from one or from two kinds of pyrroles. They supplied either carboxyl- or methyl-labeled acetate to suspensions of avian erythrocytes, isolated the radioactive heme which was formed, degraded the porphyrin carbon atom by carbon atom, and measured the radioactivity of the carbon atoms at each position. It was found (Fig. 4) that the terminal carbon atom of each vinyl group (one each on rings A and B) was about as radioactive as the center carbon atom of each of the two propionic acid residues on rings C and D and that the proximal carbon atoms of the propionic and vinyl side chains were about equally radioactive. This suggested that each of the four pyrroles had originally had a propionic acid substituent and that the vinyl groups were derived from propionic acid residues. Further examination revealed that on rings A and B the following pairs of carbon atoms were approximately equally radioactive (Fig. 4): the carbon atom of the methyl substituent and the terminal carbon atom of the vinyl group; the proximal carbon atom of the vinyl group and the ring carbon atom adjacent to the methyl substituent; and the β-ring carbon atom of the "propionic side" of the pyrrole and the α-ring carbon atom of the "methyl side" of the ring. By an extension of the argument regarding the origin of the vinyl groups from propionic acid residues, one could propose that the methyl side chains were derived from acetic acid side chains. This led Shemin and Wittenberg (1951) to propose that all four rings of Proto IX are derived from a common pyrrole which has an acetic and a propionic acid side chain.

Uro- and coproporphyrins (Figs. 1 and 2) had been identified as constituents of the urine of persons afflicted with congenital porphyria; these compounds were considered to be by-products of abnormal heme synthesis (Lemberg and Legge, 1949). At about the time that the tracer work was in progress, renewed serious attention was being given to the nature of the porphyrins excreted by porphyria patients. The development of a paper chromatographic method for the separation of porphyrins on the basis of the number of carboxyl groups per molecule (Nicholas and Rimington, 1949) permitted the qualitative analysis of mixtures of urinary porphyrins on a microscale. Using this method, porphyrins with 7, 6, 5, and 3 carboxyl groups per molecule, as well as the

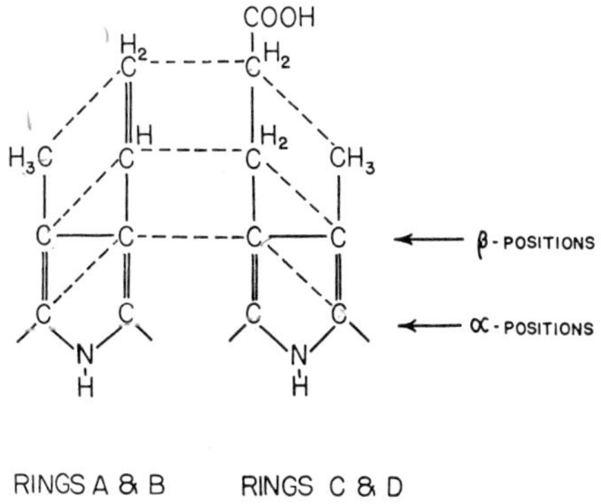

FIG. 4. Labeling patterns of pyrrole rings of protoporphyrin from experiments by Shemin and Wittenberg (1951) using C^{14}-acetate. Carbon atoms which were found to be equally radioactive are connected by broken lines. The carboxyl carbon atom of rings C and D was radioactive only in experiments with $CH_3C^{14}OOH$. One of the carbon atoms adjacent to the nitrogen atom did not become labeled in these experiments; this carbon atom is derived from glycine.

already familiar octa- and tetracarboxyllic Uro and Copro, were found in porphyria urines (Nicholas and Rimington, 1951; McSwiney et al., 1950). These observations, together with a changing point of view, brought to this field the conviction (e.g., Granick, 1951) that the biosynthesis of Proto IX proceeded approximately as follows:

Glycine + Acetate → Common pyrrolic precursor → Uro III → Heptacarboxy porphyrin → Hexacarboxy porphyrin → Copro III → Tricarboxy porphyrin(s) → Proto IX

Additional support for such a sequence came from the discovery of a *Chlorella* mutant which accumulates highly carboxylated porphyrins (Bogorad and Granick, 1953a).

Against this background came information on steps in the span from glycine plus succinate to Proto IX. New intermediates were identified; new observations relegated Uro, Copro, and other compounds to side tracks; and, naturally, new problems were revealed.

Porphobilinogen Utilization and Formation

Freshly passed urine of individuals with the hereditary disease, acute porphyria, generally looks fairly normal but on standing at room temperature, particularly in the light, it comes to resemble heavy port wine in color. The wine color is partly that of uroporphyrin but is mainly attributable to *porphobilin,* an as yet uncharacterized compound. Another peculiarity of urine from this source is that when fresh it gives a strong red color upon being mixed with Erlich's reagent, *p*-dimethylaminobenzaldehyde in acid solution. This reaction is characteristic of certain

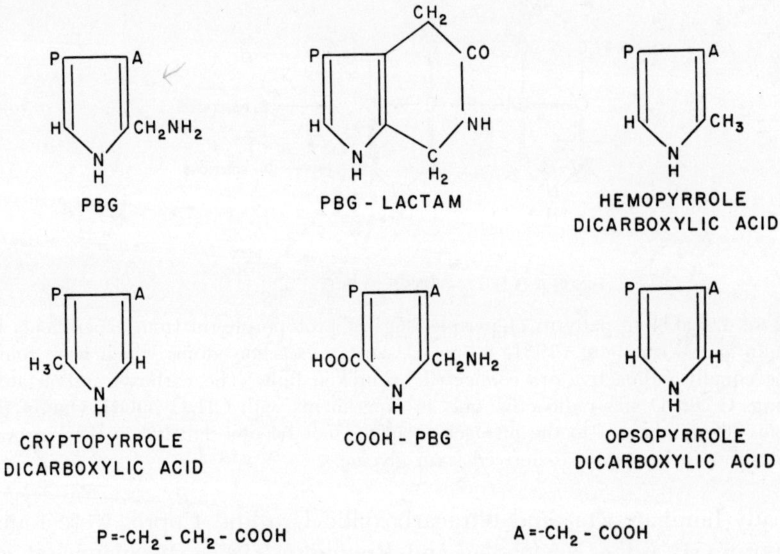

FIG. 5. Porphobilinogen and porphobilinogen analogs.

pyrroles and polypyrroles. Reddened porphyria urine, on the other hand, gives a weak or negative Ehrlich reaction. Thus, the increase in porphobilin is accompanied by a decrease in Ehrlich reactivity. The Ehrlich-reacting material, first observed by Sachs (1931), was named *porphobilinogen* (PBG). PBG was studied for many years but the PBG-era began only after Westall (1952) had crystallized the compound (Fig. 5) and its structure was established (Cookson and Rimington, 1953; Granick and Bogorad, 1953). It was soon demonstrated that PBG could serve as a substrate for the production of a variety of porphyrins, including Proto IX. The array of porphyrins produced from PBG by a frozen and thawed preparation of *Chlorella* is shown in Table I (Bogorad and Granick, 1953a). Broken-cell preparations from a number of other biological sources also catalyze these reactions. Avian erythrocytes (Falk et al., 1953) have been widely used to study porphyrin biosynthesis from PBG.

TABLE I

PORPHYRINS FORMED FROM PBG BY FROZEN AND THAWED Chlorella PREPARATIONS[a]

Porphyrins	Per cent of total porphyrin synthesized
Uroporphyrin	28%
Protoporphyrin	14%
Other porphyrins with 2 to 7 carboxyl groups per molecule (including coproporphyrin)	58%

[a] From Bogorad and Granick (1953b).

The nature of the immediate precursor of PBG was resolved by Shemin and Russel (1953) at about the same time that PBG was shown to be a substrate for porphyrin formation. δ-Aminolevulinic acid (ALA) labeled with carbon-14 was synthesized and was shown to be used for the synthesis of heme by avian erythrocytes. A report by Neuberger and Scott (1953), who had been working along the same lines, confirmed the findings of Shemin and Russel. Lascelles (1955) demonstrated the synthesis of porphyrins and bacteriochlorophyll from ALA by *Rhodopseudomonas spheroides.*

This information on the utilization of PBG and ALA opened the way for detailed enzymological investigations of porphyrin biosynthesis.

ALA-dehydrase, the enzyme which catalyzes the condensation of two molecules of ALA to make PBG, has been partially purified from avian erythrocytes (Granick, 1954; Schmid and Shemin, 1955), rabbit reticulocytes (Granick, 1958), and liver from various animals (Gibson et al., 1955; Iodice et al., 1958). Table II contains a summary of the properties of various ALA-dehydrase preparations. Gibson et al. report that their preparation is homogeneous on electrophoresis and is about 270-fold purer than the aqueous extracts of ox liver acetone powder with which they start. An interesting addition to the information in Table II is the report by Iodice et al. (1958) that they found 0.1% Cu and traces of Mg in purified ALA dehydrase from beef liver.

Thus, these observations showed that PBG is used for the enzymatic synthesis of porphyrins, that this pyrrole is formed from two molecules of ALA (Fig. 6), and that modifications in porphyrin side chains probably occur after the first tetrapyrrole is formed.

TABLE II
ALA-DEHYDRASE

Enzyme preparation	Activator	Inhibitor	pH (Optimum)	K_m
From ox liver acetone powder, (Gibson et al., 1958)	Cysteine or glutathione	$Cu^{++} = Hg^{++} =$ $Ag^+ > Zn^{++} > Pb^{++}$ (at 10^{-3} M), EDTA	6.7	1.4×10^{-4} M
From chicken erythrocytes	Glutathione	Metals and EDTA	6.3	5×10^{-4} M
From rabbit reticulocytes (Granick, 1954, 1958)	Glutathione	Metals and EDTA	6.7	5×10^{-4} M
From duck erythrocytes (Schmid and Shemin, 1955)	Glutathione	—	—	—

ALA Synthesis

Before examining some details of the utilization of PBG the course of ALA synthesis will be considered. Prior to the discovery of the role of ALA in porphyrin biosynthesis Shemin and Kumin (1952) had obtained evidence for the incorporation of succinate without rearrangement into porphyrin; this supported an earlier suggestion made by Shemin and

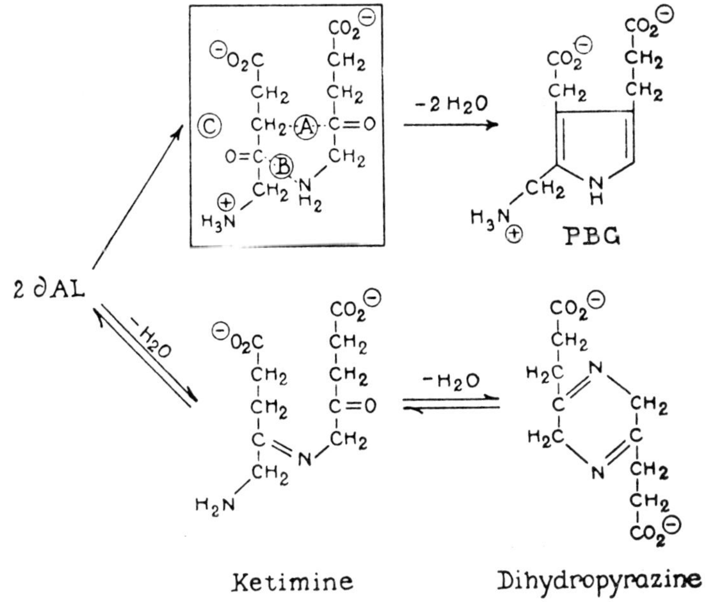

Fig. 6. Mechanism of action of ALA-ase (Granick and Mauzerall, 1958).

Wittenberg (1951). The recognition of the participation of succinate, probably as succinyl-CoA, taken with the earlier evidence for the incorporation of the methylene carbon and the nitrogen atoms of glycine, led Shemin to consider ALA a probable intermediate. Furthermore, in the first report of the utilization of ALA for porphyrin biosynthesis, the succinate-glycine cycle (Fig. 7) was proposed (Shemin and Russel, 1953). Evidence has been obtained for some segments of this scheme (e.g., Shemin, 1955), but for the present discussion the production of ALA is of greatest interest. The results of some studies of systems which catalyze ALA formation are summarized in Table III. The more highly purified preparations can use succinyl-CoA and glycine as substrates and pyridoxal phosphate is the only cofactor required. Isolated cell particulates have generally been found to be incapable of utilizing succinyl-

CoA, presumably because of permeability barriers, but do make ALA when supplied with glycine and α-ketoglutarate or succinate. Cofactors required for preparations of this sort vary but generally include Co-

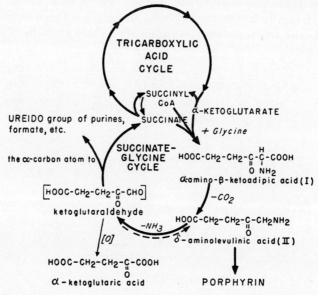

Fig. 7. The succinate-glycine cycle (Shemin and Russel, 1953).

enzyme A, pyridoxal phosphate, and DPN or ATP. Shemin and Kikuchi (1958) have outlined some possible routes of ALA synthesis from succinyl-CoA and glycine (Fig. 8).

The earliest indications that pyridoxal phosphate and Coenzyme A might participate in ALA biosynthesis came from Lascelles' (1957) work on *Tetrahymena vorax* and Schulman and Rickert's (1957) studies on porphyrin synthesis by preparations of red cells from vitamin-B_6-deficient and pantothenate-deficient ducks.

Tetrapyrrole Formation

UROPORPHYRINOGEN I

As already mentioned, PBG is the key intermediate in porphyrin formation in the sense that carbon and nitrogen atoms in PBG are committed for use in porphyrin biosynthesis unless the pyrrole is excreted. During the initial investigations of the formation of porphyrins from PBG by frozen and thawed preparations of *Chlorella* (Bogorad and Granick, 1953b), the effect of heating the preparations prior to incubation with

TABLE III
SYSTEMS FOR ALA SYNTHESIS

Enzyme preparation	Substrates	Cofactors	Inhibitors
Freeze-dried particles from chicken erythrocytes (Gibson et al., 1958)	Glycine + α-Ketoglutarate	DPN and pyridoxal phosphate	p-Chloromercuribenzoate, L-penicillamine, l-cysteine
	Glycine + Succinyl-CoA		
Particle-free extracts of R. spheroides and R. rubrum. (Kikuchi et al., 1958)	Succinate + Glycine	CoA, Pyridoxal phosphate, ATP, and Mg^{++}	—
	Succinyl-CoA + Glycine	Pyridoxal phosphate	—
Hemolyzed chicken erythrocytes (Granick, 1958)	Glycine + α-Ketoglutarate Glycine + Succinate	CoA, pyridoxal phosphate, and DPN	DNP, azaserine

PBG was studied and extremely interesting, though unexpected, results were obtained. Preheating the preparations at 55° for 30 minutes resulted in no alteration in their capacity to catalyze the consumption of substrate (Fig. 9).

The enzyme catalyzing this process was apparently stable at 55°C although it was destroyed at 100°C. Preheating did, however, have a striking effect upon the kinds of porphyrins produced; instead of the array of porphyrins produced by unheated preparations (Table I) only

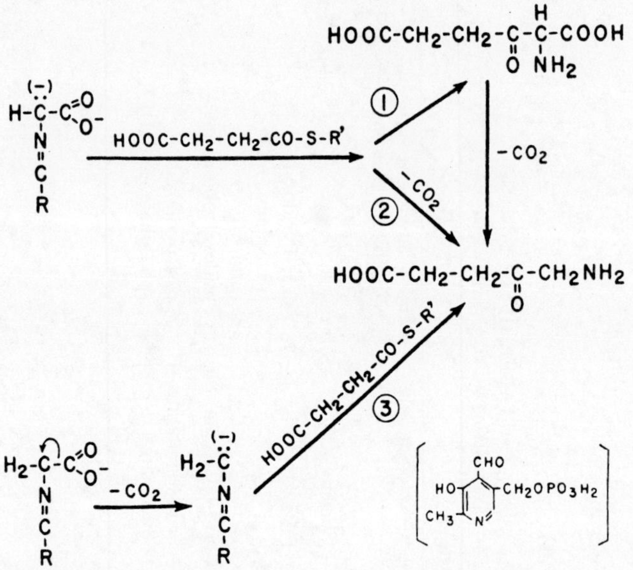

Fig. 8. Mechanisms of δ-aminolevulinic acid synthesis from succinyl coenzyme A and pyridoxal phosphate derivative of glycine (Shemin and Kikuchi, 1958).

Uro accumulated. Thus, the enzymes responsible for the production of Copro, etc. were apparently much more susceptible to heat inactivation than the PBG-consuming enzyme. The most interesting consequence of the preheating was revealed when the Uro which accumulated was found to be type I. Uro I is found normally in just a few organisms but is excreted in relatively large amounts by humans and other animals with the hereditary disease, congenital porphyria (Lemberg and Legge, 1949).

Preheating, in a sense, converted the *Chlorella* preparations from the "normal" to the "congenital" type; this kind of alteration in the behavior of porphyrin-making systems by heating has been confirmed with avian red cell and human erythrocyte preparations (Rimington and Booij, 1957; Booij and Rimington, 1957).

The synthesis of Uro I from PBG is easily visualized to proceed by the linear condensation of four molecules of PBG followed by the head to tail cyclization of the tetrapyrrole. The production of the III isomer from PBG as the sole subtrate is more difficult to understand since ring D appears to be in "backwards" (Fig. 1). But, as already discussed, the chlorophylls and Proto IX are structurally related to Uro III, not Uro I, and consequently the course of synthesis of the III isomer is of para-

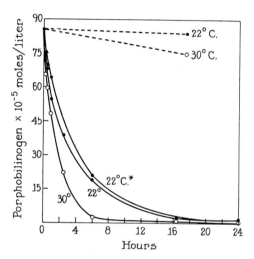

FIG. 9. The consumption of PBG in the presence (solid lines) and absence (broken lines) of frozen and thawed *Chlorella* preparation. Incubation temperatures are shown. *Preparation heated at 55°C for 60 minutes prior to addition of PBG and incubation (Bogorad and Granick, 1953b).

mount importance biologically. The alteration (by preheating), of a system which makes Uro III and isomerically related porphyrins into a system which can make only Uro I from PBG provides some interesting suggestions about the relationship between the biosynthesis of Uroporphyrins I and III. The simplest interpretation of the conversion phenomenon appeared to be the following:

(a) *At least* two enzymes participate in the synthesis of Uro III from PBG.

(b) These enzymes differ in their susceptibility to heat inactivation. The enzyme which controls the rate of consumption of PBG is more stable than the second enzyme, which is required for Uro III formation.

(c) The enzyme which controls the rate of PBG consumption catalyzes the linear condensation of the pyrrole molecules and, in the absence of the second enzyme, Uro I is formed.

(d) The second enzyme somehow acts with the first one, or on some of its products, to produce Uro III.

Working from this interpretation, attempts were made to isolate individual enzymes in order to study the reaction mechanisms.

Aqueous extracts of acetone powders of spinach leaf tissue proved to be an excellent source of the Uro I-making enzyme, porphobilinogen deaminase (PBG-D) (Bogorad, 1955, 1958b). When crude enzyme preparations are used, porphyrin production frequently parallels PBG

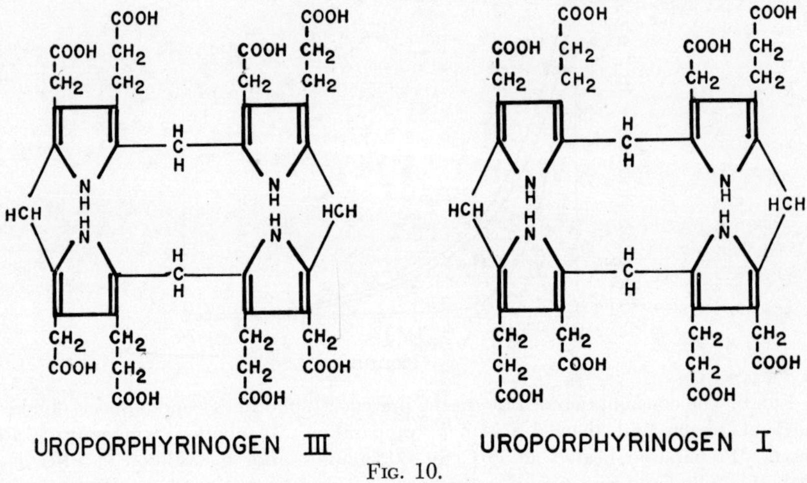

UROPORPHYRINOGEN III   UROPORPHYRINOGEN I

Fig. 10.

consumption; with purified preparations, especially when incubated anaerobically, PBG is consumed but the reaction mixtures remain virtually colorless until oxidized by oxygen or iodine. The colorless product is uroporphyrinogen (Urogen) I (Fig. 10) and the over-all reaction is:

$$4 \text{ PBG} \rightarrow 1 \text{ Urogen I} + 4 \text{ NH}_3$$

The yields approach 100% when the reaction proceeds anaerobically. The enzyme is inhibited by p-chloromercuribenzoate and the reversal of this inhibition by cysteine suggests the presence of a functional —SH group.

The specificity of PBG-D was investigated by incubating a series of synthetic analogs of PBG (most of them kindly supplied by Dr. S. F. MacDonald of the Division of Pure Chemistry, National Research Council, Ottawa, Canada) with PBG and PBG-D. With the exception of opsopyrrole dicaboxylic acid (OPD), none of the analogs shown in

Fig. 5 had any effect on the reaction. OPD was found to inhibit the reaction competitively (Fig. 11) (Bogorad, 1957a). Inhibition of PBG consumption by OPD has also been shown to occur in avian erythrocyte systems (Carpenter and Scott, 1959).

In as much as the completely symmetrical Urogen I is produced from PBG by PBG-D, it seems reasonable to conclude that PBG-D catalyzes either (a) the sequential condensation of PBG molecules to produce linear and then cyclic tetrapyrroles, or (b) the production of dipyrroles

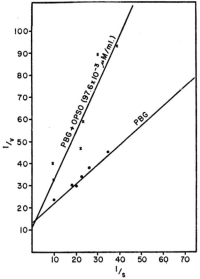

Fig. 11. Kinetics of PBG consumption by PBG-D in the presence and absence of OPD.

and their condensation with one another. This straightforward interpretation has been difficult to check because intermediates could not be detected. Experiences of this kind could almost lead one to conclude that all four pyrroles condense at once — a most inconvenient proposition for reasons which should become obvious later.

Very recently it has become possible, by including appropriate concentrations of hydroxylamine or ammonium ions in the reaction mixtures, to force the accumulation of a compound or compounds which react with Ehrlich's reagent differently from PBG but in a manner similar to that in which some dipyrrylmethanes behave. (Unfortunately the analogous tri- and tetrapyrrylmethanes are not available and it has not been possible to determine from the reaction with the Ehrlich reagent whether the accumulated material is a di-, tri-, or tetrapyrrole or a mixture.) It is especially interesting that the compound or compounds which accumulate appear to be consumed when incubation of the reaction mixture is continued; thus, these compounds appear to be intermediates in Urogen I formation. This work has progressed to the spot stage; using C^{14}-PBG as the substrate, a radioactive material which moves slightly slower than Uro I but faster than PBG has been detected upon paper electrophoresis of reaction mixtures containing hydroxylamine. Attempts are now in progress to accumulate and isolate this material for further study.

Thus, three classes of PBG-D inhibitors are known: noncompetitive inhibitors like p-chloromercuribenzoate; the competitive inhibitor OPD; and a group of inhibitors, including hydroxylamine, which have little or no effect on the rate of consumption of PBG but seem to arrest the utilization of some intermediates. The last group appears to be the most useful as an experimental tool. The action of this group of inhibitors, however, raises the possibility that PBG-D preparations contain more than one kind of enzyme.

These, then, are some of the things we know about Urogen I formation, but to understand the synthesis of chlorophylls and heme we must know about the formation of the III isomer.

Uroporphyrinogen III

Aqueous extracts of wheat germ proved to be an excellent source of an enzyme which, when incubated with PBG-D and PBG, brings about the production of Urogen III instead of Urogen I (Bogorad, 1958c); the rate of PBG consumption is essentially controlled by the concentration of PBG-D. The best preparations of uroporphyrinogen isomerase (U-Is), at concentrations which lead to Urogen III production in the presence of PBG-D, show no capacity for catalyzing PBG consumption when incubated alone with the pyrrole. U-Is thus appears to be the second enzyme — the one destroyed by heating frozen and thawed *Chlorella* preparations. Again, both PBG-D and U-Is preparations may contain more than one enzyme.

As to the mechanism of Urogen III formation: we know more about how it does not proceed than how it does. These are some of the facts which must be considered in any speculation on the subject:

(1) Urogen III is not produced when Urogen I is incubated with U-Is (Bogorad, 1958c) or systems which catalyze the formation of Urogen III from PBG (Granick and Mauzerall, 1958). Consequently it is clear that the cyclized tetrapyrrylmethane is not the substrate for U-Is; the isomerase does not open the Urogen I ring, slip out one pyrrole, flip it over, and make Urogen III in

this manner. (However, a mixture of isomers including mainly Urogen III can be made nonenzymically by heating Urogen I with formaldehyde (Mauzerall and Granick, 1958b).

(2) As far as can be determined, the isomerase has no effect, such as shifting the aminomethyl group from one α-position to the other of PBG, when incubated alone with the pyrrole (Bogorad, 1957b, 1958c).

(3) Kinetic studies reveal that there is probably some interaction between PBG and U-Is when PBG-D is active (Bogorad, 1958c).

These observations, plus the virtual certainty that PBG-D catalyzes the linear condensation of PBG molecules, make it appear reasonable that the substrates for U-Is are a di- or tripyrrylmethane plus PBG. It is of considerable interest that hydroxylamine, whose effect on the Urogen I-synthesizing system has been discussed, is a powerful inhibitor of Urogen III production (Bogorad, 1958c). It is not clear yet whether hydroxylamine affects Urogen III synthesis directly by attacking U-Is or indirectly by preventing the production of a substrate for this enzyme. The latter would be the more interesting.

The detailed mechanism of the formation of Urogen III from PBG has been the subject of extensive speculation. The more than half-dozen hypotheses which have been published can be grouped into a few classes (Table IV).

TABLE IV
CLASSES OF HYPOTHESES OF UROPHORPHYRINOGEN III BIOSYNTHESIS

I. T-PYRRYLMETHANE INTERMEDIATES (PYRROLE TRANSFERS)	
Bogorad and Granick (1953b)	4 PBG → 1 Urogen III
Shemin et al. (1955)	5 PBG → 1 Urogen III + 1 OPD + 1 CH_2O
Jackson and MacDonald (1957)	4 PBG + 1 OPD → 1 Urogen III + OPD
II. "TRANSPYRRYLASE" (TRANSFER VIA ENZYME-PYRROLE COMPLEX)	
Bogorad (1958a)	3 PBG → Tripyrrylmethane Tripyrrylmethane + PBG + Enz → Dipyrrylmethanes I and II → Urogen III
III. "GROUP TRANSFERS" (DISPLACEMENTS)	
Cookson and Rimington (1954)	PBG → OPD + CH_2O 3 PBG + OPD + CH_2O → Urogen III
Bullock et al., (1958)	Displacement and shift of $-CH^+_2$ group to opposite α-position. Shift via N.

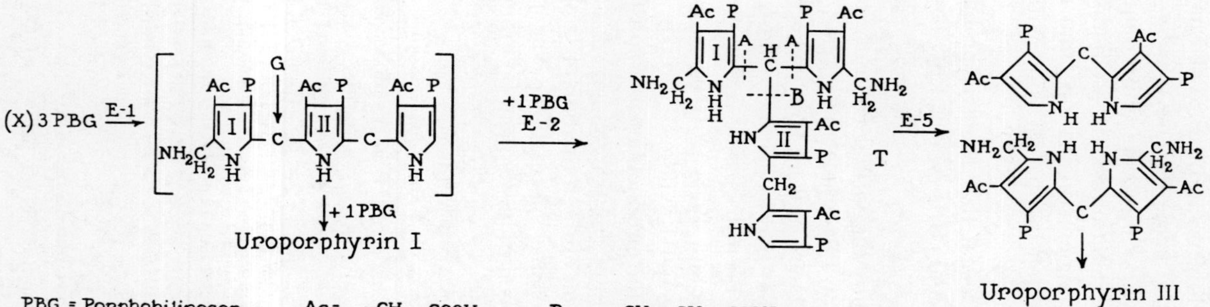

PBG = Porphobilinogen   Ac = -CH$_2$-COOH   P = -CH$_2$-CH$_2$-COOH

Fig. 12. Hypothetical scheme for the biosynthesis of Urogen III from PBG (Bogorad and Granick, 1953b).

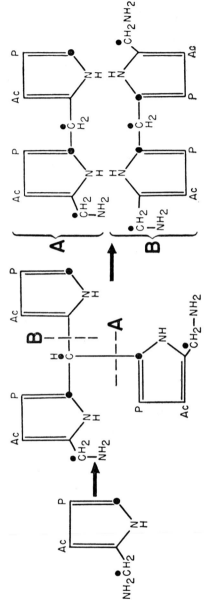

FIG. 13. Hypothetical scheme for the biosynthesis of Urogen III from PBG. KEY: *Ac*, acetic acid side chain; *P*, propionic acid side chain; ●, α-carbon atom of glycine and δ-carbon atom of δ-aminolevulinic acid (Shemin et al., 1955).

One group of hypotheses [Bogorad and Granick, 1953b (Fig. 12); Shemin *et al.*, 1955 (Fig. 13)] includes a T-pyrrylmethane of some kind as an intermediate. It is characteristic of these proposals that the bridge between the pyrroles which will be rings C and D of the porphyrin is formed early and from the aminomethyl group of another molecule of PBG. The pyrrole which has contributed its α-substituent is then split off and either discarded as OPD (Shemin *et al.*, 1955) or used to make Urogen III as part of a dipyrrole formed before the split (Bogorad and Granick, 1953b). A pertinent criticism of this kind of scheme is that, organochemically, a pyrrole condenses with a dipyrrylmethene (an oxidized dipyrrylmethane) to produce a tri-pyrrylmethane but Urogen III biosynthesis appears to occur without a dipyrrylmethene intermediate. This, however, would not be an insurmountable objection if the enzymes involved could temporarily oxidize the di- or tripyrrole (Bogorad, 1958a).

A slightly different kind of hypothesis, but one involving transfers of pyrrole rings and therefore semirationally included in this category,

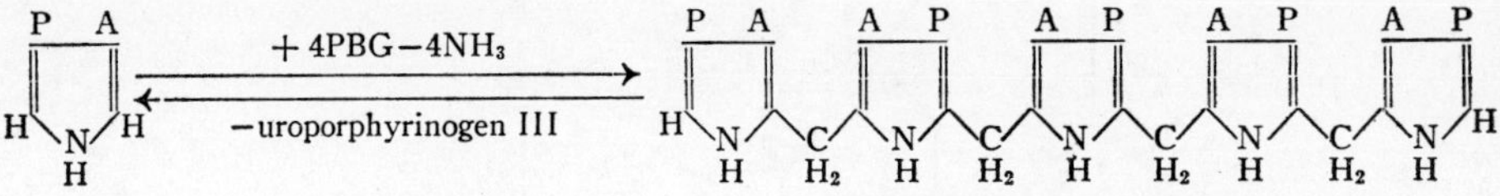

$$A = CH_2.COOH, \quad P = CH_2CH_2.COOH$$

Fig. 14. Hypothetical scheme for the biosynthesis of Urogen III from PBG with OPD turnover and via a pentapyrrane intermediate (Jackson and MacDonald, 1957).

is that of Jackson and MacDonald (1957) (Fig. 14) in which OPD serves as a cofactor.

Another type of hypothesis which has been suggested (Fig. 15) invokes a transpyrrylase mechanism (Bogorad, 1958a). According to this hypothesis a linear tripyrrole is formed and then a monopyrrolic residue is "picked off" of the chain by an enzyme, which catalyzes the condensation of this unit with a molecule of PBG. In this way two different dipyrrylmethanes are formed. Urogen III is formed when pairs of the different dipyrroles condense. Studies on the enzymatic synthesis of Urogen III from PBG in the presence of synthetic methylene di-PBG (Bogorad and Marks, unpublished) suggest that the scheme as illustrated in Fig. 15 may be incorrect but a mechanism of this general type, differing only in the kinds of dipyrroles formed, may be correct.

Hypotheses of a third kind involve transfers of groups from one pyrrole to another or from one α-position to another of the same pyrrole. The most elaborate scheme of this kind (Fig. 16) has been proposed by Bullock *et al.* (1958) as an extension of a proposal by Robinson (1955).

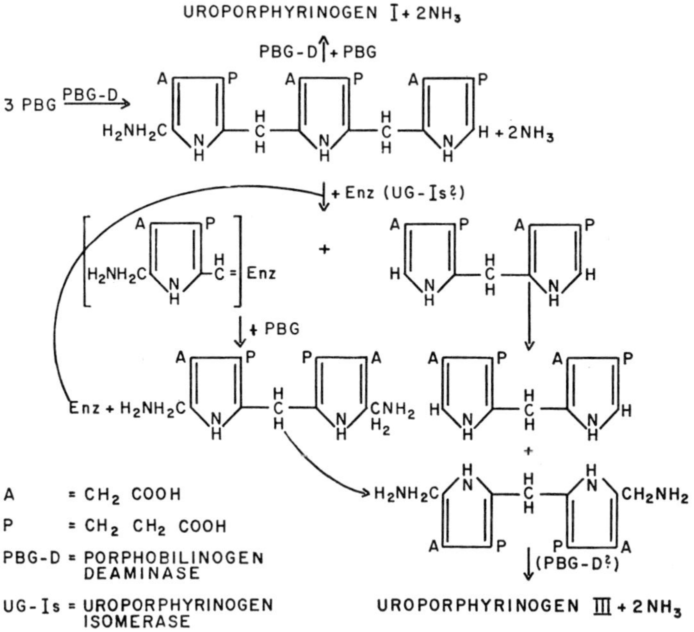

15. BIOSYNTHESIS OF PROTOCHLOROPHYLL 247

FIG. 15. Hypothetical scheme for the biosynthesis of Urogen III from PBG (Bogorad, 1958a).

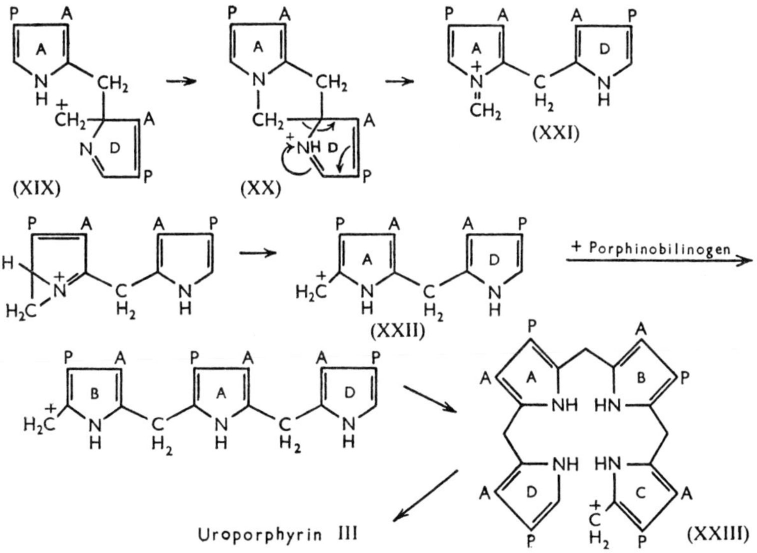

FIG. 16. Hypothetical scheme for the biosynthesis of Uro III from PBG involving reactions on the substituted α-position of PBG and shifts of —CH$^+_2$ groups (Bullock et al., 1958). KEY: A, acetic acid side chain; P, propionic acid side chain.

The basic assumption here is that the substituted α-position of PBG is more reactive than the unsubstituted one. The $-CH^+_2$ group, originally at the α-position which has been attacked, will migrate first to the pyrrole nitrogen and then to the previously unsubstituted α-position of the attacking pyrrole. Repetitions of this behavior would lead to the formation of a linear tetrapyrrole which would cyclize to Urogen III.

Table IV summarizes the over-all reactions, as well as a few intermediate steps, of hypotheses of Urogen III biosynthesis. Data are available to evaluate some of the schemes. Thus, for example, the hypothesis of Shemin et al. (1955) is inconsistent with the observations that (a) yields of Uro III from PBG approach 100% (Dresel and Falk, 1956; Bogorad, 1958c); (b) only about 10% of the predicted amount of $C^{14}H_2O$ accumulates when aminomethyl-C^{14}-PBG is used as the substrate for the enzymatic synthesis of Urogen III and, furthermore, approximately the same amount of C^{14}-formaldehyde is produced during the enzymatic synthesis of Urogen I (Bogorad and Marks, 1960); and (c) no detectable OPD accumulates in the reaction mixture during the enzymatic synthesis of Urogen III (Bogorad, unpublished) although this hypothesis requires that one molecule of OPD should be produced for every five molecules of PBG consumed.

The hypothesis of Jackson and MacDonald (1957) is inconsistent with the observation that H^3-OPD is not incorporated into Urogen III during its enzymatic synthesis from PBG (Bogorad and Marks, unpublished). [Similar observations, using C^{14}-OPD, have been made by Carpenter and Scott (1959)].

The proposal of Cookson and Rimington (1954) is inconsistent with the observations that neither H^3-OPD nor C^{14}-formaldehyde are incorporated into Urogen III during its enzymatic synthesis *in vitro* from PBG (Marks and Bogorad, unpublished).

The remaining hypotheses are not necessarily correct — merely untested or untestable with currently available synthetic substrates.*

---

* *Note added in proof:* J. Wittenberg [*Nature* **184**, 876 (1959)] has recently advanced the interesting proposal that Urogen III might be formed by the interchange of PBG residues between two open-chain linear tetrapyrroles. In Wittenberg's scheme an octapyrrole, consisting of a ring of six pyrroles with two additional pyrrole residues extending from opposite sides of the ring, is an intermediate between the linear tetrapyrroles and Urogen III. The interacting tetrapyrroles are of the type which would be formed by the linear condensation of four molecules of PBG and, if an interchange of pyrrole residues did not occur, a head-to-tail condensation of each original tetrapyrrole would result in the formation of one molecule of uroporphyrinogen I. The mechanism of interchange of pyrrole residues resembles elements of the T-polypyrrole mechanisms which have been suggested.

Decarboxylation of Uroporphyrinogens

Despite the widely held view that Copro III and Proto IX were derived from Uro III, numerous attempts to obtain evidence for such conversions provided no support for this opinion. Finally it was found that porphyrinogens were precursors and intermediates (Neve et al., 1956; Bogorad, 1955, 1957b, 1958d; Granick and Mauzerall, 1958). Table V shows the results of a study of the utilization of Uro I, PBG, and Urogens I and III by frozen and thawed preparations of Chlorella. Urogen was utilized for the formation of Proto, Copro, etc.; Uro was not.

TABLE V
PORPHYRIN CONVERSIONS BY FROZEN AND THAWED Chlorella PREPARATIONS[a]

Substrate	Porphyrins recovered (%)	
	Uro	Ether-soluble porphyrins[b]
PBG	22.0	78.0
Urogen I	24.5	75.5
Uro I	100.0	—
		Proto
Urogen III		63.0

[a] From Bogorad (1958d).
[b] Mostly Copro.

Mauzerall and Granick (1958a) report that a single enzyme or enzyme system, Urogen decarboxylase, catalyzes the reactions in the span from Urogen to Coprogen. Some results of their investigations of this enzyme are summarized in Table VI.

TABLE VI
UROPORPHYRINOGEN DECARBOXYLASE[a]

Specificity: No effect on PBG, ALA, or Uro. Rates of decarboxylation: Urogen III > IV > II > I

Inhibited by: Hg^{++}, Cu^{++}, Mn^{++}, iodoacetate, p-chloromercuribenzoate. Inhibition prevented by excess glutathione.

pH optimum: 6.8.

K_m: Less than 5×10^{-6}.

[a] From Mauzerall and Granick (1958a).

Biosynthesis of Protoporphyrin IX and Heme

An understanding of the steps between Coprogen III and Proto IX (or Protogen IX) has proved to be very elusive. These are the meager clues: (a) oxygen is required for Proto synthesis (Falk et al., 1953); (b) hematoporphyrin IX, monovinyl-monohydroxyethyl deuteroporphyrin IX, and small amounts of Proto IX and other porphyrins accumulate in cultures of Chlorella mutant W_5B-17 (Bogorad and Granick, 1953a; Granick et al., 1953).

A reasonable scheme, roughly analogous to fatty acid metabolism, can be written with hematoporphyrin as one of the intermediates (Fig. 17).

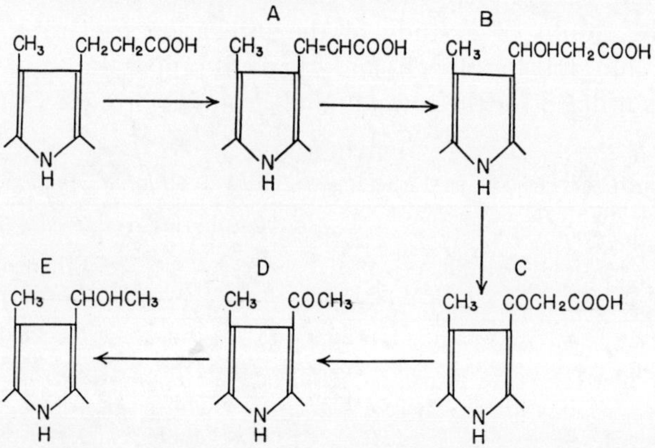

FIG. 17. A possible sequence of intermediates between Copro (or Coprogen) III and Proto (or Protogen) IX.

However, frozen and thawed *Chlorella* preparations which can catalyze the formation of Proto IX from PBG fail to utilize hematoporphyrin IX, hematoporphyrinogen IX, diacetyl deuteroporphyrin IX, or diacetyl deuteroporphyrinogen IX for the synthesis of Proto IX (Marks and Bogorad, 1958). Thus, there is no direct experimental evidence to support this scheme. Hematoporphyrin appears to be a stabilized derivative of the true intermediate which might be, for example, a phosphorylated hematoporphyrin. Another possible course could be via an acrylic intermediate or a concerted oxidation and decarboxylation of Copro III (Granick, 1955; Mauzerall and Granick, personal communication).

Another problem which arises at this point in the biosynthetic chain of porphyrins is: how far up on the path are the enzymes capable of handling only porphyrinogens? The best clue to the absolute limit is that porphyrinogens cannot chelate metals while the rigidly planar porphyrins are excellent chelators. Thus, the porphyrinogens must be oxidized, at the latest, immediately prior to coordination with the metal. It is tempting to guess that in the synthesis of cytochromes and hemoglobin, for example, first the iron-protein complex is formed and then the porphyrinogen is oxidized by the iron-protein. The result would be that the porphyrin, immediately as it is formed from the porphyrinogen, is in place and chelates with the iron; covalent bonds with the protein would be formed later. This type of mechanism would also provide a ready control for matching the rates of protein (e.g., globin) and Proto production. The main objection to this proposal is that under some circumstances free *heme* is formed from PBG or Proto in *in vitro* systems (e.g., Nishida and Labbe, 1959). This could be explained as occurring by the nonenzymatic coupling of protoporphyrin and iron in the manner described by Heikel *et al.* (1958), although Labbe's (1959) evidence argues against this, or by the splitting of the protein-to-iron linkage in the course of extracting the heme for assay.

A biosynthetic scheme for the formation of Proto IX, revised from the "primitive" one given earlier (page 231), is given in Fig. 18.

Protochlorophyll Formation

The biosynthetic relation between Proto IX and chlorophyll was firmly established by Granick's discovery (1948a) of a *Chlorella* mutant which produces no chlorophyll but accumulates Proto IX instead. The subsequent isolation of a *Chlorella* mutant which produces Mg-Proto IX (Granick, 1948b) led Granick to suggest that Proto was the last common member of the biosynthetic chains of heme and chlorophyll — insertion of iron committing the porphyrin to use in cytochromes, hemoglobin, or other heme enzymes; insertion of magnesium leading to the use of the porphyrin for chlorophyll formation.

Dr. Holt's report during these meetings that the *Chlorobium* chlorophyll which he has studied may lack the cyclopentanone ring requires that we place its point of origin apart from the main line of porphyrin biosynthesis somewhere after Mg-protoporphyrin but before Mg-vinyl pheoporphyrin a_5 (protochlorophyll lacking phytol), a compound first described in nature (Granick, 1950) as occurring in a *Chlorella* mutant. Mg-vinyl pheoporphyrin a_5 has also been identified in etiolated barley seedlings (Loeffler, 1955).

The most recent report on the production of Mg-vinyl pheoporphyrin a_5 and related compounds comes from Granick (1959). He has found that cuttings of etiolated barley seedlings accumulate the pheoporphyrin and Proto IX when supplied ALA. The further finding that when α-α'-dipyridyl is supplied with the ALA the cuttings accumulate Mg-Proto and an unspecified Mg-Proto monoester may be pertinent to the problem of iron deficiency chlorosis in plants as well as perhaps providing a new clue to the reactions which lead to the formation of the cyclopentanone ring. It may be that it is the propionic acid side chain on ring C of Mg-Proto IX which is esterified prior to formation of the cyclopentanone

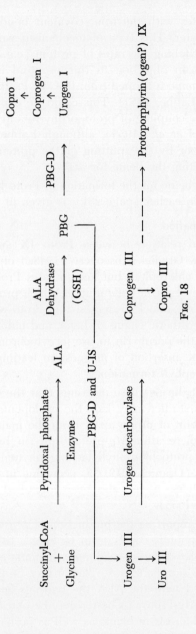

Fig. 18

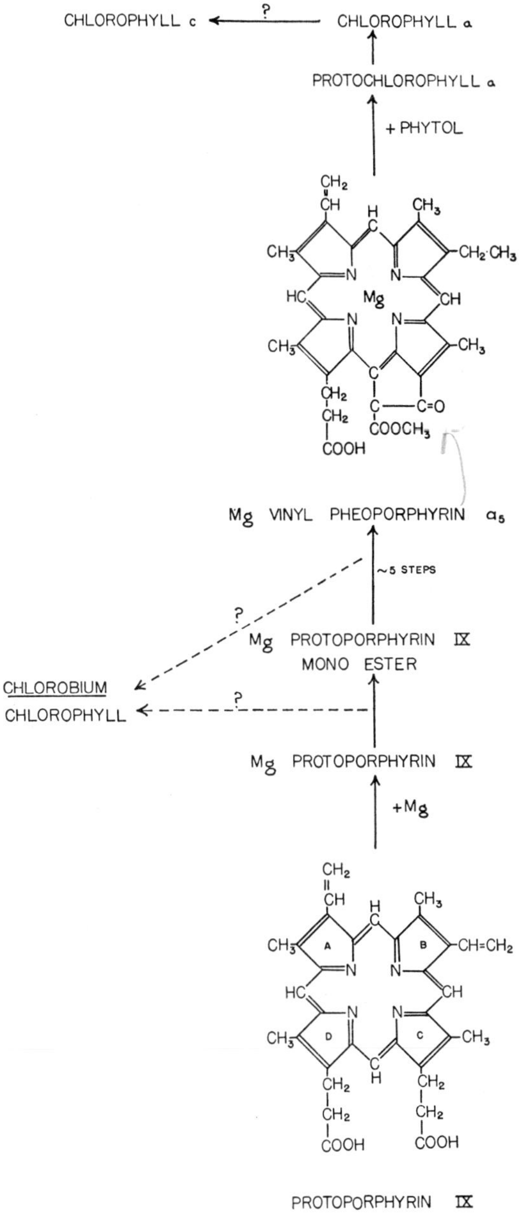

15. BIOSYNTHESIS OF PROTOCHLOROPHYLL 253

FIG. 19. A scheme for the biosynthesis of protochlorophyll from Proto IX. Reactions based primarily on speculation are shown with broken lines.

ring. This would serve to prevent the decarboxylation of the propionic acid residue during or after the oxidation to the carbonyl level of the carbon atom adacent to the ring. After the formation of the carbonyl group, the middle carbon atom of this side chain would become more reactive than before and be more likely to form a bond with the carbon atom which joins rings C and D. Investigations of the sequence of events here may provide valuable clues to the mechanisms of conversion of propionic to vinyl side chains in the production of Proto IX and the production of the acetyl substituent in bacteriochlorophyll. If this kind of speculation resembles the facts, bacteriochlorophyll may be derived not from chlorophyll, but might depart from the "main line of porphyrin biosynthesis" at some point between Copro III and Proto IX; the key intermediate in this case could be, for example, Copro III with the propionic acid side chain on ring A esterified. Figure 19 provides a hypothetical scheme of the steps between Proto X and protochlorophyll a; the most speculative sequences are connected by dotted lines.

A number of major problems in the biosynthetic chain between protoporphyrin and the chlorophylls remain to be solved. The mechanism of the incorporation of magnesium is an area which has hardly been touched, except in some early work by Smith (1947). Other problems cannot be attacked satisfactorily until the details of the structure of some of the chlorophylls have been established. Finally, the relationships between chloroplast development and chlorophyll formation are just beginning to be explored (e.g., von Wettstein, 1959; Sager, 1959; Bogorad et al., 1959).

REFERENCES

Bogorad, L. (1955). *Science* **121**, 878.
Bogorad, L. (1957a). *Plant Physiol.* **32**, xli.
Bogorad, L. (1957b). *In* "Research in Photosynthesis" (H. Gaffron, ed.), p. 475. Interscience, New York.
Bogorad, L. (1958a). *Natl. Acad. Sci. Natl. Research Council Publ.* **557**, 74.
Bogorad, L. (1958b). *J. Biol. Chem.* **233**, 501.
Bogorad, L. (1958c). *J. Biol. Chem.* **233**, 510.
Bogorad, L. (1958d). *J. Biol. Chem.* **233**, 516.
Bogorad, L., and Granick, S. (1953a). *J. Biol. Chem.* **202**, 793.
Bogorad, L., and Granick, S. (1953b). *Proc. Natl. Acad. Sci. U. S.* **39**, 1176.
Bogorad, L., and Marks, G. S. (1960). *J. Biol. Chem.* In press.
Bogorad, L., Pires, G., Swift, H., and McIlrath, W. J. (1959). *Brookhaven Symposia in Biol.* No. **11**, 132.
Booij, H. L., and Rimington, C. (1957). *Biochem. J.* **65**, 4P.
Bullock, E., Johnson, A. W., Markham, E., and Shaw, K. B. (1958). *J. Chem. Soc.* **287**, 1430.
Carpenter, A. T., and Scott, J. J. (1959). *Biochem. J.* **71**, 325.
Comfort, A. (1950). *Science* **112**, 279.

Cookson, G. H., and Rimington, C. (1953). *Nature* **171**, 875.
Cookson, G. H., and Rimington, C. (1954). *Biochem. J.* **57**, 476.
Dresel, E. I. B., and Falk, J. E. (1956). *Biochem. J.* **63**, 80.
Falk, J. E., Dresel, E. L. B., and Rimington, C. (1953). *Nature* **172**, 292.
Fischer, H. (1915a). *Z. Physiol. Chem.* **95**, 34.
Fischer, H. (1915b). *Z. Physiol. Chem.* **96**, 148.
Gibson, K. D., Neuberger, A., and Scott, J. J. (1955). *Biochem. J.* **61**, 618.
Gibson, K. D., Laver, W. G., and Neuberger, A. (1958). *Biochem. J.* **70**, 71.
Granick, S. (1948a). *J. Biol. Chem.* **172**, 717.
Granick, S. (1948b). *J. Biol. Chem.* **175**, 333.
Granick, S. (1950). *J. Biol. Chem.* **183**, 713.
Granick, S. (1951). *Ann. Rev. Plant Physiol.* **2**, 115.
Granick, S. (1954). *Science* **120**, 1105.
Granick, S. (1955). *Ciba Foundation Symposium on Porphyrin Biosynthesis and Metabolism* p. 143.
Granick, S. (1958). *J. Biol. Chem.* **232**, 1101.
Granick, S. (1959). *Plant Physiol.* **34**, xviii.
Granick, S., and Bogorad, L. (1953). *J. Am. Chem. Soc.* **75**, 3610.
Granick, S., and Mauzerall, D. (1958). *J. Biol. Chem.* **232**, 1119.
Granick, S., Bogorad, L., and Jaffe, H. (1953). *J. Biol. Chem.* **202**, 801.
Grinstein, M., Kamen, M. D., and Moore, C. V. (1949). *J. Biol. Chem.* **179**, 359.
Heikel, T., Lockwood, W. H., and Rimington, C. (1958). *Nature* **182**, 313.
Iodice, A. A., Richert, D. A., and Schulman, M. P. (1958). *Federation Proc.* **17**, 248.
Jackson, A. H., and MacDonald, S. F. (1957). *Can. J. Chem.* **35**, 715.
Kennedy, G. Y., and Vevers, H. G. (1954). *J. Marine Biol. Assoc. United Kingdom* **33**, 663.
Kikuchi, G., Shemin, D., and Bachmann, B. (1958). *Biochem. et Biophys. Acta* **28**, 219.
Labbe, R. F. (1959). *Biochim. et Biophys. Acta* **31**, 589.
Lascelles, J. (1955). *Ciba Foundation Symposium on Porphyrin Biosynthesis and Metabolism* p. 265.
Lascelles, J. (1957). *Biochem. J.* **66**, 65.
Lemberg, R., and Legge, J. W. (1949). "Hematin Compounds and Bile Pigments," Interscience, New York.
Loeffler, J. E. (1955). *Carnegie Inst. Wash. Year Book* **54**, 159.
McSwiney, R. R., Nicholas, R. E. H., and Prunty, F. T. G. (1950). *Biochem. J.* **46**, 147.
Marks, G. S., and Bogorad, L. (1958). Unpublished.
Mauzerall, D., and Granick, S. (1958a). *J. Biol. Chem.* **232**, 1141.
Mauzerall, D., and Granick, S. (1958b). *Intern. Congr. Biochem. 4th Congr., Vienna, 1958 Abstr. Communs.* No. 1-28 p. 4.
Muir, H. M., and Neuberger, A. (1949). *Biochem. J.* **45**, 163.
Muir, H. M., and Neuberger, A. (1950). *Biochem. J.* **47**, 97.
Neuberger, A., and Scott, J. J. (1953). *Nature* **172**, 1093.
Neve, R. A., Labbe, R. F., and Aldrich, R. A. (1956). *J. Am. Chem. Soc.* **78**, 691.
Nicholas, R. E. H., and Rimington, C. (1949). *Scand. J. Clin. & Lab. Invest.* **1**, 12.
Nicholas, R. E. H., and Rimington, C. (1951). *Biochem. J.* **48**, 306.
Nishida, G., and Labbe, R. F. (1959). *Biochim. et Biophys. Acta* **31**, 519.
Radin, N. S., Rittenberg, D., and Shemin, D. (1950). *J. Biol. Chem.* **184**, 745.

Rimington, C., and Booij, H. L. (1957). *Biochem. J.* **65**, 3P.
Robinson, R. (1955). "The Structural Relations of Natural Products." Oxford Univ. Press, London and New York.
Sachs, P. (1931). *Klin. Wochschr.* **10**, 1123.
Sager, R. (1959). *Brookhaven Symposium in Biol.* No. **11**, 101.
Schmid, R., and Shemin, D. (1955). *J. Am. Chem. Soc.* **77**, 506.
Schulman, M. P., and Richert, D. A. (1957). *J. Biol. Chem.* **226**, 181.
Shemin, D. (1955). *Ciba Foundation Symposium on Porphyrin Biosynthesis and Metabolism* p. 4.
Shemin, D., and Kikuchi, G. (1958). *Ann. N. Y. Acad. Sci.* **75**, 122.
Shemin, D., and Kumin, S. (1952). *J. Biol. Chem.* **189**, 827.
Shemin, D., and Rittenberg, D. (1946). *J. Biol. Chem.* **166**, 621.
Shemin, D., and Russel, C. S. (1953). *J. Am. Chem. Soc.* **76**, 4873.
Shemin, D., and Wittenberg, J. (1951). *J. Biol. Chem.* **192**, 315.
Shemin, D., Russel, C. S., and Abramsky, T. (1955). *J. Biol. Chem.* **215**, 613.
Smith, J. H. C. (1947). *J. Am. Chem. Soc.* **69**, 1492.
Westall, R. G. (1952). *Nature* **170**, 614.
von Wettstein, D. (1959). *Brookhaven Symposia in Biol.* No. **11**, 138.
Wittenberg, J., and Shemin, D. (1949). *J. Biol. Chem.* **178**, 47.
Wittenberg, J., and Shemin, D. (1950). *J. Biol. Chem.* **185**, 103.

Discussion

NEILANDS: Your suggestion that the metal ion goes in by first combining with the enzyme protein and then later with the porphyrin is attractive in several respects because of the concentration of heme that you find in nature. Nature has not much respect for the porphyrin but great respect for the porphyrin-metal complex and it very seldom seems to make any excess heme. On the other hand, I think that the *in vitro* experiment has been done in which you mix globin, iron, and protoporphyrin; the yields of hemoglobin are pretty small, not very convincing. I was going to mention one other point about how this metal may get in. I had a student, by the name of Orlando, who treated coproporphyrin and several other porphyrins with sodium amalgam. Then he added ferric ion and got very good yields of the iron complex, but only if he added the iron shortly after treating the porphyrin with sodium amalgam, after it goes through that greenish-yellow stage.

BOGORAD: When you reoxidize this kind of thing, you get a whole series of colored intermediates. The first thing you see is a strong 500 mμ band; it looks like kind of an orange solution, and the guess would be that this is probably two dipyrryl methenes isolated from one another. Now I do not know at what state metal incorporation is most likely. It would not pick up the iron when it was completely reduced; it would have to be on its way back up to being oxidized.

NEILANDS: The picture which seemed most attractive was, say, two reduced bridges and then the two, or three, pyrrole nuclei with the double bonds between them lying in a plane. Now the structure, even with one reduced bridge in it, must be quite unstable and there must be a large decrease in free energy accompanying the change to the much more stable planar configuration; Orlando pictured this reaction as being one in which the metal ion was squeezed in by using the energy from the porphyrinogen stage to the porphyrin.

16

Protochlorophyll Transformations

JAMES H. C. SMITH

Department of Plant Biology, Carnegie Institution of Washington, Stanford, California

Protochlorophyll Biosynthesis

In the preceding paper, Dr. Bogorad has summarized in a most interesting and effective way the path of porphyrin biosynthesis as it relates to chlorophyll formation. His last diagram showed the series of steps proposed several years ago by Granick which outlines the path of chlorophyll synthesis fundamentally as we conceive it today. This being the case, it is necessary to begin by referring to the same diagram.

Granick's (1950a) scheme is shown in Fig. 1. Dr. Bogorad has discussed the steps from "glycine + acetic acid" to "protoporphyrin IX." The next step in the scheme is the introduction of magnesium to form magnesium protoporphyrin IX. This compound was isolated from a chlorophyll mutant of *Chlorella* and forms a rational step in the biosynthetic scheme. Although the introduction of magnesium could be a simple chelation, it is more probably mediated by enzyme action, even as iron is enzymatically incorporated into protoporphyrin IX to form hemin (Labbe, 1959). The next steps involve the closure of the cyclopentanone ring (ring V) and methylation of the carboxyl group attached. In the absence of any experimental evidence on the mechanism of these processes, only a formal presentation of a possible chain of events can be hypothesized. Formally, the β-carbon atom of the propionic acid group becomes oxidized to a keto-group, the α-carbon of the same acid group undergoes oxidative condensation with the γ-carbon of the porphyrin ring, the carboxyl group is methylated, and the vinyl group at C-4 is reduced to ethyl. The compound resulting from these changes is magnesium vinyl pheoporphyrin a_5 or protochlorophyllide. This compound is well-known and has not only been identified in a mutant strain of *Chlorella* (Granick, 1950b) but has been found in etiolated leaves where it is an intermediate in the normal chlorophyll-forming process (Loeffler, 1955).

Granick originally postulated his scheme on the basis of intermediates obtained largely from *Chlorella* mutants. Recently, however, he has iden-

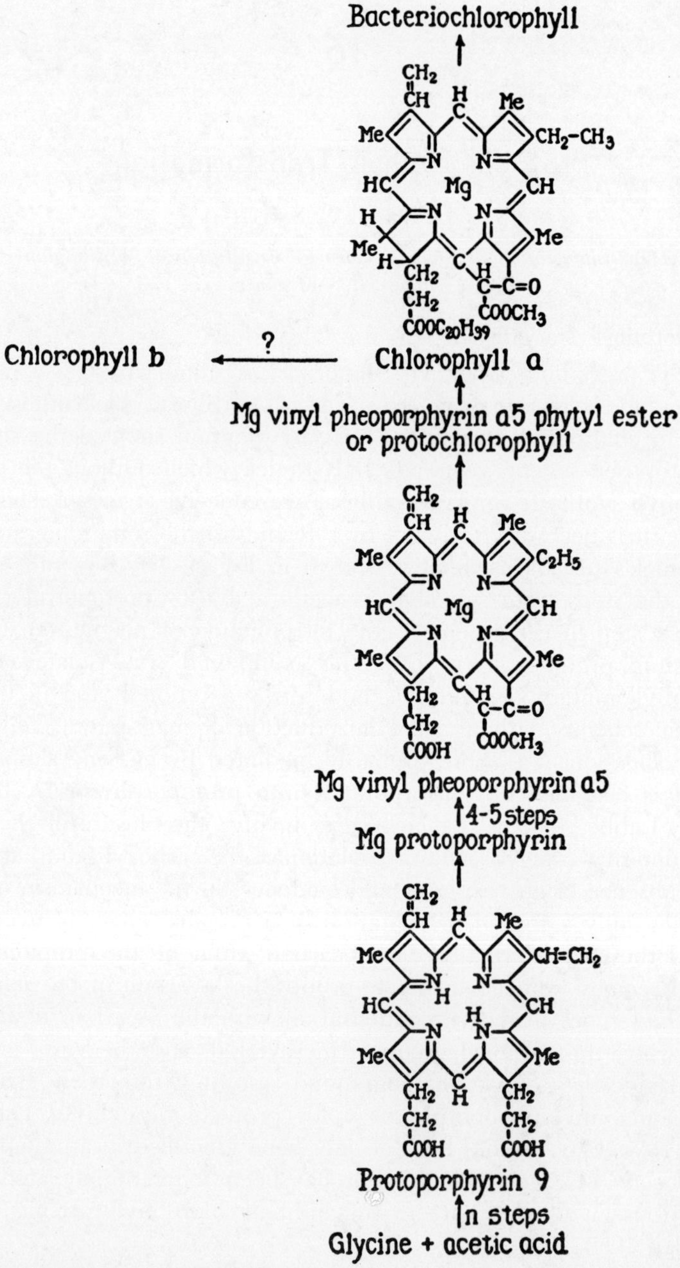

Bacteriochlorophyll
↑

Chlorophyll b ←—?—— Chlorophyll a
↑
Mg vinyl pheoporphyrin a5 phytyl ester
or protochlorophyll
↑

Mg vinyl pheoporphyrin a5
↑ 4-5 steps
Mg protoporphyrin
↑

Protoporphyrin 9
↑ n steps
Glycine + acetic acid

FIG. 1. Abridgment of Granick's scheme for the biosynthesis of chlorophylls (Granick, 1950a).

tified the same intermediates—protoporphyrin, magnesium protoporphyrin, magnesium protoporphyrin methyl ester, and magnesium vinyl pheoporphyrin a_5—in excised etiolated barley leaves fed δ-aminolevulinic acid (Granick, 1959).

In normal etiolated leaves, magnesium vinylpheoporphyrin a_5 coexists with its phytylated derivative, protochlorophyll. Both these compounds are converted to their corresponding chlorophyll derivatives when the leaves containing them are illuminated: protochlorophyll is converted directly to chlorophyll a, and protochlorophyllide to chlorophyllide a. The latter is subsequently esterified with phytol, through thermal processes, to form chlorophyll a. Thus two alternative pathways exist between protochlorophyllide and chlorophyll a (Smith et al., 1957a; Wolff and Price, 1957). Since the conversion of protochlorophyll to chlorophyll requires the addition of hydrogen atoms at carbons 7 and 8, this reaction is a photohydrogenation (Fischer, 1940).

Certain other steps in Granick's scheme, the formation of chlorophyll b and of bacteriochlorophyll, will be mentioned later.

Phototransformation of Protochlorophyll

This paper deals chiefly with the phototransformation of protochlorophyll, or its phytyl-free derivative, to the corresponding chlorophyll a derivative. Since the photoreaction is the same for both, the term protochlorophyll will be used to cover both unless specifically designated otherwise.

PROTOCHLOROPHYLL AS CONTINUOUS PRECURSOR TO CHLOROPHYLL

Doubt no longer exists that the protochlorophyll of dark-grown seedlings gives rise to the chlorophyll initially formed therein by action of light (cf. Smith and Young, 1956). Doubt has been expressed, however, that protochlorophyll is the precursor of chlorophyll a throughout its continuous accumulation (Virgin, 1955). Nevertheless, several lines of evidence suggest that it performs this function: First, there is regeneration of protochlorophyll in briefly illuminated dark-grown seedlings when they are returned to the dark at room temperature and this new-formed protochlorophyll is transformed by light to chlorophyll. A combination of these two processes could well account for the accumulation of chlorophyll. A necessary corollary to this is that the chlorophyll formed does not revert to protochlorophyll in the dark. This has been confirmed by Virgin (1955) who demonstrated that no decrease in chlorophyll occurred even though a considerable quantity of protochlorophyll was formed. There is no experimental evidence for a dynamic equilibrium between chlorophyll and protochlorophyll.

Next, in seedlings kept at different temperatures, the initial rate of protochlorophyll formation in the dark parallels the rate of chlorophyll formation in the light (Virgin, 1955). When the temperature of seedlings is decreased below room temperature, the rate of regeneration of protochlorophyll is slower until at 0° C it is zero (Virgin, 1955). The initial rate of this protochlorophyll regeneration corresponds to the rate of accumulation of chlorophyll in the light which is slower the lower the temperature and ceases near 0° C (Virgin, 1955; Smith, 1949a, b; Smith and Koski, 1948).

The experiments just cited were made on leaves that had accumulated only relatively little chlorophyll. They demonstrated that protochlorophyll could be and probably is the precursor of chlorophyll in the early stages of chlorophyll accumulation. They did not answer the question, however, whether protochlorophyll is the possible or probable precursor of chlorophyll in the later stages of its accumulation.

The first evidence bearing on this question was obtained by Koski (cf. Smith and Koski, 1948) who isolated protochlorophyll by chromatography from green mature barley leaves harvested from the field. This evidence implied that protochlorophyll was a possible precursor of chlorophyll even under these conditions. Very recently Krasnovskii and co-workers (Litvin et al., 1959) demonstrated the formation and transformation of protochlorophyll in fully green leaves. They held the leaves in the dark for some time, then illuminated part of them at −150° C and part at room temperature. Those illuminated at −150° C, where transformation is inhibited, showed protochlorophyll fluorescence whereas the others did not. From this experiment they concluded that:

> "Comparison of our data on the rate of accumulation of protochlorophyll with data on the rate of renewal of chlorophyll, obtained by isotopic methods . . . gives the basis to conclude that the biosynthesis of chlorophyll in green leaves also is accomplished with the participation of an active form of protochlorophyll." (Translation by H. W. Milner.)

By use of the C^{14}-labeled porphyrin intermediate, δ-aminolevulinic acid, Roux and his associates (Duranton et al., 1958) induced the formation of labeled protochlorophyll in fully green tobacco leaves. The labeled protochlorophyll was formed in the dark, whereas labeled chlorophyll a was formed in the light. The presumption was that the protochlorophyll failed to accumulate in the light because of its rapid phototransformation to chlorophyll. These experiments bring convincing evidence that protochlorophyll is the precursor of chlorophyll a throughout the chlorophyll accumulation process.

Origin of Chlorophyll b

Only chlorophyll a is formed by photoconversion of the protochlorophyll initially present in dark-grown leaves. In most of the higher plants, however, as chlorophyll content increases, both chlorophylls a and b accumulate. Chlorophyll a comes from protochlorophyll, but what is the source of chlorophyll b? Two postulates concerning its origin have been made: that it comes either (i) from a protochlorophyll b, or (ii) from chlorophyll a.

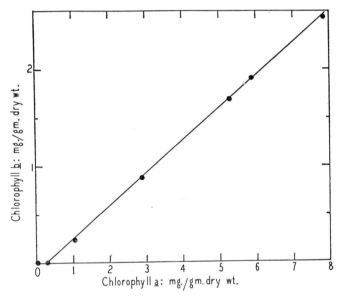

Fig. 2. The relation between the accumulation of chlorophylls a and b at 17.1° C under continuous illumination of dark-grown seedlings of barley (cf. Smith, 1949b).

The first of these postulates seems untenable because protochlorophyll b has never been identified in leaves. Certainly such a compound does not exist originally in dark-grown leaves, else chlorophyll b would arise from the initial illumination of the leaves. Furthermore, it is doubtful if it is formed from δ-aminolevulinic acid as is protochlorophyll a; otherwise, Roux and his associates (Duranton et al., 1958) would have found it, and they specifically state, "*Nous n'avons pas trouvé de protochlorophylle b.*"

Also, it seems equally unlikely that chlorophyll b comes from chlorophyll a. In the experiments of Koski and Smith (cf. Smith, 1949b) the quantities of chlorophylls a and b in barley leaves were followed during greening. These are plotted against each other in Fig. 2. At first only chlorophyll *a* was present, but after chlorophyll *b* appeared its content was directly proportional to the content of chlorophyll *a*. This straight-line relation means that the rate of formation of chlorophyll *b* was directly proportional to the rate of formation of chlorophyll *a* and suggests that both chlorophylls had their origin in a common process or common precursor. This relation seems to preclude any possibility that chlorophyll *b* arose directly from chlorophyll *a;* otherwise, its rate of formation would have been proportional to the quantity of chlorophyll *a* present and would have given a curve convex to the abscissa.

Experiments with labeled carbon also support the contention that chlorophyll *b* does not come from chlorophyll *a*. From experiments with labeled glycine and acetate, Della Rosa *et al.* (1953) concluded that "... chlorophylls *a* and *b* are derived from a common precursor but are not interconvertible." And Roux and his associates believed that at most only a very small fraction of labeled chlorophyll *a* could be converted directly into chlorophyll *b*, a conclusion derived from their experiments with labeled δ-aminolevulinic acid in tobacco leaves (Duranton *et al.*, 1958).

These facts, therefore, make it highly improbable that chlorophyll *b* arises either from a preformed protochlorophyll *b* or from chlorophyll *a*. Its formation still remains a puzzle.

Protochlorophyll and Bacteriochlorophyll

Very little is known about the steps in the biosynthetic chain leading to other chlorophylls. It is significant, therefore, that R. Y. Stanier (Stanier and Smith, 1959) has recently isolated a pigment with spectral absorption properties similar to protochlorophyll from the "tan mutant" of *Rhodopseudomonas spheroides,* a purple bacterium whose normal photosynthetic pigment is bacteriochlorophyll. The question arose whether this pigment was identical to protochlorophyll either from leaves or from inner seed coats of squash.

A comparison of the absorption curve of the bacterial pigment with the absorption curves of the protochlorophylls from the other sources is shown in Fig. 3. All pigments were chromatographically purified and dissolved in ether for spectroscopic measurement. The shape of the whole spectral curve of the bacterial pigment resembles more closely the curve of the seed-coat protochlorophyll than of the bean-leaf protochlorophyll. The most conspicuous differences lie in the Soret bands where the maximum of the bacterial pigment's absorption lies at 440.5 mμ, that of squash inner seed-coat protochlorophyll at 438 mμ, and that of bean-leaf protochlorophyll at 433 mμ. The spectroscopic results indicate the bacterial pigment to be a protochlorophyll and to be more closely akin to seed-coat than to leaf protochlorophyll. Proof of its function as a precursor of bacteriochlorophyll is still needed. If such proof is found it might well demonstrate the unity of the biosynthetic pathways of chlorophyll a and bacteriochlorophyll for a long way along the course of their synthesis.

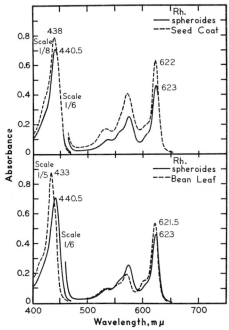

Fig. 3. Comparison of the spectrum of a protochlorophyll-like pigment from *Rhodopseudomonas spheroides* with known protochlorophylls: *top*, protochlorophyll from inner seed coats of squash; *bottom*, protochlorophyllide from etiolated bean leaves (Stanier and Smith, 1959).

External Factors Affecting Protochlorophyll Transformation

And now let us examine the effects of various external factors on the photoconversion of the protochlorophyll initially present in etiolated leaves.

Light Intensity

First let us look at the effect of light intensity. Koski (1949) showed by the use of polychromatic light that the initial rate of transformation of protochlorophyll to chlorophyll a was about proportional to the incident intensity. Later Smith and Benitez (1954) found that in etiolated barley leaves exposed to monochromatic light (589 mμ) of constant intensity the conversion followed a second-order rate law with respect to protochlorophyll concentration. Under these conditions the rate constant was directly proportional to the incident intensity. For example, at 50 and 17 foot-candles the corresponding rate constants were 0.0190 and 0.0066. The ratio of intensities was 2.94 and of rate constants 2.88, values that agree very well.

Wavelength

Next, let us consider the effect of wavelength on the transformation. The second-order rate law for the conversion is obeyed extremely well at 589, 579/577, and 546 mμ. At 436 mμ, although there is fair adherence to this law, the deviation is greater than the experimental error warrants (Smith and Benitez, 1954). No cause is known for this deviation at the present time, but it is suggestive that the lack of strict conformity to the second-order law comes in the spectral region of strongest carotenoid absorption.

That carotenoids alter the effectiveness of light in causing the transformation is clearly evident from the action-spectrum measurements of Koski et al. (1951). These workers compared the action spectra of etiolated carotenoid-rich normal and carotenoid-free albino corn leaves. The effectiveness of red light was the same in both, whereas the effectiveness of blue light was much less in the normal leaves. The lowered effectiveness in blue light was attributed to the screening action of the yellow pigments.

The action spectrum of albino leaves resembles the absorption spectrum of protochlorophyll within the leaf (Smith, 1959a), and of pure protochlorophyll dissolved in methanol when the usual holochromatic shift is discounted (Koski et al., 1951). These similarities leave no room for doubting that protochlorophyll is the photoreceptor for its own conversion.

Temperature

Another external factor that affects the conversion of protochlorophyll in leaves is temperature. The work of Koski (1949; cf. also Smith and Young, 1956) showed that at temperatures of 5° and 18° C the rates of transformation were equal. At very low temperatures, however, the rates were much reduced (Smith and Benitez, 1954). At 20.2°, —40.0°, and —70.0° C, the second-order velocity constants were respectively 0.000132, 0.0000735, and 0.0000138. Furthermore, the limit of conversion varied with temperature. For the same three temperatures, the limits of transformation were 95.3, 71.3, and 58.4%. At —195° C no transformation occurred, a fact utilized by Litvin et al. (1959) to demonstrate fluoroscopically the presence of protochlorophyll in mature green leaves.

Approximately 40° C was the upper threshold at which barley leaves began to lose their ability to transform protochlorophyll. As the temperature was raised above this, their transforming power was progressively reduced until only 4% remained after heating for 5 minutes at 55.3° C. The rates at which the conversion ability was destroyed at different temperatures suggested that the destruction was caused by the denaturation of a protein. This protein could be either an enzyme catalyst or a carrier for the pigment and necessary for its conversion. That it is an enzyme seems unlikely because of the low temperatures at which the transformation can occur. That it is a carrier protein is highly probable because of protochlorophyll's spectral properties in the natural state and because of the physical and chemical properties of the isolated holochromatic complex.

Internal Changes Following Transformation

We have been discussing the effects of external factors on the protochlorophyll-chlorophyll conversion in an attempt to get a clearer picture of the system undergoing phototransformation within the leaf. We have considered chiefly what happened to the protochlorophyll. Now let us examine the chlorophyll. This product of the reaction undergoes certain changes that may also be useful for the analysis of the reacting system.

PHYTYLATION

One change is the rapid phytylation of the chlorophyllide a produced by the photoconversion. Only a minor part of the pigment possessing the protochlorophyll spectrum is true protochlorophyll; the rest is protochlorophyllide. The latter is converted by light to chlorophyllide a which is rapidly phytylated to chlorophyll a at room temperature (Loeffler, 1955; Wolff and Price, 1957) but not at temperatures near 0° C. This makes it clear that phytylation is a thermochemical rather than a photochemical reaction. These results raise two important questions, however; namely, (i) why is only part of the protochlorophyll-type pigment originally phytylated, and (ii) from where does the phytol come for the esterification of chlorophyllide a. These results hint that one acid group of protochlorophyllide is originally bound so as to be inaccessible for phytylation but that photoconversion to chlorophyllide a makes it available for esterification through chlorophyllase action. The source of the phytol is undetermined.

SPECTRAL SHIFT

Another remarkable change following the conversion is the transposition of the chlorophyll spectrum first produced. Shibata (1957) found that the chlorophyll first formed by transformation of protochlorophyll has its absorption maximum at about 684 mμ. On standing at room temperature for a period of 14 minutes following the conversion, the position of this maximum gradually shifts to near 673 mμ and the height of the band increases appreciably. Since the two chlorophyll spectra have an isobestic point, the mechanism undoubtedly involves the direct transformation of one molecular species into another. When the leaf is illuminated and held at low temperature (below 5° C) after being illuminated, the spectral shift is either stopped or greatly inhibited. The large temperature coefficient of the shift indicates a reaction with large heat of activation. Speculation concerning the cause and nature of the shift has suggested several possibilities: (i) that the conversion changes the configuration of the pigment so as to require its rearrangement or reorientation on the carrier protein; (ii) that the pigment is transferred from one carrier to another with consequent change of spectrum; or (iii) that phytylation occurs which causes a realignment of the pigment molecules in the interfaces of the holochrome. The last hypothesis has some justification from observations made (Smith et al., 1959) concerning the phytylating ability and spectral shift in etiolated leaves of several corn mutants. Those mutants in which the phytylating ability was low or absent gave very small shifts in the absorption spectrum of the new-formed chlorophyll, whereas a mutant that had normal phytylating ability gave a spectral shift about the same as that of normal corn. The relation between these two values in the different samples was not quantitative but showed some degree of correlation.

In regard to the other two hypotheses, however, it is well to recall an old experiment of Smith and Benitez (1953) in which etiolated barley leaves were homogenized in glycerine and the homogenate freed of debris by centrifugation. The cold supernatant showed the characteristic absorption band of native protochlorophyll at 650 mμ. When the supernatant was warmed to room temperature this absorption band was shifted to 635 mμ. The chlorophyll derived from the 650-mμ form had an absorption maximum near 680 mμ, whereas that derived from the 630–635-mμ form had a maximum close to 672 mμ (cf. Smith et al., 1957a). The conclusion from these experiments is that rearrangements or reorientations can take place before the conversion and without disrupting the convertible structure; they have nothing to do with the change of shape of the pigment molecule nor the transferring of the pigment from one carrier molecule to another resulting from the photoconversion. That the configuration of the carrier and its relation to the pigment is affected both by the photochemical action and by the aqueous glycerine is not an improbable cause of these spectral shifts.

Properties of Isolated Holochrome

In the preceding discussion, the transformation of protochlorophyll inside the leaf has been chiefly considered. Although a great deal can be learned from such studies, the ideal approach would be to isolate the active system from the leaf and to examine its physical and chemical properties in detail. An attempt has been made to do this and what follows is a survey of the results obtained.

Isolation

In our laboratory, the first active extrafoliar systems were extracted by braying etiolated leaves in glycerine (Smith and Benitez, 1953). The extracts contained active holochrome but concentration and purification of active material from the viscous glycerine solutions were impossible. Krasnovskii and Kosobutskaya (1952) found, however, that an active extract of bean leaves could be obtained in a phosphate buffer solution. Thanks to this discovery the way was opened for further purification of the active component.

Smith and Kupke (1956; cf. Smith et al., 1957a) showed that a glycine buffer extract, pH $\sim$ 9.5, of etiolated bean leaves, when dialyzed and subjected to ultracentrifugal analysis, exhibited two peaks in the sedimentation diagram. These peaks corresponded to boundaries having sedimentation coefficients of 3 to 4 and of 16 to 17 S respectively. By a combination of preparative centrifugation and spectrophotometric analysis, it was clearly demonstrated that the particle showing the sedimentation coefficient of 16 to 17 S was the active protochlorophyll-containing component. This particle could be sedimented as a pellet by centrifugation at 91,000 g for 3 hours and resuspended in buffer in active transformable condition.

Purification of the protochlorophyll holochrome was carried out chiefly by repeated fractional precipitation with ammonium sulfate. Kupke (cf. Smith et al., 1956) found that slightly ammoniacal 40 per cent ammonium sulfate, pH $\sim$ 8.1, separated the holochrome from most of the other components in the extracts. The precipitated holochrome was readily soluble in glycine buffer in active condition. Dialysis against distilled water followed by lyophilization gave a yellow fluffy powder that was easily soluble in buffer. The protochlorophyll contained in the powder was relatively stable in the cold and dark and was partially convertible to chlorophyll by light in both the solid and dissolved states.

The purification procedure most recently adopted includes a dialysis of the original extract followed by precipitation of certain contaminants with a small quantity of barium chloride, extensive ammonium sulfate fractionation, and centrifugation by swinging bucket in a density gradient of sucrose. The sucrose zone containing the pigment was removed and its properties determined.

ABSORPTION SPECTRUM OF PROTOCHLOROPHYLL HOLOCHROME

The absorption spectrum of the protochlorophyll holochrome is shown in Fig. 4 (Smith and Coomber, 1959). This spectrum has four

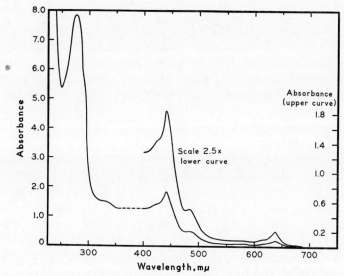

FIG. 4. Absorption spectrum of the protochlorophyll holochrome purified by density gradient high-speed centrifugation (Smith and Coomber, 1959).

clear-cut absorption regions: (i) the region at about 280 mµ, typical protein absorption; (ii) the prominent Soret band near 441 mµ, mostly attributable to protochlorophyll; (iii) the region about 485 mµ, characteristic of carotenoids; and (iv) the long wavelength absorption band of protochlorophyll at 637.5 mµ.

The protein absorption at 280 mµ is undoubtedly chiefly caused by tryptophan and tyrosine. Estimation of the molecular ratio of protochlorophyll to tryptophan is about 1 to 11, and to tyrosine approximately 1 to 16. The molecular species responsible for the small absorption band at 335 mµ has not yet been identified. The ratio of the absorbancies at 280 and 260 mµ, viz., 1.45, indicates less than 1% impurity of nucleic acid (Warburg and Christian, 1942).

Particle Weight

The ultraviolet absorption at 280 mµ gave a measure of the protein present in a given volume of holochrome solution. The values found agreed well with those determined by the biuret method. By comparing the ratio of protein to protochlorophyll, determined by spectral absorption, and by assuming a molecular ratio for protein to protochlorophyll of unity, an average "molecular weight" of 0.96×10^6 was estimated for the pigment complex. This value is in accord with the particle weight derived from ultracentrifugal measurements, 0.7×10^6. Previously Smith and Kupke (1956) had estimated the particle weight as 0.4×10^6 based on an assumed density of 1.333 for the particle and an observed sedimentation coefficient of 16 to 17 S. Recent density determinations by use of a swinging bucket centrifuge with solutions of the holochrome suspended in density gradients of sucrose gave a density for the particle of 1.16. This value combined with the sedimentation coefficient used previously led to the revised particle weight of 0.7×10^6 gm. These two analyses support the concept that each holochrome particle contains only one protochlorophyll molecule.

Carotenoids

So far, it has been impossible to obtain the holochrome free of carotenoid. This does not necessarily mean that carotenoids are required for the protochlorophyll-chlorophyll conversion, because albino leaves, although they possess little or no carotenoid pigment, carry out this phototransformation (Koski and Smith, 1951). The carotenoids of the holochrome have not yet been identified.

Absorption Spectrum of Chlorophyll-*a* Holochrome

The protochlorophyll holochrome just described is capable of being converted to chlorophyll *a* holochrome. This is shown in Fig. 5. Here the curve for the protochlorophyll holochrome before being illuminated is reproduced from Fig. 4 for comparison with the curve of the illuminated holochrome—the chlorophyll holochrome. The characteristic spectral changes resulting from the conversion are evident: the large shift in the position of the main absorption band in the red end of the spectrum, and the great decrease in height of protochlorophyll's Soret peak. No change was noted in either the carotenoid or protein absorption bands.

The shape of the chlorophyll holochrome's absorption band in the blue-violet end of the spectrum is remarkable. It differs radically from the corresponding absorption band of chlorophyll in organic solvents [cf. Fig. 6; (Trurnit and Colmano, 1959)]. In the chlorophyll holo-

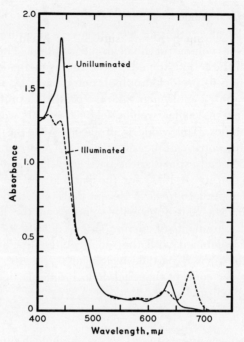

Fig. 5. Comparison of the illuminated and unilluminated holochrome which had been purified by density-gradient high-speed centrifugation (Smith and Coomber, 1959).

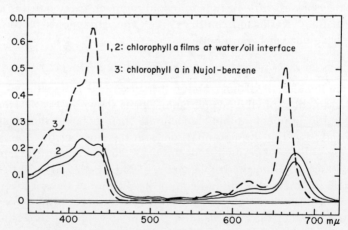

Fig. 6. Comparison of the absorption spectra of chlorophyll *a* in monolayers at water-oil interfaces (curves *1* and *2*) and in a Nujol-benzene solution (curve *3*) (Trurnit and Colmano, 1959).

chrome, the peak near 410 mµ is slightly higher than that near 440 mµ, whereas in chlorophyll solutions this peak is relatively much lower. Chlorophyll *a*, either when adsorbed on filter paper (Smith *et al.*, 1957b) or when adsorbed at an oil-water interface (Fig. 6), has an absorption spectrum like that of chlorophyll in the holochrome. This suggests that in the holochrome chlorophyll *a* may be similarly disposed.

Fluorescence Polarization

Further deductions concerning the state of chlorophyll in the holochrome can be made from measurements of fluorescence polarization. The polarization of fluorescence occurs when, under certain circumstances, fluorescence is excited with plane-polarized light. The degree of the polarization depends, among other things, upon the size of the fluorescing molecule, its concentration, and the viscosity of the medium containing it. In principle, if conditions are such that the molecule cannot rotate between its absorption of the plane-polarized exciting light and its emission of fluorescence, the fluorescence will be plane-polarized except when a sufficient number of energy transfers occur between randomly oriented excited and unexcited molecules (Goedheer, 1957).

Previous work had shown that the fluorescence polarization of chlorophyll in mature plants is very low (Latimer and Smith, 1958), whereas that from newly formed holochromatic chlorophyll is relatively high. These facts indicated that worthwhile information concerning the disposition of chlorophyll during its accumulation might be obtained by following the change in fluorescence polarization of the chlorophyll holochrome as chlorophyll increased.

Solutions of the chlorophyll holochrome were obtained by the methods used to extract protochlorophyll holochrome. Dark-grown bean leaves were exposed to light for various lengths of time. At known intervals, leaf samples of equal weights were removed and ground in a glycine-glycerine buffer mixture of pH $\sim$ 9.5. The extracts were centrifuged at 10,000 g for 15 minutes to remove debris. The fluorescence polarization of each supernatant was measured at 644 mµ with an apparatus like that formerly used by Goedheer (1957); the chlorophyll content was estimated from measurements of absorbance at 670 mµ.

Within the error of measurement, the reciprocals of the fluorescence polarizations obtained by the method described when plotted against the chlorophyll absorbancies yield a straight line as shown in Fig. 7 (Goedheer and Smith, 1959). This conforms to the equation commonly used to relate these quantities (Förster, 1951), namely

$$1/P = 1/P_0 + AC\,\tau$$

In this equation, P represents the fluorescence polarization observed; P_0, the polarization when C, the concentration of the pigment, is zero; τ, the lifetime of the activated state; and A is a constant.

The results of these experiments show that the fluorescence polarization decreases as the chlorophyll content increases. This would be expected from the energy-transfer hypothesis if the chlorophyll as it accumulates would continue to distribute itself on carrier particles already occupied by chlorophyll rather than to go to particles not so occupied. In the first condition, the compactness of the pigment would favor energy transfer; in the second, the dispersion would hinder it. The conclusion

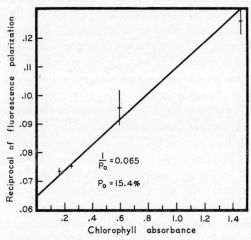

FIG. 7. A plot showing the influence of chlorophyll accumulation on the depolarization of chlorophyll fluorescence from holochrome extracts of bean seedlings. In this plot $1/P$ is taken as a measure of the fluorescence depolarization (Goedheer and Smith, 1959).

seems justified, therefore, that chlorophyll as it increases tends to build up on particles already containing chlorophyll. Whether the chlorophyll accumulates on the same particles that originally contained the protochlorophyll, or whether it is transferred to a different particle following the transformation are questions under consideration at the present time.

The extrapolated value of the fluorescence polarization of the holochrome at zero chlorophyll concentration is 15.4%. This is much lower than the 29% found by the same set-up for chlorophyll dissolved in castor oil where the pigment's rotation is inhibited. Therefore, this lowering of the fluorescence polarization shows that the holochromatic chlorophyll has some rotational freedom. The rotation cannot be a property of the entire holochrome since it could not rotate fast enough to lower the fluorescence polarization appreciably because of its large particle weight. In consequence, the rotation must be due to the chlorophyll itself. The chlorophyll could rotate independently of the holochrome if it were not rigidly bound to but contained within the holochrome in an oil- or fat-water interface. Additional evidence for this disposition of chlorophyll comes from the shape of its absorption curves in the holochrome which in the blue-violet region of the spectrum (cf. Fig. 5) is very much like that of pure chlorophyll in an oil-water interface (Fig. 6).

Quantum Yield of Conversion

Another property of the protochlorophyll-chlorophyll transformation that is of fundamental significance is the quantum yield of the reaction. Measurements made by Smith and French (1958; cf. also Smith, 1959b) on clear solutions of the holochrome gave values which averaged 0.6 molecule per quantum. Two wavelengths of light were used: 642 mµ, obtained from a monochromator, and 644 mµ, from a cadmium arc. Both beams were monochromatized with appropriate filters. From the results obtained, it is difficult to say whether the reaction has an intrinsic quantum yield of 1.0 or of 0.5. Further work is needed to distinguish between the two. Since the quantum yield is near 0.5, it is tempting to speculate that two quanta are required to transfer the two hydrogen atoms or the two electrons necessary to the hydrogenation of the double bond. But in view of what is known of the reaction, this possibility seems remote. It appears much more likely, considering the probable bimolecular nature of the process and the direct dependence of rate on light intensity, that the transformation is a one-quantum process with an over-all efficiency of 0.6. Our present concept is that a single photoactivation step is necessary for the protochlorophyll-chlorophyll transformation.

From the foregoing analysis, we have come to a tentative picture of the protochlorophyll holochrome. It is a particle of the order of a million "molecular weight" and contains only one protochlorophyll molecule. This pigment is embedded in an interface of fat-water-protein in which it has considerable freedom of rotation. Photochemically, the protochlorophyll-chlorophyll transformation is very efficient. As the leaves become green, the chlorophyll accumulates on particles already containing chlorophyll rather than going to particles unoccupied by this pigment. Throughout this accumulation the chlorophyll *a* is derived from a photochemical transformation of protochlorophyll.

Acknowledgments

The writer gratefully acknowledges the kind permission of the authors and publishers named to copy the following figures from other publications for use in this article: Fig. 1, Dr. Sam Granick (author) and The Harvey Society and Charles C Thomas (publishers); Fig. 6, Dr. H. J. Trurnit (author) and The Elsevier Publishing Company (publisher); and Figs. 3, 4, 5, and 7, Carnegie Institution of Washington (publisher).

References

Della Rosa, R. J., Altman, K. I., and Salomon, K. (1953). *J. Biol. Chem.* **202**, 771-779.
Duranton, J., Galmiche, J. M., and Roux, E. (1958). *Compt. rend. acad. sci.* **246**, 992-995.
Fischer, H. (1940). *Naturwissenschaften* **28**, 401-405.
Förster, T. (1951). "Fluoreszenz organischer Verbindungen," p. 175. Vandenhoeck and Ruprecht, Göttingen.
Goedheer, J. C. (1957). "Optical Properties and In Vivo Orientation of Photosynthetic Pigments," pp. x and 90. Drukkerij Gebr. Janssen, Nijmegen, Netherlands.
Goedheer, J. C., and Smith, J. H. C. (1959). *Carnegie Inst. Wash. Year Book* **58**, 334-336.
Granick, S. (1950a). *Harvey Lectures Ser.* **44**, 220-245.
Granick, S. (1950b). *J. Biol. Chem.* **183**, 713-730.
Granick, S. (1959). *Plant Physiol.* **34**, xviii.
Koski, V. M. (1949). Ph.D. Thesis, University of Minnesota, Minneapolis, Minnesota.
Koski, V. M., and Smith, J. H. C. (1951). *Arch. Biochem. Biophys.* **34**, 189-195.
Koski, V. M., French, C. S., and Smith, J. H. C. (1951). *Arch. Biochem. Biophys.* **31**, 1-17.
Krasnovskii, A. A., and Kosobutskaya, L. M. (1952). *Doklady Akad. Nauk S. S. S. R.* **85**, 177-180.
Labbe, R. F. (1959). *Biochim. et Biophys. Acta* **31**, 589-590.
Latimer, P., and Smith, J. H. C. (1958). *Carnegie Inst. Wash. Year Book* **57**, 293-295.
Litvin, F. F., Krasnovskii, A. A., and Rikhireva, G. T. (1959). *Doklady Akad. Nauk. S. S. S. R.* **127**, 699-701.
Loeffler, J. E. (1955). *Carnegie Inst. Wash. Year Book* **54**, 159-160.
Shibata, K. (1957). *J. Biochem. (Tokyo)* **44**, 147-173.
Smith, J. H. C. (1949a). *In* "Photosynthesis in Plants" (J. Franck and W. E. Loomis, eds.), pp. 209-217. Iowa State College Press, Ames, Iowa.
Smith, J. H. C. (1949b). *J. Chem. Educ.* **26**, 631-638.
Smith, J. H. C. (1959a). *Proc. Intern. Photobiol. Congr. 2nd Congr., Torino, Italy, 1957* pp. 333-342.
Smith, J. H. C. (1959b). *Brookhaven Symposia in Biol.* **11**, 296-302.
Smith, J. H. C., and Benitez, A. (1953). *Carnegie Inst. Wash. Year Book* **52**, 149-153.
Smith, J. H. C., and Benitez, A. (1954). *Plant Physiol.* **29**, 135-143.
Smith, J. H. C., and Coomber, J. (1959). *Carnegie Inst. Wash. Year Book* **58**, 331-334.

Smith, J. H. C., and French, C. S. (1958). *Carnegie Inst. Wash. Year Book* **57**, 290-293.
Smith, J. H. C., and Koski, V. M. (1948). *Carnegie Inst. Wash. Year Book* **47**, 95.
Smith, J. H. C., and Kupke, D. W. (1956). *Nature* **178**, 751-752.
Smith, J. H. C., and Young, V. M. K. (1956). In "Radiation Biology" (A. Hollaender, ed.), Vol. III, pp. 393-442. McGraw-Hill, New York.
Smith, J. H. C., Kupke, D. W., and Giese, A. T. (1956). *Carnegie Inst. Wash. Year Book* **55**, 243-248.
Smith, J. H. C., Kupke, D. W., Loeffler, J. E., Benitez, A., Ahrne, I., and Giese, A. T. (1957a). In "Research in Photosynthesis" (H. Gaffron *et al.*, eds.), p. 471. Interscience, New York.
Smith, J. H. C., Shibata, K., and Hart, R. W. (1957b). *Arch. Biochem. Biophys.* **72**, 457-464.
Smith, J. H. C., Durham, L. J., and Wurster, C. F. (1959). *Plant Physiol.* **34**, 340-345.
Stanier, R. Y., and Smith, J. H. C. (1959). *Carnegie Inst. Wash. Year Book* **58**, 336-338.
Trurnit, H. J., and Colmano, G. (1959). *Biochim. et Biophys. Acta* **31**, 434-447.
Virgin, H. I. (1955). *Physiol. Plantarum* **8**, 630-643.
Warburg, O., and Christian, W. (1942). *Biochem. Z.* **310**, 384-421.
Wolff, J. B., and Price, L. (1957). *Arch. Biochem. Biophys.* **72**, 293-301.

Discussion

KAMEN: What is the spectrum of the protein with the chlorophyll removed?

SMITH: I do not know. We have not determined that yet. It would be a very interesting thing to do.

ARNON: I understand that you have looked for oxygen evolution in the photochemical transformation of protochlorophyll and did not find any. Could you tell us how you looked for it?

SMITH: We have not done it with this particular protochlorophyll holochrome. We did it in leaves, using the phosphorescence of trypaflavine absorbed on silica gel to detect the oxygen. We analyzed the transformation of protochlorophyll to chlorophyll, under hydrogen. We transformed about 80 to 90% of the protochlorophyll to chlorophyll in the hydrogen and obtained less than 2% of the amount of oxygen that should have come off.

WEBER: Did I understand correctly that you have a molecular weight of the order of a million and one solitary chlorophyll?

SMITH: This is what our analysis has shown, yes.

HENDRICKS: It means then that the chlorophyll gets turned over, that finally the amount of chlorophyll in the system is high compared with this million molecular weight.

SMITH: Oh, yes.

HENDRICKS: Is it possible that the association of chlorophyll to the protein is much like that of an enzyme intermediate? A hydrogen transfer process usually has to go on with things in close association, but I am surprised that the bimolecular kinetics would keep up with it at low temperatures.

SMITH: Well, they do not quite keep up, you see. What actually happens is that at $-70°$ you only get 60% transformation. At $22°$ you get 95% transformation. In other words, as you go lower and lower you get less and less of the protochlorophyll that can be transformed. I looked on this in this way. That you activate a pigment, or some molecule, by light, and it can maintain its activated state for only so long. But you have a very viscous medium which becomes more and more viscous at low temperatures so the molecule cannot get to the thing that it has to react with before it loses its activation at low temperature as well as it can with high temperature.

WEBER: Those figures there could give you the energy of activation that you must have in the excited state to get your reaction. You have rates at $-70°$, $-20°$, and so on. From those figures can you calculate an energy of activation for the excited state?

SMITH: I did this some time ago, and it did not come out to be a constant figure.

STANIER: There was a structural problem that I wonder if you could clarify for us a bit. Presumably, when you do the *in vitro* transformation of the holochrome, you end up with a chlorophyll *a* holochrome, which would have essentially similar physical properties to the protochlorophyll holochrome you began with. Is this correct?

SMITH: Frankly, we have not done this on the analytical centrifuge. This is something that we should do, but we have not done it.

STANIER: Because it seems to me that if this is the case, which one would really suspect, it would be extremely interesting to see how long that thing persists *in vivo* after the primary transformation has taken place. You must presumably get very rapid structural changes which then go on to make the mature chloroplast.

SMITH: Of course there are some very important changes that take place here. Going back to our work on the building up of photosynthesis, you can illuminate and get the chlorophyll and get no photosynthesis. Then you set the leaves aside in the dark and bring them back into the hydrogen apparatus, the phosphorescence indicator, and you get just a little bit of oxygen. But if you irradiate for 5 minutes after being in the dark for 2 hours, and then put it in the apparatus, the oxygen just streams out. In other words, something happens during that 2 hours and during that second irradiation, but we have no conception of what it is.

ARNON: Well, what happens during those 2 hours in the dark may be unrelated to your photochemical changes. This may be the rate, let us say, of the formation of enzymes needed for CO_2 assimilation. We know that some of them are light-induced —TPN-triosephosphate dehydrogenase, for example. You may see one thing which affects your pigment and then coupled with that you may need another period of activation for enzymes needed for CO_2 assimilation.

FOX: Are there not some fairly gentle methods of dissociating these chromogens from the proteins? I am thinking of some of the reversible changes with slight alteration of pH and temperature.

SMITH: I do not know, but about 20% alcohol, for 20 or 30 minutes, destroys the transformed holochrome. So do detergents, except digitonin.

KAMEN: You know if you take some bean leaves grown in the dark and extract with laundry soap, you know, the old-fashioned laundry soap, you get this very strange transformation and I just wonder if you are not emulsifying whatever the stuff is that it is going to transform. Maybe some of the difficulties lie in that you still have not found the right system to really do all the things you would like it to do.

SMITH: We can extract at pH 7; you do not have to do it at pH 9.5. It just seemed more stable at 9.5.

KAMEN: How stable is this preparation after you have made it? Leaving it in the dark.

SMITH: Well, it is very stable unless bugs get into it. That can cause trouble.

WEBER: Have you done flashing light kinetics on the transformation?

SMITH: We have done flashing light kinetics, or Virgin did, down to milliseconds, and could find no change. We had hoped we could find out whether there was a second reaction but of course the millisecond is a fairly long time, even so.

WEBER: You said you found no change, what did you mean?

SMITH: Well, I mean you get the same amount of transformation for a given amount of light as you do with continuous light.

BENDIX: I am interested in the carotenoid component. I was wondering if there was any possibility that the reason you cannot get the carotenoid component off is that it is an integral part of the thing you are dealing with and you take it apart in the process of taking it off.

SMITH: We keep thinking we might oxidize the carotenoid and still get a colorless protein without destroying the other color, but we have not been able to get it off. It is not necessary, you see, because the transformation goes in a carotenoidless mutant.

BENDIX: I was wondering if you have looked for oxygen evolution in the carotenoidless mutants.

SMITH: Yes, we did, but it was not satisfactory. We used corn that had gotten a little bit too old. I think we tried it twice and neither time did we have any indication of oxygen evolution. But this is not to be taken as a serious try, really.

WOLKEN: Dr. Smith, what is your best source of protochlorophyll? Bean seedlings?

SMITH: We had hoped sunflower cotyledons were going to be, this big green cotyledon, but it turned out to be another pigment. But the best source is beans, because it is the only one that we can use with aqueous solutions and get the holochrome. We can use anything with glycerin and get it, but beans are the only source that we can use for the holochrome to purify.

ARNON: Apropos beans, have you tried *Aspidistra*? If you could raid some bar and get some *Aspidistra* plants it might be worth trying because *Aspidistra* is one of the plants that Lubimenko reported he could get water-soluble chlorophyll out of. Beans was added to that list by Hansen, who had a special variety of beans that did it better. But Lubimenko maintained that *Aspidistra* was the best. So I suggest that you try that.

SMITH: We found one leaf that is a wonderful leaf if you need a lot of leaf surface, and that was taro. You can grow big etiolated leaves that have quite a bit of chlorophyll, whereas most etiolated leaves are about the size of your little finger. This is marvelous if you need a lot of surface for photochemical work. I have never tried extracting the holochrome. It is not a practical plant for preparation because it is a bulb and takes a long time to sprout and mechanisms are built in so that it will sprout only when it wants to.

17

The Photosensitivity of Sea Urchins

N. MILLOTT

Department of Zoology, Bedford College, University of London, London, England

Many sea urchins respond to light. Some are orientated by it, others are stimulated to perform complex reactions which result in taking up and holding opaque covering, while some show movements of their spines, podia, and pedicellariae. Many of these reactions follow changes in intensity and they are particularly striking in the tropical genus *Diadema*. These reactions were studied originally by von Uexküll (1900) and later by Millott and Yoshida, who studied the species *Diadema antillarum* Philippi. Most of the succeeding account is based on these studies.

The reactions are interesting for several reasons; in the first place, the urchins are eyeless and comparatively little is known of the intimate mechanisms involved in the photosensitivity of such animals. Again, many of the responses are shadow reactions, and despite the widespread occurrence of such reactions, especially among invertebrates, the mechanisms have received very little attention, though the work of von Buddenbrock (1930) on *Balanus* and of Föh (1932) on *Helix* should be mentioned.

The reactions of the spines of *Diadema* are particularly striking and we have concentrated attention on these. When the animal is suddenly illuminated or shaded, the spines respond by a sharp swing which may be repeated a greater or lesser number of times so that, in response to a strong stimulus, prolonged waving occurs. To be brief we call these the "on" and "off" reactions, noting that the "off" is not only much more vigorous and prolonged, but also much easier to elicit and requires for its inception only about one thousandth of the intensity change necessary to call forth the "on" reaction. It is thus not only more interesting, but also easier to study, so that we have devoted most of our attention to it.

To make what we know of the mechanism clear, it is necessary to emphasize certain features of echinoid anatomy. The pentamerous radiate symmetry characteristic of echinoderms is expressed on the surface as 5 meridians extending between oral and aboral poles. These are the radii or ambulacra and they bear the podia. The intervening areas, the interadii or interambulacra, are clothed by the large poisonous spines whose movement has been studied.

A section of a radius (Fig. 1) shows the salient features of the nervous system. It is very diffuse and superficial, extending over the entire surface in the epidermis, at the base of which it forms a felt of very fine fibers and scattered cells, often densely packed. There is some centralization, each ambulacrum carrying a radial nerve within the shell (or test), giving off numerous small branches at regular intervals on either side, which traverse the test to emerge and fan out in the thin, delicate skin. The radial nerves are linked by a nerve ring in the oral region. The spines are moved by muscles confined to their bases which are innervated from the superficial nerve layer.

From such an organization "operational units" can be isolated in the form of sectors and, provided that they include the radial nerve, their spines will show active movements in response to photic stimuli for several hours.

Preparations of the kind described above can be mounted in a dark tank (Fig. 2) and the spine movements can be recorded photographically. Stimulation is accomplished by admitting or interrupting light beams of varying width. Where small spots of light are required, the beam can be projected through the objective lens of a compound microscope (Fig. 3). The internal and external surfaces can thus be "scanned" by spots.

The need for the radial nerve in preparations shows that the responses are reflexes. That it is involved in the "off" response has been known since the early work of von Uexküll, who maintained a sharp distinction between this and the "on" reaction, in that the latter involves conduction only through superficial skin pathways. We have long suspected this distinction, and experiments conducted last year (Millott and Yoshida, 1959) invalidated it (Plate I, 1–3), for removal of the radial nerve abolishes the photic response, just as in the case of the "off" reaction,

FIG. 1. Transverse section through a radius of a young individual of *Diadema antillarum*, showing the branches of the radial nerve which supply the tube feet, integument, and spines. The section is composite and somewhat diagrammatic. Since the section is thin, only the lateral edges of the integumentary nerve appear; the intervening portion, which is outside the plane of section, is indicated by mechanical stippling. KEY: *ep.s.*, epineural sinus; *h.c.ep.*, hillock in covering epithelium; *i.m.n.*, inner (radial) margin of the integumentary nerve; *m.*, melanophore; *m.p.*, heavy deposit of melanin in tube foot; *n.r.*, portion of nerve ring around spine base; *o.m.n.*, outer (interradial) margin of the integumentary nerve; *p.n.*, podial nerve; *p.s.*, prolongation of epineural sinus; *r.n.c.*, radial nerve; *r.p.c.*, radial perihaemal canal (subneural sinus); *r.w.v.c.*, radial water vascular canal; *s.n.l.*, superficial nerve layer; *t.*, decalcified test. (From N. Millott, *Phil. Trans. Roy. Soc. London* **B238** (1954). Reproduced by permission of the Royal Society of London.)

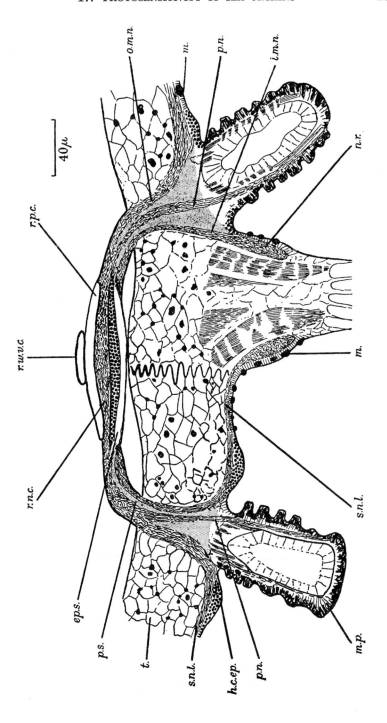

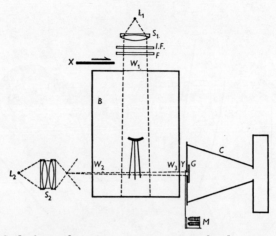

Fig. 2. Method of recording spine movements in isolated pieces of test. B, tank; C, camera; F, neutral density filter; I.F., interference filter (used in determining spectral sensitivity); G, ground glass screen; L_1 and L_2, light sources; M, time and stimulus signalling apparatus; S_1 and S_2, lens systems; W_1, W_2, W_3, windows in opaque sides of tank; X, shutter; Y, slit. [From N. Millott and M. Yoshida, J. Exptl. Biol. **34**, 394 (1957).]

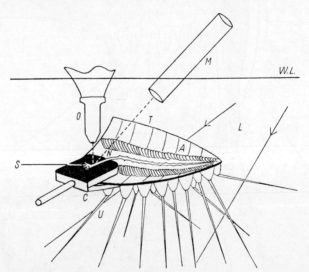

Fig. 3. The method of producing small spots of light. A, ampullae of podia; C, clamps; L, light beam passing beneath preparation so as to cast a shadow on the spines, movements of which are photographed; M, microscope for viewing light spot; N, radial nerve; O, objective lens of compound microscope; S, position of light spot; T, piece of test; W.L., water level. [From M. Yoshida and N. Millott, Experentia **15**, 13 (1959).]

but it does not prevent spines reacting to electrical stimulation. Von Uexküll's distinction was unfortunate in that it served to discourage the inevitable comparison between "on" and "off" reactions and the similarly named effects in retinae.

When preparations are scanned internally and externally, the whole skin is shown to be sensitive but, more than this, the radial nerve itself can be stimulated by shading, even though the obliterated light spot is only about 10 μ in diameter (Plate I, 4A). There is no doubt that internal sensitivity is localized in the nerve, for shading a spot just outside the margin of the nerve (Plate I, 4B) elicits no response (Yoshida and Millott, 1959).

There is thus an exceedingly diffuse photosensitivity, which corresponds in distribution with that of the nervous system, and at first sight the animal appears to afford just another instance of the widespread and somewhat mysterious dermal light sense, sometimes described as "dermatoptic." But it is more than this, for the sensitive outside surface of an echinoid is more than just "skin," it embodies a prominent felt of nerve and in some respects resembles a vast spreading retina. This exaggerated, loose, and somewhat brash comparison is of course limited, but as will be seen later there are some valid similarities.

With this in mind, the question of the character of the receptive elements arises. Apart from the fact that they must be diffuse, and moreover present in the internally placed radial nerve, as well as in the superficial body wall, we know little. We have yet failed to find morphologically defined receptors at the organ level. They have been described in an allied species, but the organs that appear to correspond with them in *Diadema antillarum* are iridophores, and play no direct part in the spine response (Millott, 1953). Being scattered over the outside receptive surface, they may perhaps earn their keep by diffusing light like a tapetum (Plate I, 5). It is all the more tempting to suggest this, because studies with Dr. Manly (unpublished), have shown that by virtue of their colloidal contents, they scatter light that looks predominantly blue (though its spectral quality has not yet been determined) and there are indications that the animal is most sensitive to blue light.

The quest for areas that are specially photosensitive, shows that the most sensitive areas are associated with the ambulacra. Because the animal is relatively insensitive to increases in illumination, it was not possible to localize specially receptive areas for the "on" reaction more precisely, but for the "off" reaction the most sensitive areas proved to be the ambulacral margins at approximately the point where the branches of the radial nerves emerge on the surface. It may be noted in passing that this sensitivity distribution was revealed by two approaches. In the earlier one, the relative threshold for the shadow response was determined; in the one more recently used, the places where inhibition of the shadow response was most effectively produced by illumination were determined. The distribution coincided, showing that it is easiest to elicit the shading response at the same place as it is easiest to inhibit it. This hints at the importance of inhibition in the response (see below).

One is therefore driven to the conclusion that nerve elements themselves are excited by light. This is not an isolated observation, other instances having been described by Prosser (1934), Welsh (1934), Arvanitaki and Chalazontis (1949), and most recently by Kennedy (1958). Though this constitutes an apparently unorthodox variant of the classical concept of a reflex, where the receptive and conducting elements are differentiated morphologically as well as physiologically, there is no fundamental divergence, for certain of the nerve elements could be differentiated physiologically for photoreception at the cellular level, and perhaps also morphologically. Only further study of the nervous system can decide this.

Since structures absorbing visible light are involved, it is pertinent to consider pigmentation. Here we have nothing to add to what we have already published, except an important reservation. The pigmentary system of *Diadema* is very complex, the most prominent colors are black,

---

PLATE I

(*1, 2, and 3*). Photographic records of spine movements of one preparation. The time scale for each record appears below fig. 3, one white and one adjacent black area marking a one second interval. Each record should be read from left to right.
(*1*) Typical response: A, light admitted at arrow; B, light extinguished at arrow.
(*2*) Effect of removing the radial nerve: A, light admitted at arrow; B, light extinguished at arrow. (*3*) Effect of stimulating electrically the outer surface after removal of the radial nerve. The stimulus was continued between the two arrows. [From N. Millott and M. Yoshida. *Proc. Zool. Soc. London* **133**, 67 (1959).]

(*4*) The effect of extinguishing a small spot of light focused onto the radial nerve. The position and size of the spot is shown by the black circle superimposed on the photograph of a portion of the radial nerve (white area) in the left hand section of the figure. The scale represents 0.5 mm. The effect of obliterating the spot is shown in tracing A. Tracing B is a control experiment in which the light spot was focused onto a position just outside the margin of the radial nerve, as shown by the white circle. The time scale (in seconds) appears above each tracing. The interruption of the light beam is shown by the disappearance of the black band below each tracing. [From M. Yoshida and N. Millott. *Experientia* **15**, 13 (1959).]

(*5*) Portion of skin from an inter-ambulacrum of *Diadema antillarum* showing the melanophores (center) and iridophores (sides). Approximately 50 ×. (Photograph by Dr. Brenda Manly and Mr. M. Gross.)

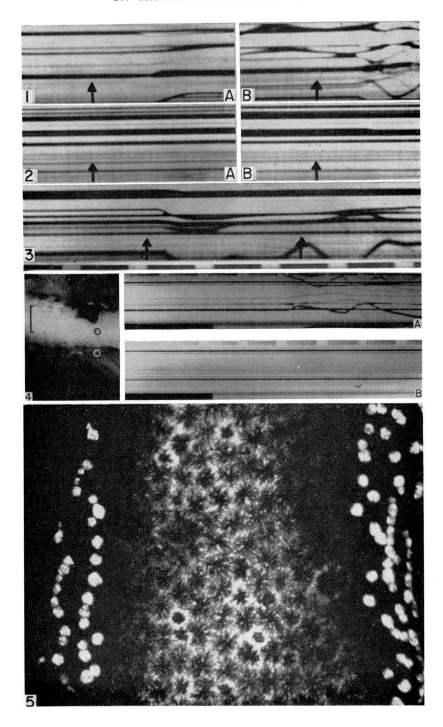

purple, and red. All are widely distributed, the red particularly so. Some of the pigment occurs in large chromatophores, which recent studies with Dr. Manly (unpublished), show to form a vast mobile network in the skin immediately overlying the felt of nerves. The chromatophores show active photomechanical movement, so that here again there is an analogy with certain retinas.

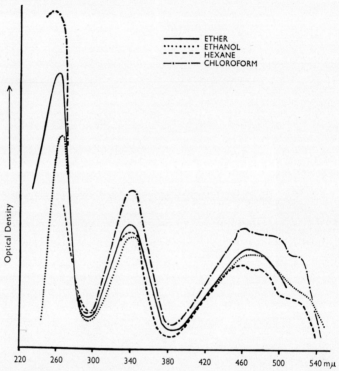

Fig. 4. Absorption spectrum in various solvents, of the purified hydroxynaphthoquinone pigment from *Diadema antillarum*. [From N. Millott, *Proc. Zool. Soc. London* **129**, 263 (1957).]

Some of the black pigment is an ill-defined "melanin," the red and purple is a hydroxynaphthoquinone resembling echinochrome A; the spectral absorption of which is shown in Fig. 4. In the visible range, it absorbs maximally in ethanol, at 467 mμ and in acid ether at 462/3 mμ (Millott, 1957). Such pigments are common in echinoids. This pigment is particularly interesting because it occurs in the radial nerves, and there is sometimes enough of it to make them look pink. However, much of it is not in nerve elements but inside the ubiquitous amebocytes. Not all of it is found there, but whether it is in the nerve elements or merely the interstices between them has not yet been determined with certainty.

The spectral sensitivity, originally determined on a basis of the light energy necessary to elicit the responses to a standard shadow (Millott and Yoshida, 1957), showed a maximum at 465 mμ (Fig. 5). This has been recently redetermined by an improved method (Fig. 6), in which

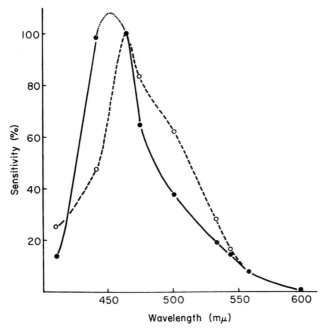

Fig. 5. The action spectrum of the shadow reaction of *Diadema antillarum* as recently redetermined (solid line), compared with that previously obtained (broken line). [From M. Yoshida and N. Millott, *J. Exptl. Biol.* **37**, 390 (1960).]

isolated pieces of test bearing single spines were subjected to shading by changing instantaneously from a constant intensity of white light, to that of various colors, the relative effectiveness of which is measured by the degree of adjustment in their intensity necessary to produce a threshold shadow response (Yoshida and Millott, 1960). Sensitivity was maximal between 455 and 460 mμ (Fig. 5).

These values approach the absorption maximum for the ubiquitous red pigment when extracted and the point of maximum sensitivity of the chromatophores (Yoshida, 1957), in the skin of *Diadema setosum* (Leske), which contain the red pigment. The significance of this is not yet clear because the color of echinochrome varies with several factors including pH and the state of ionization (Ball, 1936; Kuhn and Wallenfels, 1941; Millott, 1957), and the spectral absorption of the pigment *in situ* has not yet been determined.

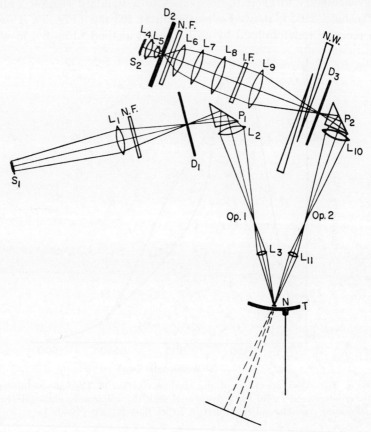

Fig. 6. Method used to redetermine the spectral sensitivity. D_1, D_2, and D_3, diaphragms; I.F., interference filter; L_1–L_{11}, lenses; N, radial nerve; N.F., neutral filter; N.W., neutral wedge; Op. 1, beam of white light; Op. 2, beam of colored light; P_1 and P_2 prisms; S_1 and S_2, tungsten filament lamps; T, test with spine. [From M. Yoshida and N. Millott, *J. Exptl. Biol.* **37**, 390 (1960).]

The recent study of the relationship between stimulus and response in preparations with a single spine has yielded interesting results concerning the events in the shadow reflex which follow photoreception.

In the "off" reaction there are two environmental agents, light and a decrease in intensity. Are both stimuli? Here we face the intriguing problem of the action of the shadow, and whether it acts as a stimulus by influencing some balanced reversible photosensitive system in a way opposite to the action of light, or whether it merely marks an interruption, permitting a rebound from a suppressive or inhibitory action due to the light. The shadow would appear to be the stimulus because the response follows so closely after it. Föh regarded it as such and claimed that both its duration and intensity exerted significant effects, comparable in their relationships with the duration and intensity of lighting. Further he fitted them in to the theoretical scheme proposed by Hecht to explain photoreception based on photochemical considerations.

Our recent results have led us to reject Föh's approach, because if the shadow is a stimulus in the strict sense, the reaction which follows it should be affected significantly and in the same way by its duration and intensity. This is not so.

The reaction of the single spine preparation to a standard stimulus is sufficiently constant (Fig. 7), to enable us to see what effects occur on

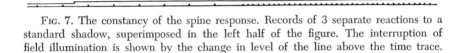

FIG. 7. The constancy of the spine response. Records of 3 separate reactions to a standard shadow, superimposed in the left half of the figure. The interruption of field illumination is shown by the change in level of the line above the time trace. Time in seconds. [From N. Millott and M. Yoshida, *J. Exptl. Biol.* **37**, 363 (1960).]

reaction time, size, frequency and duration of the contractions when the light and shade are varied.

Changing the intensity of the shadow (measured as per cent decrease in field illumination), affects the *whole* reaction; the reaction time decreases steadily, whereas the frequency of the contractions and the duration of the reaction increase as the shading is increased. The effect of duration is very different; it does not affect the whole reaction but mostly the later part, so that the amplitude of the first contraction is unaffected, and the reaction time is affected only near the threshold, and then erratically. The amplitude and frequency in the later reaction and its duration, all increase with the shadow, though the increase falls off. In all its features, with the notable exception of the reaction time and initial amplitude, the reaction is reduced when light is readmitted, as can be seen by comparing the effects of limited shadows and those that are prolonged.

The differing effects of duration and intensity of shading can be contrasted with the similar effects of duration and intensity of lighting. In the case of the latter, as they are increased, the reaction time decreases while the amplitude (initial as well as later), frequency of beat, and duration of the reaction increase. The different effect on the reaction time is shown graphically in Fig. 8.

The fact that only the later part of the reaction is susceptible to the effect of varying the duration of shading hints at an additional mechanism coming into play at this time. Again, the lack of an obvious effect of light (unless very bright) until its intensity is decreased, suggests that it may be exerting an inhibitory effect, which persists until shading releases a

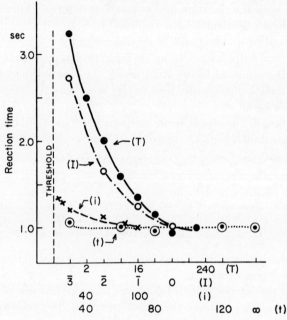

Fig. 8. The different effects of lighting and shading on reaction time. Ordinates, reaction time in seconds.
    Curve T (●—●), duration of lighting;
    Curve I (O–·–O), intensity of lighting;
    Curve i (×---×), intensity of shading;
    Curve t (⊙····⊙), duration of shading.
Abscissae: T, in seconds; I, in arbitrary logarithmic units; i, percentage decrease in field intensity; t, in milliseconds. [From N. Millott and M. Yoshida, *J. Exptl. Biol.* **37**, 363 (1960).]

reaction whose duration and vigor are in proportion to the preceding illumination. If shading is not total, the reaction will be inhibited to a degree depending on the light remaining. When more light is readmitted, the additional mechanism hinted at above comes into play to suppress whatever remains of the reaction. The failure to affect the first part of the reaction could be due to the latency of inhibition.

To substantiate this idea it is necessary to show that light inhibits the spine response. The simplest way of doing this is to vary the intensity of the light readmitted after shading. Fig. 9 shows the increased suppressing effect on the duration of the shadow reaction, which results from increasing the intensity of readmitted light.

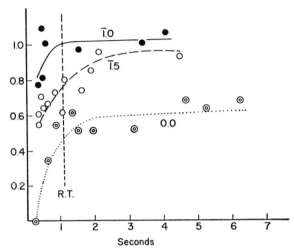

FIG. 9. The effect of the intensity of light readmitted after shading on the duration of the shadow reaction. Abscissae: duration of shading in seconds. Ordinates: ratio of the duration of the response to that of a control response during which no light was readmitted. The vertical dotted line shows the reaction time. The figures alongside each curve show the relative intensity of readmitted light, expressed in arbitrary logarithmic units. [From N. Millott and M. Yoshida, *J. Exptl. Biol.* **37**, 376 (1960).]

The inhibition due to light and the interaction between the effects produced by light at different places are shown by shading one point and illuminating another. This was done on the radial nerve and on the outside surface. When the two light spots are near together or coincide, there is marked inhibition, and as they are moved apart, the inhibition lessens, (Fig. 10), so that there is an inhibitory gradient. In some cases when the illuminated spot had been moved to a point where it failed to inhibit, moving it still further away reversed the effect, so that the response increased in size, but only the later part of it.

The interaction appears to occur both centrally and peripherally. That interaction is peripheral is shown by the fact that cutting the radial nerve between the two light spots projected on to it does not stop interaction between them. On the other hand, cutting the branches of the radial nerve on either side does not impair the inhibiting gradient existing between light spots projected within the denervated area, whatever the position of the spine in respect to it. This suggests interaction in the radial nerve itself.

The effect of making various cuts reveals a bewildering variety of effects, indicating complex patterns of interaction and there are not only

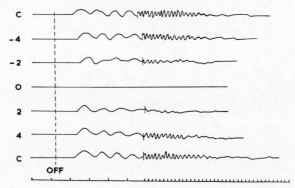

Fig. 10. The effect on inhibition of the position at which light is readmitted. Typical records taken from one preparation. The numbers alongside each record show the position of the inhibitory light spot on the radial nerve; position 0 is approximately midway between the ambitus and the periproct. The degree of linear separation from this in millimeters is indicated by the numbers: positive values being toward the oral pole; negative toward the aboral. *C* marks control reactions during which no light was readmitted after shading, *before* (upper tracing) and *after* (lower tracing) the intervening records were taken. Time in seconds. The vertical dotted line shows the time at which shading occurred. In each case (other than the controls) light was readmitted after a standard interval. [From N. Millott and M. Yoshida, *J. Exptl. Biol.* **37**, 376 (1960).]

indications of tonic inhibition, but also of interaction between "on" and "off" responses. This might explain the high threshold of the "on" response, for if inhibition preceded it by virtue of a shorter latency, it is conceivable that only relatively high intensities would overcome it.

It is therefore clear that the events behind photoreception in *Diadema* are complicated, involving interplay between excitation and inhibition that is reminiscent, in a general way, of that associated with complex photoreceptors. The inhibiting gradients in the skin may perhaps serve to emphasize contrast (resulting from changing patterns of shade cast on it) in a way that could prove advantageous in signalling the presence of moving objects such as predators.

If this were so, there is still less reason to regard the dermal light sense of echinoids as being primitive.

References

Arvanitaki, A., and Chalazontis, N. (1949). *Arch. sci. physiol.* **3**, 27-44.
Ball, E. G. (1936). *J. Biol. Chem.* **114**, vi.
Föh, H. (1932). *Zool. Jahrb.* **52**, 1-78.
Kennedy, D. (1958). *Ann. N. Y. Acad. Sci.* **74**, 329-336.
Kuhn, R., and Wallenfels, K. (1941). *Ber. deut. chem. Ges.* **74B**, 1594-1598; cited from Fox, D. L. (1953). "Animal Biochromes," p. 201. Cambridge Univ. Press, London and New York.
Millott, N. (1953). *Nature* **171**, 973.
Millott, N. (1957). *Proc. Zool. Soc. London* **129**, 263-272.
Millott, N., and Yoshida, M. (1957). *J. Exptl. Biol.* **34**, 394-401.
Millott, N., and Yoshida, M. (1959). *Proc. Zool. Soc. London* **133**, 67-71.
Prosser, C. L. (1934). *J. Cellular Comp. Physiol.* **4**, 363.
von Buddenbrock, W. (1930). *Z. vergleich. Physiol.* **13**, 164-213.
von Uexküll, J. (1900). *Z. Biol.* **40**, 447-476.
Welsh, J. H. (1934). *J. Cellular Comp. Physiol.* **4**, 379-388.
Yoshida, M. (1957). *J. Exptl. Biol.* **34**, 222-225.
Yoshida, M., and Millott, N. (1959). *Experientia* **15**, 13.
Yoshida, M., and Millott, N. (1960). *J. Exptl. Biol.* **37**, 363.

Discussion

Fox: Have any tests been made on the iridocytes to show whether those may be guanidine or something of that sort?

Millott: No, it is not that. The iridophore is a bump on the surface with beautiful laminae, reminiscent, on a very much bigger scale, of Dr. Wolken's beautiful preparation that he described yesterday. Now, these laminae have between them a gel; you can cut them with a microdissection apparatus, and colloidal matter flows out, leaving a beautiful blue trail. The blue color is due to a Tyndall effect of the colloidal material.

18

Sensitivity to Light in the Sea Anemone *Metridium senile* (L.): Duration of the Sensitized State

WHEELER J. NORTH

Division of Marine Biochemistry and Institute of Marine Resources, Scripps Institution of Oceanography, University of California, La Jolla, California

When light is allowed to fall upon one side only of the sea anemone *Metridium senile*, a local contraction of the parietal musculature usually occurs beneath the illuminated area and the anemone bends toward the source of the light (Bohn, 1906). The entire body surface of *Metridium* is apparently photosensitive but certain wavelengths are more effective in eliciting the bending response. Maximum sensitivity generally occurs in the blue-green at about 500 mµ; often more than a thousand times as much energy is required to elicit a response at 650 mµ (North and Pantin, 1958). We were unable to find in the animals either macroscopic or microscopic units suggestive of photoreceptors. North (1957) found that at low levels of light intensity a bending traverse of a millimeter or less occurred, and that the magnitude of the movement was independent of the intensity. He found no significant effect of temperature upon the reaction time (the period elapsing from the instant of illumination until a response was observed) over a 17° C range, but the reaction time varied inversely with the intensity of illumination, and it was concluded that, within the ranges of intensity studied, the photosensitive response obeyed the Bunsen-Roscoe law. Stimulation of *Metridium* required roughly 5×10^9 incident quanta/cm^2 of blue-green light; it apparently made no difference whether the radiant energy was all delivered in 1 second at a relatively high intensity or in 10 seconds at 1/10 the high intensity. It was not known how many of the 5 billion quanta necessary for stimulation were absorbed by the photosensitive system but it was considered entirely possible that less than 10 quanta were responsible. Neither the identity nor the concentration of the photosensitive pigment or pigments are known, and until such information becomes available conclusions about the number of quanta required by the pigment molecules must rest upon indirect evidence.

Methods and Material

The apparatus described by North (1957) was utilized throughout the experiments herein reported. The power supply of the calibrated light source was carefully regulated to avoid errors arising from fluctuations in line voltage. The light was focused in a rectangular strip, 1.5×8 cm., up the side of an animal. An Ilford 603 gelatin filter was used to supply blue-green light, which was the only fraction employed in the present study. This filter had a maximum transmission of 12% at 493 mμ and the values dropped to 1% at 473 and 520 mμ. Intensities were regulated with Ilford neutral density gelatin filters.

Moderate sized specimens of *Metridium* (about 3 cm. diameter and 10 cm. high when expanded) were utilized. During an experiment the animal was maintained in a system of recirculating seawater with a temperature regulation of $\pm \frac{1}{4}°$ C. At other times the anemones were kept in fresh running seawater refrigerated to 13° C. Only the white color phase of *Metridium* was utilized in the experiments although most of the other color phases described by Fox and Pantin (1941) were available. North and Pantin (1958) showed that the pigments causing coloration probably do not act as specific photosensitive substances but rather as inert filters. By using white anemones any variations in sensitivity arising from different concentrations of insensitive body pigments were eliminated.

During an experiment a close check was maintained on the dimensions of the anemone to ensure that the animal maintained approximately the same shape and volume, since any change might alter the number of photosensitive units lying within the illuminated strip.

Results

The time elapsing from the initiation of illumination until the onset of bending is of considerable importance, since at any given intensity this reflects the total number of photoeffective quanta delivered to the animal. In several photosensitive invertebrates this reaction time has been shown to consist of 2 parts: (i) a sensitization period during which time the light must be left on to elicit response, and (ii) a following latent period during which no illumination is necessary. Both these periods can occupy several seconds. North (1957) was unable to find a latent period in *Metridium* and concluded that it was either absent or too short to be detected by the methods used. The reaction time in *Metridium*, then, is of about the same duration as the sensitization period, and it has been found that if the light is extinguished before the onset of bending, an anemone will remain in a sensitized state which persists for about 10 seconds. In order to study this phenomenon it is necessary to take the average of many observations because the length of the reaction time under normal conditions exhibits as much as a sevenfold fluctuation.

To illustrate, Fig. 1 shows a typical record obtained from an animal maintained in total darkness except for short light stimuli of less than 25 seconds duration at intervals of 5 minutes or more. The length of the reaction time is seen to be quite variable, although it oscillates randomly around an average value of 5.2 quanta/cm²/second. Observations in-

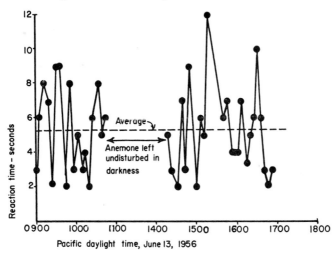

Fig. 1. Variation of reaction time for a *Metridium* observed over an 8 hour period at an average temperature of 11.4° C. The points were arbitrarily connected by straight lines. Energy of the stimulus was 8.6×10^8 quanta/cm²/second of blue-green light.

dicated that this average value remained constant for at least a week. North (1957) studied large numbers of reaction times obtained from 8 anemones and found that they formed a consistent pattern which in every case could be closely approximated by a Poisson distribution. This very usefully enabled the application of ordinary statistical procedures to the data.

It is possible, then, to stimulate an anemone a number of times at a given intensity and, from the results, to predict successfully an illumination period which is insufficient to cause a response but which would nonetheless comprise a substantial portion of the sensitization period. For the sake of convenience we shall designate this as the presensitization period. One can then stimulate the anemone for a given presensitization period, turn off the light for a few seconds, then restore illumination for a postsensitization period until a bending response is obtained. By varying the intermediate dark period one can measure the length of time the photosensitive system can retain the effects of the presensitization light.

A number of experiments of this kind revealed that an intermediate dark period of up to 10 seconds caused very little change in the total time of illumination required to elicit response. If the dark period is

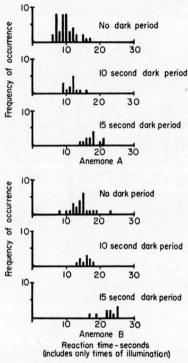

Fig. 2. Results of 2 experiments to determine the effect on the sensitization state of an anemone by introducing an intermediate dark period during photic stimulation of the animal.

increased to 15 seconds, however, the length of the postsensitization period increases to values corresponding to those obtained when the anemone is stimulated with continuous illumination. Figure 2 shows the results of 2 experiments on different animals tested at the critical periods of 10 and 15 seconds. Table I shows the average values of total illumination time required by the animals and it can be seen that there was a slight increase for an intermediate dark period of 10 seconds in both instances, but that the much greater increase after 15 seconds of intermediate darkness amounted to the length of the presensitization period.

TABLE I
AVERAGE VALUES OF TOTAL ILLUMINATION TIME, IN SECONDS, REQUIRED FOR
RESPONSE BY 2 ANEMONES UNDER DESIGNATED REGIMES OF STIMULATION[a]

	Total illumination time (seconds)	
	Anemone A	Anemone B
No intermediate dark period	9.9 (40)	14.4 (23)
10 seconds intermediate dark period	11.6 (14)	15.6 (10)
15 seconds intermediate dark period	17.6 (13)	22.5 (10)

[a] Figures in parentheses indicate the number of observations. The presensitization period for anemone A was 6 seconds and that for anemone B was 8 seconds of illumination.

This suggests that the effects of the presensitization stimulus were absent after 15 seconds of darkness.

The Mann-Whitney U test (Siegel, 1956) was used to examine the probability that the results of Fig. 2 could have occurred by chance. For each anemone, values were sought for the probability that the 10-second intermediate dark period data were not different from the no dark period data and for the probability that the 10-second data were the same as the 15-second data. The results are presented in Table II. If we set the

TABLE II
PROBABILITIES DETERMINED BY THE MANN-WHITNEY U TEST (ONE-TAILED) THAT
TWO GROUPS OF DATA COULD BE DERIVED FROM A SINGLE DISTRIBUTION BY CHANCE

Data groups compared	Anemone A	Anemone B
No dark period vs. 10 seconds	0.009	0.042
10 seconds vs. 15 seconds	$\ll 0.001$	$\ll 0.001$

level of significance at $p = 0.05$, it is readily seen that the test rejects the hypothesis that any of the groups could be derived from the same distribution by chance. A total of 5 anemones was used in studying the effects of an intermediate dark period and no evidence appeared which was contradictory to the above results. It is concluded that sensitization induced by a subthreshold light stimulus begins to disappear after about 10 seconds of darkness and has vanished by 15 seconds.

Discussion

If total decay of sensitization occurs within 15 seconds, what happens when the intensity of the stimulus is made so dim that it is necessary to illuminate the animal for longer than 15 seconds? It might be expected that considerable difficulty in eliciting a response would occur because the effects produced in the initial phases of the sensitizing period might disappear before the necessary quanta for response were absorbed by the photosensitive system and thus the threshold level might never be reached. Such, however, is not the case. It is little trouble, for example, to obtain verification of the Bunsen-Roscoe law at very dim intensities where the average reaction time is 60 seconds.

It is possible that facilitation phenomena could play a part in the photosensitive response and herein may lie the explanation for the paradox outlined above. Further work, however, needs to be done at extremely dim intensities before conclusions can be drawn. Studies of this kind are difficult because the chances of a spontaneous movement unrelated to the light stimulus increase as the length of the reaction time increases. This factor could be circumvented by lowering the temperature, which decreases the rate of spontaneous movements, but the intertidal anemones we have thus far used do not reliably maintain an expanded state below 11° C. Deep water animals accustomed to 6° C or lower are needed but would, of course, need to be tested first to ensure that a different environment had not caused fundamental changes in the photosensitive machinery. Studies of this kind might contribute valuable information on the hypothesis of quantal uncertainty (North, 1957), which ascribes the variation in reaction time to the infrequency of quantum capture by the photosensitive machinery of *Metridium*. One of the conclusions resulting from the hypothesis is that these anemones are able to respond to less than 10 absorbed quanta. If this is true the photosensitive response should cease to follow the Bunsen-Roscoe law at longer average reaction times (say 2 to 4 minutes) because of the decay of the sensitization period demonstrated above. If an absolute threshold were found it would be strong evidence in favor of the hypothesis of quantal uncertainty.

Acknowledgments

It is a pleasure to acknowledge the helpful advice of Professors Carl Eckart, Denis L. Fox, and C. F. A. Pantin, and the aid and advice provided by Raymond J. Ghelardi and Conrad Limbaugh. Support from the George Allen Bisbee Donation and the Rockefeller Foundation Donation for research in Marine Biology is gratefully acknowledged.

References

Bohn, G. M. (1906). *Compt. rend. soc. biol.* **58**, 420-422.
Fox, D. L., and Pantin, C. F. A. (1941). *Phil. Trans. Roy. Soc. London* **B230**, 415-450.
North, W. J. (1957). *J. Gen. Physiol.* **40**, 715-733.
North, W. J., and Pantin, C. F. A. (1958). *Proc. Roy. Soc.* **B148**, 385-396.
Siegel, S. (1956). "Nonparametric Statistics," pp. 116-127. McGraw-Hill, New York.

Discussion

Fox: I am taking advantage of Dr. North because I know something about his work, but I should like to ask him to say a word or two about the grounds on which he can suspect the nature of the pigments involved. He has some data which give at least a few clues from action spectra as to what the pigments could be.

North: Well, you have, as I pointed out, a rather complex action spectrum with peaks in the yellow and the blue-green. I did not go into adaptation phenomena at all, but when you illuminate the anemone constantly with dim light, his threshold becomes higher; that is, it takes a stronger stimulus to cause him to bend. If you illuminate him with dim blue-green light, you will find that you have to give him a much stronger blue-green stimulus to cause the bending. However, you can still give him about the same stimulus of yellow light and get a bending movement, which suggests that there is a different pigment absorbing in this region than there is in the blue-green. Since you are adapting him with blue-green light, the yellow pigment is not being affected. If you adapt him to yellow light, you find that the blue-green region is not affected, but the region down around 420–440 mμ is affected.

French: That effect sounds a little like the enhancement effect in photosynthesis, where one wavelength has an effect on the sensitivity to another wavelength.

North: What I want to do, now that I have got this experiment that I described of turning on a light, giving him a dark period, and then giving him another light, is to turn on a light of one color, then switch over to another color, to see how many seconds it would take to cause the bending response.

19

The Photoreactions Controlling Photoperiodism and Related Responses

STERLING B. HENDRICKS

Mineral Nutrition Laboratory, Soil and Water Conservation Research Division, Agricultural Research Service, U. S. Department of Agriculture, Beltsville, Maryland

Many evident changes in plants are induced by a single light-responsive system with maxima of effectiveness in the red and blue parts of the spectrum and at the limit of visibility in the near infrared. The changes include the photoperiodic control of flowering, germination of light-responsive seed, control of etiolation, induction of anthocyanin formation, and many other growth responses. The proofs that only one pigment system is involved, that it is photoreversible, and that it can act in two very different manners, allowing responses from the threshold of visibility to full sunlight, are of interest. A single control point is indicated for many features of plant development and the nature of the control is becoming apparent.

A small group has had the good fortune to have developed the entire scope of the subject, from its foundation in response to the threshold of its biochemical mechanism.[*] This has differed from the usual experience in these times in allowing full play of experimental design and logical deduction leading to an ever widening contact with varied aspects of plant development. Significant developments have been described as they were encountered and general treatments have been published at intervals. The most recent developments, however, have come so rapidly and have opened such broad possibilities as to be challenging to describe, both as to how they were made and to where they led and might lead.

The first step is a philosophical one best stated in translation from the Latin of Francis Bacon's *Novum Organum*: ". . . many new experiments may be discovered tending to benefit society and mankind, by what we

---

[*] This group has variously been W. W. Garner and H. A. Allard, the discoverers of photoperiodism, followed by H. A. Borthwick, R. J. Downs, M. W. Parker, A. A. Piringer, N. J. Scully, and H. W. Siegelman, with W. L. Butler, P. H. Heinze, S. B. Hendricks, K. H. Norris, E. H. Toole, and V. K. Toole associated at times in their efforts. The group encompasses basic and applied aspects of plant biology, chemistry, and physics.

call literate experiences; yet comparatively insignificant results are to be expected thence, whilst the more important are to be derived from the new light of axioms deduced by certain methods and rules from the above particulars and pointing out and defining new particulars in turn. Our road is not a long plain, but rises and falls, ascending to axioms, and descending to effects.

"Now it is the greatest proof of want of skill, to investigate the nature of any object in itself alone; for that same nature which seems concealed and hidden in some instances, is manifest and almost palpable in others, and excites wonder in the former, while it hardly attracts attention in the latter."

In studies of flowering, seed germination, or a lengthening stem, the particular phenomenon often is described and its "mechanism" examined, starting not uncommonly from the final display. But the pathway from causation to display is often so devious and involved as to defeat hope of following it backwards to initial causation. A greater hope when a stimulus is involved is to start from causation. While it is equally difficult to reach an understanding of final display, the region of great bifurcation of possible pathways can more readily be approached. Because the phenomena under consideration are induced by light, it is possible to examine the nature of the initial stimulus which, in essence, was the approach followed with striking success in the study of vision both from physical and biochemical points of view.

Action Spectra: General Considerations

Progress in understanding the phenomena rests almost entirely upon discovery by direct observation, as used so effectively by Garner and Allard in their discoveries of photoperiodism, and on measurement of action spectra. Action spectra express the incident energy per unit area required to produce a given level of response as a function of the spectral region. The spectral regions can be obtained with spectrographs employing prisms or gratings or by use of interference filters. Attention is given to dispersion, resolving power, spectral band widths, spectral purity, radiation intensity, and source constancy. To avoid confounding the action of the stimulus with steps leading to its display, the time of irradiation should be as short as possible, which implies high radiation intensities. This requirement necessitates compromises on dispersion and resolving power which, in turn, fix the band widths occupied by the experimental object. The band widths should be less than the spectral separation of differences in responses. After these several conditions are met, great care must be exercised over spectral purity because the object might be a thousandfold more sensitive in one region than in another, as can be readily illustrated by etiolation response. Finally, attention should be paid to absorption and scattering of radiation by the plant.

The energy versus time of irradiation curve for a given response, such as 50% seed germination, is first established to allow, if possible, use of varying times of irradiation for constant response. This fixes the limits for reciprocity which generally exceeds 10 minutes for the particular phenomenon and can be several hours, particularly for germination.

The most significant analysis would require a knowledge of response as a function of absorbed energy rather than of incident energy. It is of value and not trivial that the absorbed energy in a given spectral region increases in direct proportion with the incident energy, but one region is not directly comparable with another because of varying absorption with spectral position and scattering of radiation by the object. The nature of scattering by heterogenous media and the nature of absorption in polypigmented systems are borne in mind, but inadequate knowledge of these factors is not a serious drawback provided suitable controls are maintained.

Probable errors arising from physical procedures can readily be kept within limits of an energy variation of $\pm$ 10%. The probable errors resulting from biological variability often exceed $\pm$ 25% and can be as great as 100%. Neither source of error is particularly troublesome.

Action Spectra: Examples for Control of Flowering

Inhibition of floral initiation of the short-day plant, soybeans, and of the long-day plant, barley, illustrates some features of the initial response. Action is evident throughout the visible spectrum with minimal energy requirements for a given action in the region of 4500 Å and 6400 to 6600 Å. The least response is in the region of 4800 to 5100 Å, and it might be that floral initiation of barley is not promoted in this region. The action spectra are sufficiently similar to suggest common causation, even though there are significant differences. The energy requirement at 6500 to 6600 Å for 50% response is the order of 5×10^{-3} joules, which does not vary by more than fivefold for many responses. In later discussions energies of $< 10^{-2}$ joules are called "low" in contradiction of "high" which is taken as > 1 joules.

Action Spectra: Examples for Seed Germination

These are illustrated in Fig. 1 in the region of 5800 to 8000 Å for germination of *Lactuca sativa* and *Lepidium virginicum* seeds. The seeds are potentiated to germinate with a minimum energy requirement near

6500 Å and are inhibited with a minimum near 7350 Å. Action spectra for floral initiation of the short-day plant, cocklebur, and germination of *Lactuca sativa* seeds are shown in Fig. 2. The two are the same within the limits of experimental error on an absolute incident energy scale. The responses of photoperiodic induction and seed germination are thereby seen to have the same causal activation.

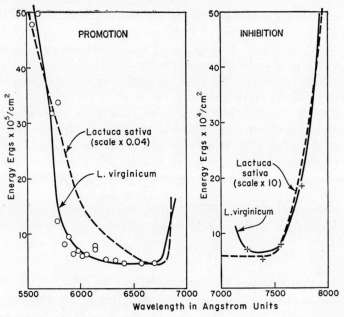

Fig. 1. Action spectra for germination of *Lactuca sativa*, var. Grand Rapids, and *Lepidium virginicum* seeds.

Equality of diverse responses on equal scale of incident energy per unit area when the objects differ so greatly in area as does a leaf and a seed is particularly significant. It indicates that the control is reduced to an equal cellular basis or, in other words, equal responses are obtained for equal absorbed energies per cell, even though one response might require cooperation of a thousand times more cells than the other. This is rather beautifully shown in another and unrelated way by the precision of measurements. To obtain the same precision it is necessary to use several hundred times as many seeds as leaves. Equal precision is obtained by use of that number of seeds having the same area as the leaf.

Responses are evident in the region of 4000 to 5800 Å. In the region of 4000 to 5000 Å, germination cannot be driven to completion, either for inhibition or promotion, which implies that both effects are present.

Reversibility and Order of the Reaction

The actions potentiated in the region of 6550 Å can be reversed by irradiation in the region of 7350 Å before response has essentially progressed. Often the cycle can be repeated many times, as illustrated by seed germination, with the resulting action dependent on the final irradiation. This reversibility lends great power to the physiological work

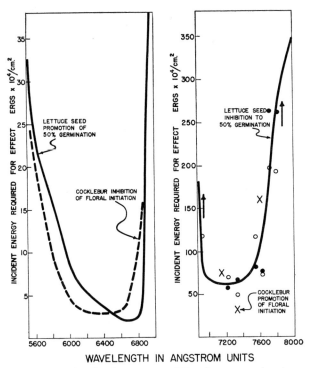

Fig. 2. Action spectra for floral initiation of *Zanthium pensylvanicum* and germination of *Lactuca sativa*, var. Grand Rapids, seeds.

permitting a high probability for establishing the identity of the stimulus for various responses and allowing detailed examination of the photoreaction.

The reversibility permits examination of the initial action by a null method which eliminates the biological response. The biological system is first potentiated for a given action by one type of radiation. The radiant energy of the other type required to eliminate the potentiation before the response takes place is then measured.

The radiant energy required for a given response, such as 40% seed germination, is independent of the temperature between 2° and 23° during short irradiation periods. This indicates that the reversible reaction is first order, depending in rate on the concentration of only one substance in a *given* instance.

Variation of Response with Degree of Pigment Conversion and the Calculation of Absorptivities

The reversibility of the potentiated responses implies a reversible photoreaction which, being first order, allows calculation of $E_a \lambda P_a S_a$ and $E_b \lambda P_b S_b$ where E is the molar absorptivity in the region $\lambda + \Delta \lambda$; P is the quantum efficiency; S is the relative amount of the incident energy reaching the site of action; λ is the wavelength; and a and b refer to the two pigment forms implied by the reversibility. The principle involved is that used by Warburg and Negelein in calculating the absorptivities of CO cytochrome oxidase.

Use of the method requires measurement of a response versus irradiance curve, with response expressed in any arbitrary units such as length of internode or percentage of germination. The calculation fixes both the value of EPS and the degree of pigment conversion required for a given response. The absorptivities and the absolute scale for response are established without having a discernible amount of pigment.

The maximum values of both $E_a \lambda P_a S_a$ and $E_b \lambda P_b S_b$ at 6550 and 7350 Å, respectively, exceed 3.0×10^7 cm²/mole for some objects. This value is near the maximum possible value for E, corresponding to an absorptivity of one electron which implies that P_a and P_b as well as S_a and S_b approach unity.

The absorptivities indicate that the effective pigment is highly colored and in one form is intense blue or green. Despite this, the color is not apparent in albino barley tissue which is responsive to stimulation as indicated by its etiolation, which indicates a very low concentration of the pigment. Its failure to be evident indicates that the concentration is $< 10^{-6} M$ in a responsive tissue.

The Dark Reaction and the Nature of Pigment Action

The pigment form with its absorption maximum at 7350 Å changes in darkness to the other form with a half-life of the order of 24 hours. This change has only been measured for lettuce seed held at 30 or 35°, which blocks germination and which can be tested for the pigment conversion by reducing the temperature to 20° when germination proceeds, depending on the degree of conversion of the pigment to the form with the absorption maximum at 6550 Å. The greater stability of this last form is also shown by the qualitative observation that it is predominant in dormant seed requiring light for germination.

The low concentration and the nature of the responses suggest that one form of the pigment is an enzyme. The pigment must amplify its action to display control of response. The active form is probably the one with the absorption maximum at 7350Å, as shown by the marked action such as germination accompanying small percentage changes to this form where the enzyme would be limiting for action. Its presence enhances germination of imbibed seed that have long been dormant, implying a positive action rather than release from an inhibition. The symmetry of an inhibitor or an activation system, though, is difficult to analyze by physiological methods and a possible flaw in logic had to be repeatedly examined at this point in the early stages of the work.

The Photoreaction — Oxidation and Reduction

The values of $E_a \lambda P_a S_a$ and $E_b \lambda P_b S_b$ vary as much as 25-fold in various objects. If the one value is low, the other is high—as is illustrated by germination results in Fig. 1. This behavior indicates that some further factor is involved in the reaction. Such a factor is an additional reactant.

The reactants are probably an oxidant and a reductant, present in considerably higher concentrations than the pigment forms and having association constants sufficiently great to assure association with either the oxidant or reductant, which satisfies the requirement for a first order reaction. Either hydrogen or an electron might be transferred in the photoreaction because the time spent in the activated state is unlikely to be adequate for transfer of multiatom groups. The probability of transfer will depend on the ratio of the concentration of the oxidant to the reductant and on the relative association constants for the four possible combinations: (i) oxidant-reduced pigment, (ii) reductant-reduced pigment, (iii) oxidant-oxidized pigment, and (iv) reductant-oxidized pigment. Transfer can take place only for the first and fourth of these associations and radiation absorbed by the other two combinations is dissipated for the reversible reaction, thereby lowering the value of EPS.

The ratio of the concentrations of the oxidant to the reductant is determined by reactions with other components and can be varied by holding the object at different temperatures, as illustrated by germination responses of *Lepidium virginicum* seeds, where an initially $E_{7350} P_b > E_{6550} P_b$ condition can be reversed to $E_{7350} P_b < E_{6550} P_b$.

The Photoreaction Formalized

The several deductions can be summarized in the general reaction

$$PH_2 + R \xrightleftharpoons[7350 \text{ Å max.}]{6550 \text{ Å max.}} P + RH_2$$

The reduced form of the pigment is assumed to be the one absorbing at the shorter wavelength because this would correspond to the lower conjugation of double and single bonds. Again, though, the argument is not rigorous and has to be subjected to continued examination. The pigment form (**P**) acts enzymatically and PH_2 is inactive, but what **P** might act upon remains to be seen.

Progress toward a further understanding was held up at this point for 5 years and then developed in an unexpected direction to the isolation of the pigment and knowledge of its biochemical function.

". . . . a much more faint and uncertain breeze of hope were to spring up from our new continent yet we consider it necessary to make the experiment . . . For the risk attending want of success is not to be compared with that of neglecting the attempt; the former is attended with the loss of a little human labor, the later with that of an immense benefit. For these and other reasons it appears to us that there is abundant ground to hope, and to induce not only those who are sanguine to make experiments but even those who are cautious and sober to give their assent." [From Bacon's *Novum Organum*.]

Some Actions Requiring High Energies

Inhibition of axis elongation of plants reaches its most pronounced display only with high energies, as in full sunlight. Action spectra for this inhibition are shown in Fig. 3 for the hypocotyl of *Sinapis alba* and for radicle elongation of *Lactuca sativa*. The entire visible spectrum is effective with maxima for response at constant energy in the region of those observed for the several responses controlled by the reversible photoreaction. Distinctive features are the relatively great action in the region of 4400 to 4700 Å and maxima for formation of anthocyanin near both 6600 and 7350 Å.

Action spectra with high energies for anthocyanin formation are similar to those for axis elongation. These are illustrated in Fig. 4 for apple skin and seedlings of red cabbage and turnip. A distinctive feature is a single maximum for action in the region of 6300 to 7500 Å, which differs in position for the several objects.

The Basis for High Energy Action

While involvement of the low energy reversible pigment system in the high energy process was suggested by the position of action maxima in the 6000 to 7500 Å region, the mode of action was not quickly recognized. A first suspicion of common action developed from noting a small subsidiary region of enhanced spectral effectiveness in the region of 6000 Å for radicle elongation and anthocyanin formation, which is also

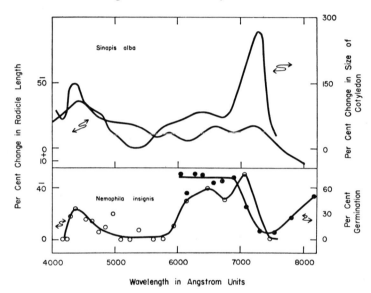

Fig. 3. Action spectra for suppression of elongation of the radicle of *Sinapis alba* (after Kohlbecker, R. (1957). Z. Botan. **45**, 507-524) and for enlargement of cotyledons (after Mohr, H., and Lunenschloss, A. (1958). *Naturwissenchaften* **23**, 578-579) in comparison with the action spectra for seed germination and radicle elongation of *Nemophila insignis*. Energies exceed 5 joules.

evident in the low energy control of seed germination. It was also noted that the high energy responses are most evident in those spectral regions where the two pigment forms both probably have appreciable absorptivities. The overlapping absorptions in the region of 5800 to 7600 Å are indicated in Fig. 5.

Consider the excitation of the reversible pigment. In the associated forms $PH_2 \cdot RH_2$ and $P \cdot R$, absorption of radiation cannot lead to hydrogen transfer but instead might be effective in photosensitization of the high energy reaction. Eventually, though, any particular molecule will be in the association $PH_2 \cdot R$ and $R \cdot PH_2$ and will be transformed upon excitation. A steady state will be set up for transformation and photosensitization dependent upon the absorptivities of the two pigment forms in the particular spectral region and upon the ratio of oxidant to reductant concentration.

The formation of anthocyanin can be followed by direct assay for moles formed per unit area per incident quantum. It can also be calculated on a relative energy scale from the absorptivities derived from low energy responses and assumed oxidant/reductant ratios. The striking agreements between observations and calculations normalized for the

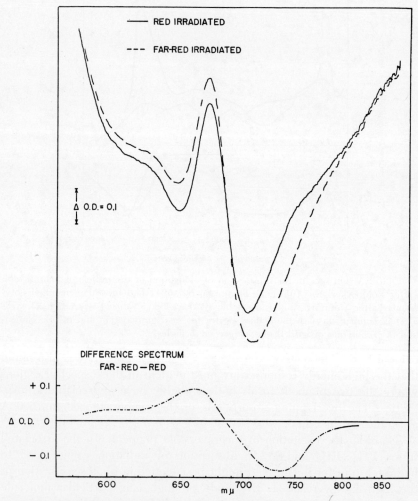

Fig. 4. Recorded optical density curves from maize shoots in the 580 to 850 mµ region after red and far-red irradiations. The difference spectrum is shown.

maximum absorbed energy (not for the spectral region) are shown in Fig. 5.

The Concentration of the Reversible Pigment

This can now be calculated from the amount of anthocyanin formed per unit of incident energy in the region of maximum rate of formation.

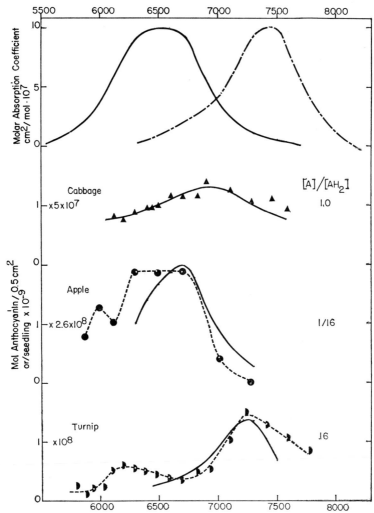

FIG. 5. Observed and calculated rates of anthocyanin synthesis in apple skins, and red cabbage and turnip seedlings in various wavelength regions at the indicated energies in ergs/cm^2 (5×10^7 etc.). The uppermost curves give the absorption coefficients for the two forms of the photomorphogenic pigment.

Consider the apple skin in which anthocyanin is formed in four to six layers of hypodermal tissue about 0.01 cm in thickness. With an incident energy of 0.6×10^{-5} einsteins (moles)/cm^2, 1.0×10^{-9} moles/cm^2 of anthocyanin are formed in the most effective region. The absorptivity of PH_2 is about 3.0×10^7 cm^2/mole and of **P** about 0.3×10^7 cm^2/mole with a ratio of PH_2 to **P** of $> 1/16$. Thus, about 0.6×10^{-11} moles/p cm^2 of pigment are present where p is the quantum efficiency for anthocyanin formation. The pigment concentration, accordingly, is $10^5 \times 0.6 \times 10^{-11} \, p \, M$ or $0.6 \times 10^{-6} \, p \, M$. If p approaches 1.0, as might be likely, the concentration is the order of $0.6 \times 10^{-6} \, M$; it is less if p is smaller.

Detection of the Pigment in Living Plants by Spectrophotometry

A pigment concentration of $10^{-7} \, M$ should be possible to detect by measurement of the difference of absorption of radiation in the region of 7350 and 6550 Å as the pigment is changed from one to the other form. Such measurements require a differential spectrometer with a noise level of less than 0.01 optical density at an O.D. of 4 and having a receiver of the order of 2π radian in solid angle. These instruments have been developed by R. K. Norris and W. L. Butler. The pigment was evident in the first test with a 1.5 cm thickness of turnip cotyledons, giving a $\Delta(\Delta \text{O.D.}) = \Delta(P_{7350} - P_{6550})$ after red radiation $- \Delta(P_{7350} - P_{6550})$ after far-red radiation $= 0.05$, with a noise level of 0.001 O.D. at an O.D. of about 3.

An initial survey indicates that the pigment is abundant in the shoot of etiolated monocotyledon seedlings. Maize seedlings were selected for further study. The difference spectrum obtained on a scanning spectrophotometer of special design (by Norris and Butler) is shown in Fig. 4.

Assay and Separation of the Pigment

A next step was to examine ground tissue with the differential spectrophotometer. The grinding had no effect on $\Delta(\Delta \text{O.D.})$ values. The solution resulting from centrifuging at 170,000 g gave an $\Delta(\Delta \text{O.D.})$ of the order of 100-fold the noise level.

The pigment is retained upon dialysis, heat-coagulates at 50° C, and is salted out of solution by 0.34 saturated $(NH_4)_2SO_4$. It is thus likely to be a protein as deduced from the physiological evidence for its enzymatic nature. Its purification and the finding of its specific enzymatic action remain to be accomplished, but seemingly offer no great obstacle.

> "For it appears at first incredible that any such discovery should be made, and when it is made, it appears incredible that it should so long have escaped men's research." [From Bacon's *Novum Organum*.]

The Nature of the Enzymatic Action and the Physiological Display

If the search for the enzymatic reaction catalyzed by P_{7350} had to be made without further guides, it would be quite hopeless and possible of success only by casual encounter. Fortunately, some guides exist. The indicated nature of the reaction, moreover, clearly points to the basic reason for the multiplicity of the physiological display. In short, a reasonable pattern can be seen for varied physiological displays sharing a common cause.

After the guides are displayed, the yet untraveled regions are entered, not in the sense of blind speculation but much as if they had been viewed from a distant vantage point. So, first, the guides.

Anthocyanin and Ethanol Formation in the Apple Peeling

Apple peels floating on a sucrose solution produce ethanol in darkness, a reaction which is blocked immediately upon adequate irradiance. This has many implications, an immediate one being the subsidiary evidence of aerobic ethanol production indicative of a relatively high degree of reduction of some coenzyme. The high degree of reduction is also required by the spectral position of the maximum for anthocyanin formation by the apple ($R/RH_2 < 1/16$). A second implication of greater pertinence as a guide is that the photocontrol is close to the utilization of a two- or three-carbon compound.

Anthocyanin formation involves two pathways of aromatic ring formation. The phloroglucinol ring incorporates three acetyl groups when the synthesizing material is supplied with acetate. The photocontrol point can be in this pathway and still serve to direct the course of ethanol production.

Anthocyanin synthesis in seedlings of *Sorghum vulgare* (variety Wheatland milo) is shown to require excitations with high energy which potentiates the synthesis. This excitation, while utilizing the reversible pigment system in the red part of the spectrum, involves another pigment with an action maximum near 4700 Å (Fig. 6). In the course of anthocyanin synthesis, which requires about 16 hours for completion, the reversible reaction is required at a point of half effectiveness 6 hours after the high energy excitation as shown in Fig. 7. The low energy reversible reaction thus enters directly into the pathway of anthocyanin synthesis.

The Control Point — A Working Hypothesis

The similarities of the many action spectra and the ubiquity of reversibility indicate that the control point is a single reaction—a bottleneck after the concepts of Krebs and Kornberg. The probable pathways

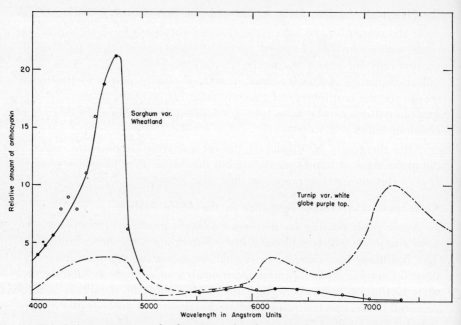

Fig. 6. Action spectrum for formation of anthocyanin formation in etiolated seedlings of *Sorghum vulgare*, var. Wheatland milo. Energy > 5 joules.

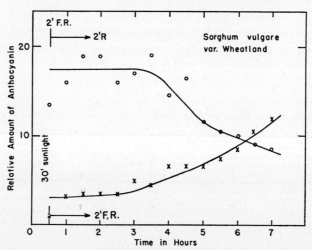

Fig. 7. Effects of red and far-red radiation on the formation of anthocyanin following irradiation of *Sorghum vulgare*, var. Wheatland milo seedlings.

of anthocyanin and alcohol synthesis suggest the nature of the controlling reaction, which for purposes of illustration—and illustration only—might be written as an acyl CoA dehydrogenation, thus

$$RH_2\text{—CO—S—CoA} + \text{Oxidant} \underset{\longleftarrow}{\overset{\text{Enzyme } P}{\longrightarrow}} R\text{—CO—S—CoA} + \text{Reductant}$$

Attention is turned as to how such a reaction might control germination, axis elongation, and flowering among the more evident responses. It matters little that what is said might later be proven wrong for the purpose is to sketch a self-consistent pattern.

A. Germination

Perhaps of greatest pertinence is that seeds requiring light to germinate are generally small ones with fat as an immediate energy source. The most primitive observed step in germination is the disappearance of fat droplets near the tip of the radicle and the appearance of starch. The expected pathway from fat to starch would involve two-carbon degradation to acetyl-S-CoA followed by formation of glucose by the glyoxalate pathway. The control point would be near the appearance of acetyl-S-CoA.

B. Elongation

The plant axis is plastic and limitation of elongation as displayed in dark-grown seedlings implies the production of a stiffening agent. An agent of this type is produced by action of the far-red absorbing form of the photoreversible pigment. An interesting feature of this action is that it does not result in a phototropic response when irradiation is restricted to one side of the plant. This implies that the stiffening agent differs from anthocyanin formation in being translocated from the original site of radiation action. It must be uniformly effective in directions transverse to the plant axis in a short time relative to the cellular generation time.

C. Flowering

This has attracted much speculation by others, which possibly has centered around the term "florigen" or hormone for flowering. The background is that the induction of flowering at the meristem is a result of perception by the leaf. It is logical that a controlling material has been transported as in control of elongation. One guess might be as good as another about this material for the pathways of synthesis from acetate or two-carbon radicals can be most diverse. A plausible pathway, though, leads from acetyl-S-CoA through mevalonic acid to sterols or carotenoids.

Specific sterols are controlling agents in the reproductive development of animals through hormone action and equivalent ones can be anticipated in plants. Sterols could be solubilized for transport by esterification with sugars as glycosides. There is the example of the agave (the century plant) in which flowering after a long vegetative interval is accompanied by a burst of sterol production.

The Day and the Length of the Night

The question of effects of the day and night lengths on plant growth can now be addressed, albeit without a completely satisfactory answer, but at least with sufficient clarity to illustrate the physiological pattern.

(a) The reversible pigment is stable under full sunlight as has been established by the immediate photoreversible responses of plants upon onset of darkness, which is illustrated by flowering of millet and *Pharbitus nil*, elongation of a number of beans, and coloration of apples.

(b) The pigment system is capable of acting during the day as determined in part by the integrated absorptions of the two pigment forms for the spectral distribution of the radiation source. Sunlight and incandescent filament sources have within 25% the same ratios of intensities/cm²/Λ ($\varkappa$ = frequency) in the region of 6550 and 7350 Å. Fluorescent sources have relatively low intensities in the region of 7350 Å. The source then sets the pigment ratio to lead to some action of P_{7350} during the day. But the photosensitization arising from the continuous excitation similar to that involved in anthocyanin formation is probably of greatest importance.

(c) The plant enters its dark period with the sugars and other energy reserves formed by photosynthesis and with the ratio of the pigment forms fixed. The ratio is of great importance for the vegetative and floral development of the plant as can be shown by maintaining the ratio against any dark reversion by extending the light period with low intensity incandescent filament or fluorescent radiation for a few hours.

(d) As darkness proceeds the reserves are utilized at a decreasing rate and the ratio of PH_2/P increases by the dark reversion of **P**. After the order of 6 hours, the pigment is chiefly in the form of PH_2 as can be shown by the effectiveness of an interruption with red radiation. The time of return of **P** to PH_2 seems short as compared with observations in seeds held dormant by high temperatures and it might be that with rapid turnover in a reaction catalyzed by **P** there is an enhanced probability of **P** reverting to PH_2, particularly if this reaction involves hydrogen transfer. At any rate, a crude temperature independence of action is possible as an increase in temperature would both hasten the action on the reserves and also hasten the decrease of the active agent **P**.

References

[These references are given for the works of Garner and Allard on photoperiodism, for the observations of Flint and McAlister and Resuhr on seed germination, and for the work of the Beltsville group, in chronological order.]

Garner, W. W., and Allard, H. A. (1920). Effect of the relative length of day and night and other factors of the environment on growth and reproduction in plants. *J. Agr. Research* **18**, 553-606.

Garner, W. W., and Allard, H. A. (1931). Effect of abnormally long and short alternations of light and darkness on growth and development of plants. *J. Agr. Research* **42**, 629-651.

Garner, W. W., and Allard, H. A. (1931). Duration of the flowerless condition of some plants in response to unfavorable lengths of day. *J. Agr. Research* **43**, 439-443.

Allard, H. A. (1932). Length of day in relation to the natural and artificial distribution of plants. *Ecology* **13**, 221-234.

Allard, H. A. (1938). Complete or partial inhibition of flowering in certain plants when days are too short or too long. *J. Agr. Research* **57**, 775-789.

Flint, L. H. (1934). Light in relation to dormancy and germination in lettuce seed. *Science* **80**, 38-40.

Flint, L. H. (1935). Sensitivity of dormant lettuce seed to light and temperature. *J. Wash. Acad. Sci.* **25**, 95-96.

Flint, L. H., and McAlister, E. D. (1935). Wave length of radiation in the visible spectrum inhibiting the germination of light sensitive lettuce seed. *Smithsonian Inst. Misc. Collections* **94**, 1-11.

Flint, L. H., and McAlister, E. D. (1937). Wave length of radiation in the visible spectrum promoting the germination of light sensitive lettuce seed. *Smithsonian Inst. Misc. Collections* **96**, 1-8.

Resuhr, B. (1939). Beitrage zur Licht keimung von *Phacelia tenacetifolia* (Benth). *Planta* **30**, 471-506.

Parker, M. W., Hendricks, S. B., Borthwick, H. A., and Scully, N. J. (1945). Action spectrum for the photoperiodic control of floral initiation in Biloxi soybean. *Science* **102**, 152-155.

Parker, M. W., Hendricks, S. B., Borthwick, H. A., and Scully, N. J. (1946). Action spectra for photoperiodic control of floral initiation in short-day plants. *Botan. Gaz.* **108**, 1-26.

Borthwick, H. A., Hendricks, S. B., and Parker, M. W. (1948). Action spectra for photoperiodic control of floral initiation of a long-day plant, Wintex barley (*Hordeum vulgare*). *Botan. Gaz.* **110**, 103-118.

Parker, M. W., Hendricks, S. B., Borthwick, H. A., and Went, F. W. (1949). Spectral sensitivities for leaf and stem growth of etiolated pea seedlings and their similarity to action spectra for photoperiodism. *Am. J. Botany* **36**, 194-204.

Parker, M. W., Hendricks, S. B., and Borthwick, H. A. (1950). Action spectrum for the photoperiodic control of floral initiation of the long-day plant *Hyoscyamus niger*. *Botan. Gaz.* **111**, 242-252.

Borthwick, H. A., Hendricks, S. B., and Parker, M. W. (1951). Action spectrum for inhibition of stem growth in dark-grown seedlings of albino and non-albino barley (*Hordeum vulgare*). *Botan. Gaz.* **113**, 95-105.

Borthwick, H. A., Hendricks, S. B., Parker, M. W., Toole, E. H., and Toole, V. K. (1952). A reversible photoreaction controlling seed germination. *Proc. Natl. Acad. Sci. U. S.* **38**, 662-666.

Borthwick, H. A., Hendricks, S. B., and Parker, M. W. (1952). The reaction controlling floral initiation. *Proc. Natl. Acad. Sci. U. S.* **38**, 929-934.

Toole, E. H., Borthwick, H. A., Hendricks, S. B., and Toole, V. K. (1953). Physiological studies of the effects of light and temperature on seed germination. *Proc. 10th Intern. Congr. Seed Testing Assoc.* **18**, 267-276.

Borthwick, H. A., Hendricks, S. B., Toole, E. H., and Toole, V. K. (1954). Action of light on lettuce seed germination. *Botan. Gaz.* **115**, 205-225.

Toole, E. H., Toole, V. K., Borthwick, H. A., and Hendricks, S. B. (1955). Photocontrol of Lepidium seed germination. *Plant Physiol.* **30**, 15-21.

Hendricks, S. B., and Borthwick, H. A. (1955). Photoresponsive growth. In "Aspects of Synthesis and Order in Growth" (D. Rudnick, ed.), pp. 149-169. Princeton Univ. Press, Princeton, New Jersey.

Toole, E. H., Toole, V. K., Borthwick, H. A., and Hendricks, S. B. (1955). Interaction of temperature and light in germination of seeds. *Plant Physiol.* **30**, 473-478.

Borthwick, H. A., Hendricks, S. B., and Parker, M. W. (1956). Photoperiodism. *Radiation Biol.* **3**, 479-517.

Toole, E. H., Hendricks, S. B., Borthwick, H. A., and Toole, V. K. (1956). Physiology of seed germination. *Ann. Rev. Plant Physiol.* **7**, 299-324.

Hendricks, S. B., Borthwick, H. A., and Downs, R. J. (1956). Pigment conversion in the formative responses of plants to radiation. *Proc. Natl. Acad. Sci. U. S.* **42**, 19-26.

Toole, E. H., Toole, V. K., Hendricks, S. B., and Borthwick, H. A. (1957). Effect of temperature on germination of light-sensitive seeds. *Proc. 11th Intern. Congr. Seed Testing Assoc.* **22**, 196-204.

Downs, R. J., Hendricks, S. B., and Borthwick, H. A. (1957). Photo-reversible control of elongation of pinto beans and other plants under normal conditions of growth. *Botan. Gaz.* **118**, 199-208.

Cathey, H. M., and Borthwick, H. A. (1957). Photoreversibility of floral initiation in chrysanthemum. *Botan. Gaz.* **119**, 71-76.

Siegelman, H. W., and Hendricks, S. B. (1957). Photocontrol of anthocyanin formation in turnip and red cabbage seedlings. *Plant Physiol.* **32**, 393-398.

Siegelman, H. W., and Hendricks, S. B. (1958). Photocontrol of anthocyanin synthesis in apple skin. *Plant Physiol.* **33**, 185-190.

Siegelman, H. W., and Hendricks, S. B. (1958). Photocontrol of alcohol, aldehyde, and anthocyanin production in apple skin. *Plant Physiol.* **33**, 409-413.

Hendricks, S. B., and Borthwick, H. A. (1959). Photocontrol of plant development by the simultaneous excitations of two interconvertible pigments. *Proc. Natl. Acad. Sci. U. S.* **45**, 344-349.

Hendricks, S. B., and Borthwick, H. A. (1959). Photocontrol of plant development by the simultaneous excitations of two interconvertible pigments. II. Theory and control of anthocyanin synthesis. *Botan. Gaz.* **120**, 187-193.

Hendricks, S. B., Toole, E. H., Toole, V. K., and Borthwick, H. A. (1959). Photocontrol of plant development by the simultaneous excitations of two interconvertible pigments. III. Control of seed germination and axis elongation. *Botan. Gaz.* **121**, 1-8.

Butler, W. L., Norris, K. H., Siegelman, H. W., and Hendricks, S. B. (1959). Detection, assay, and preliminary purification of the pigment controlling photoresponsive development of plants. *Proc. Natl. Acad. Sci. U. S.* **45**, 1703-1708.

Discussion

WOLKEN: This is a very intriguing kind of thing. Can I bring you back to the activation and inactivation mechanism? What do you now speculate is the absorbing material in the activation and what do you speculate—or you may already have information—on the inactivation?

HENDRICKS: As far as I am concerned, the inactive form is the hydrogenated form of the pigment. The hydrogen is transferred off of it, it goes to the enzyme; that is the active form.

WEBER: May I ask you what evidence you have to suggest that electrons are not transferred?

HENDRICKS: Not very much. It is just that something is transferred and in this sort of reaction, a hydrogen atom must be transferred, you have not got time to transfer anything else.

WEBER: You can have a *cis-trans* isomerism.

HENDRICKS: You can have a *cis-trans* isomerism. That is the reason I gave the evidence indicating that there is another reactant. There are, however, two arguments against the other reactant. The first is that the reaction is first order and the second is that the system, after you dialyze it, retains its reversibility. But don't forget the case of the protochlorophyll that Dr. Smith indicated to us yesterday. It has to get some hydrogen, and while it does not go first order, it still goes at a low temperature. However, it might not be an oxidation or reduction. It is just shown this way to fit the whole thing together.

ARNON: Do I understand that in some seeds germination is dependent on light, and in others it is not? Is that correct?

HENDRICKS: Yes.

ARNON: If that is the case, and in all germination you are dealing with the same reaction, how is it that some seeds do not require light and some do?

HENDRICKS: How is it that one plant flowers without the response and another one needs the response to flower? The thing is that you are apparently dealing with a bottle-neck, that the level at this point is determined by many things, and that side pathways around it can operate and control the level. Now you take the lettuce seed for instance. It is easy enough to get varieties of lettuce seed that are not controlled in their germination by this response. It is easy enough to bring those given varieties of lettuce seed into control simply by heating them to 30–35° after they imbibe, and holding them there for 24 to 48 hours. Just holding them at the high temperature at which they will not germinate will then make them such that when you reduce them to a favorable temperature, they are light responsive. This is also shown in many plants, such as the tomato. The tomato is not responsive in its flowering control to the light reaction. It is responsive to the light control in its etiolation—it etiolates if you go to dim light. It is also responsive to light control by the formation of a yellow pigment in the cuticle of the fruit. But it is not controlled in the flowering of the plant, so you see the phenomenon displayed, but not to an extent such that it controls one of the crucial reactions in the system. In other words, there are ways around. The control is not a complete one; there are always anastomosing pathways around the system. I guess the analogous illustration would be to cut a nerve and other pathways take over in the responsive system.

20

Hematin Compounds in Photosynthesis

MARTIN D. KAMEN[*]

Graduate Department of Biochemistry, Brandeis University, Waltham, Massachusetts

Introduction

Recognition of the existence and function of hematin compounds[†] in photosynthetic systems is recent. Nevertheless, sufficient data have accumulated to warrant two assertions. The first is that hematin compounds constitute an important fraction of the protein in chloroplasts and chromatophores. The second is that they are functional in mediating reactions which bridge the gap between photon absorption and the onset of biosynthesis.

Historical Development

The first suggestion that compounds analogous to respiratory catalysts might be present and functional in photosynthesis seems to have been made in 1939 by Hill, whose experimental skills and insights have supplied the basis for so many of the modern advances in photosynthesis research. In his classic memoir on the chloroplast reaction, Hill noted that the yield of molecular oxygen obtainable from illuminated chloroplasts was limited by the partial pressure of oxygen developed in the reacting chloroplast system (Hill, 1939). He suggested that this limitation in yield of oxygen arose from a back-oxidation of the photochemically produced reductant by the molecular oxygen evolved. The reduction of oxygen, he suggested, could proceed through a system of respiratory catalysts analogous to those known to function in the classic oxidase systems. Later, Hill incorporated this concept in his general theory of photosynthesis as a chemosynthetic process coupled to light absorption (Hill, 1951).

The possibility that respiratory pigments unique to chloroplasts might exist became a reality in 1943 when Scarisbrick and Hill reported the existence of a new heme protein in green plant tissues which they

---

[*] Communication No. 53 in series, "Publications, Graduate Department of Biochemistry, Brandeis University, Waltham, Massachusetts."

[†] The term "hematin compounds" is meant to include all proteins with tetrapyrrolic iron chelates as prosthetic groups.

called "cytochrome f" (Scarisbrick and Hill, 1943). Nine years later, Davenport and Hill were able to report the isolation and purification of this cytochrome and to show it was a protein variant of cytochrome c (Davenport and Hill, 1952). They also demonstrated it to exist in all chlorophyllous systems, and to be concentrated in the green parts of plants. A cytochrome of "b" type was also found somewhat later by Hill (1954).

With these facts at hand, Davenport and Hill (1952) suggested that the cytochrome system in green plants was unique to the photoactive structures of photosynthetic tissues and that it mediated a back-flow of electrons from oxygen (or its precursor) to the photoreductant in an energy-storing process, which was analogous to reversal of electron flow through cytochromes in the classic oxidase system.

An important development was the demonstration that no oxidase existed for cytochrome f (Davenport and Hill, 1952; Hill, 1958) a fact in agreement with the failure to demonstrate cytochrome-a-type proteins in chloroplasts (Hill and Scarisbrick, 1951). Hill raised the possibility that chloroplasts might be considered as similar to mitochondrial systems in which the terminal respiratory oxidase was replaced by chlorophyll (Hill, 1958).

All of this left out of account the well-known anaerobic photosyntheses of photosynthetic bacteria, in particular, the obligate anaerobic green and purple sulfur bacteria. It will be recalled that, as late as 1953, the general impression existed that cytochromes were unique to aerobic or facultative tissues. This generalization dated from the early work of Keilin and others in the 1920's and early 1930's, in which a number of clostridial and streptococcal species of anaerobes were examined (Keilin, 1925; Yaoi and Tamiya, 1928; Fujita and Kodama, 1934; Tamiya and Yamagutchi, 1933; Frei *et al.*, 1934). Hence, none of the green or purple sulfur bacteria were supposed to contain hematin compounds. However, none of the great groups of chemosynthetic and photosynthetic anaerobes had, in fact, been included in surveys of cytochrome distribution. This unfortunate circumstance delayed for many years the recognition that hematin compounds could exist and react in systems with no functional dependence on oxygen reduction.

In 1954–1955, three different laboratories, simultaneously and independently, announced the isolation and characterization of soluble c-type cytochromes in large yields from some obligate anaerobes. Postgate (1955) in England and Ishimoto and Koyama (1955) in Japan reported the existence of these hematin compounds in extracts obtained from species of the sulfate reducer, *Desulfovibrio desulfuricans*. Kamen and

Vernon isolated a c-type cytochrome from extracts of a green sulfur photoanaerobe, *Chlorobium limicola* (Kamen and Vernon, 1954), an observation extended later by Gibson and Larsen (1955) to another species of *Chlorobium*.

The existence of a c-type cytochrome in a facultative photoheterotroph, *Rhodospirillum rubrum*, had been established by Vernon and Kamen (1954; Kamen and Vernon, 1955; Vernon, 1954).* Later systematic studies from the same laboratory revealed the presence of c-type cytochromes in all species of purple photosynthetic bacteria, whether photoanaerobic or facultative (Newton and Kamen, 1956; Bartsch and Kamen, 1960). These studies completed the demonstration that heme proteins were common to all varieties of photosynthetic tissue, regardless of nutritional habit or degree of tolerance of oxygen.

Recently, a resurgence of interest in cytochrome f has been noted. Nishimura, the first worker to publish on cytochrome f outside the Cambridge school, has reported the isolation of this hematin compound from light-grown *Euglena* and shown that cytochrome f synthesis parallels that of chloroplasts and chlorophyll (Nishimura, 1959). Katoh (1959) has reported isolation of cytochrome f from a large variety of red, brown, and blue-green algae. Hulcher and Vishniac (1958) have explored techniques for extracting the porphyrin moiety of cytochrome f and other chloroplast hematin compounds, in an effort to devise a screening technique for demonstrating their presence free of uncertainties inherent in procedures dependent on protein isolation.

Another series of findings has been vital in establishing function of hematin compounds. Reference is made to elaboration of techniques for examining fast reactions both *in vivo* and *in vitro*, using methods of dynamic spectrophotometry. These procedures, which we owe mainly to efforts of such investigators as Duysens, Britton Chance, and Lündegardh, have been supplemented by Witt using flash spectrophotometric methods based in turn on the pioneering investigations of Porter and his colleagues.

At the moment a large literature dealing with such researches is developing. While it is premature to attempt an evaluation of the many papers at hand, all workers agree these researches indicate an intimate association between oxidation reactions involving the hematin compounds of functional photoactive subcellular structures and the energy-storing

---

* Elsden *et al.* (1953) described this cytochrome as the first soluble bacterial cytochrome to be isolated in substantially pure form. However, Egami *et al.* (1953) had reported at about the same time the isolation of a partially purified cytochrome from a halo-tolerant bacterial aerobe.

mechanisms activated in the earliest phases of photosynthesis (Smith and Chance, 1958). A few salient examples of this work will be discussed in a later section.

It is evident that the tempo and magnitude of researches on hematin compounds is accelerating, now that methods are at hand which enable adequate isolation and chemical investigation on the one hand, and specification of mechanisms of fast reactions in complex biochemical systems on the other. The present activity bearing on hematin compounds in photosynthesis is all the more amazing when it is recalled that only 20 years have elapsed since Hill's very tentative suggestion about photorespiratory analogs.

Many pressing questions have risen in the wake of these developments, particularly relating to new chemical capabilities of hematin compounds in photoactivated systems. I believe the purposes of this symposium will be served best if I confine discussion to certain points which have not received sufficient attention in previous articles reviewing the status of hematin compounds. The reader will find a number of references to earlier articles in the appended bibliography (Smith, 1954; Kamen, 1955, 1956). These deal with the many data available which are not included in this report.

Distribution Patterns

Present knowledge about hematin compounds known to be associated specifically with photoactive structures may be summarized briefly. (1) All photosynthetic systems contain cytochromes of the "c" type. However, as we will see later, there are considerable variations in physical properties evident in these hematin compounds depending on the source materials from which they are derived. (2) No other single category of cytochrome is present invariantly in all photosynthetic tissues. (3) Catalases and peroxidases appear to be ubiquitous and easily obtained in soluble form. (4) A new type of hematin compound, the so-called "RHP" type protein, occurs which appears to be confined to the purple photosynthetic bacteria. These compounds have been described in detail elsewhere (Bartsch and Kamen, 1958).

At least three questions arise immediately. First, how unique are these hematin compounds which occur in high concentration in photosynthetic tissues? Secondly, what correlation, if any, is there between variations in the hematin pattern of different photosynthetic systems and corresponding changes in the distribution pattern of chlorophylls and accessory pigments? Thirdly, are these hematin compounds structurally adapted so that they can participate directly in photochemical reactions?

Relative to the first question, it may be noted that hematin compounds spectrochemically similar to these found in photosynthetic tissues are also present in bacteria which possess analogous metabolic patterns, particularly the chemosynthetic bacteria. Thus, "c" cytochromes are seen in the anaerobic sulfate reducers which are chemosynthetic analogs of the obligate photoanaerobes. The "c" cytochromes of the *Athiorhodaceae* find their analogs in corresponding acidic proteins present in a number of facultative anaerobes which reduce oxides of nitrogen. The case of the green plants and algae is particularly arresting, for here there is little doubt that specialization of hematin compounds has resulted in a sharp demarcation between chloroplast cytochrome c, that is, cytochrome f, and mitochondrial cytochrome c. In *Euglena*, as remarked previously, cytochrome f is now known to be associated solely with the chloroplasts (Nishimura, 1959).

As is well known, both chloroplasts and chromatophores can couple the photochemical absorption of energy to production of ATP, the so-called "photophosphorylation" (Arnon *et al.*, 1954; Frenkel, 1954; Newton and Kamen, 1957). It seems certain that the mechanism involved is analogous to that for oxidative phosphorylation in mitochondria with the important reservation that the electron transport involved is cyclic and occurs between endogenous photo-oxidant and photoreductant, rather than between exogenous hydrogen donors and oxygen.

It may be concluded that hematin complexes are present in all biochemical systems which utilize electron transport to ultimate H acceptors in order to produce useful high-energy intermediates, such as phosphate esters, acyl intermediates, etc. Only in the case of strictly fermentative species, such as *Clostridia* and certain streptococci, are such hematin compounds missing. The H-acceptors can be inorganic, e.g., oxides of sulfur or nitrogen, or they can be the special photo-oxidants generated in the photochemical phase of photosynthesis. Thus, the uniqueness of the hematin compounds is associated, most likely, with the specificity of the ultimate hydrogen acceptor.

From this view, it may be suggested that Keilin's great generalization about cytochromes in aerobic tissues can be sharpened to the statement that a unique set of hematin compounds is required when reduction of molecular oxygen occurs. It is possible that specificity is lodged wholly in the terminal oxidase. Alternatively, one may envisage the possibility that each oxidant requires a *minimal* set of unique hematin compounds. Thus, the mitochondrial system is wholly adapted to reduction of molecular oxygen, the hematin system of *Desulfovibrio* to sulfate reduction. the cytochromes of *Micrococcus denitrificans* to nitrate reduction, etc.

This does not exclude the possibility that a given hematin system can function with more than one terminal oxidant—this appears to be true in the case of a number of facultative bacteria, such as the nitrate reducers. However, there is little doubt that the mitochondrial hematin system is unable to function except with molecular oxygen as hydrogen acceptor.

In the case of the facultative Athiorhodaceae, there remains unresolved the manner in which one and the same hematin system functions either with photo-oxidant, or with oxygen as terminal acceptor. However, space limitations preclude further discussion of this and many other questions which no doubt have occurred to the reader. The author must pass on to questions raised by the comparative biochemistry and structural analysis of hematin compounds.

Physicochemical Considerations, Including some Remarks on a Unitary Theory of Photosynthesis

The varied physicochemical characteristics of photosynthetic hematin compounds can be appreciated by examination of Table I which shows some of the physicochemical data available for the "c" cytochromes and RHP-type compounds. A similar table could be presented showing the variations encountered in c-type cytochromes isolated from a variety of nonphotosynthetic tissues.

It is evident that a great range of physicochemical variation is encountered in the single category of "cytochrome c." At present we have no certain knowledge as to the origin of these variations. Obviously, the elucidation of these data at the molecular level holds much of interest to chemists interested solely in protein structure, regardless of its bearing on photosynthetic mechanisms.

A beginning has been made in what may be called the comparative biochemistry of cytochrome c. Tuppy and Paleus (1955) have published amino acid sequence analyses of the heme-bearing peptide from a variety of "c" cytochromes, including that from *Rhodospirillum rubrum* (Paleus and Tuppy, 1959). A characteristic constant feature of these hemopeptides is the appearance of two cysteine residues, separated by two amino acid residues, and proximal to histidine and lysine (or arginine) residues. No explanation for the great changes noted in the electrochemical potentials of the iron atom is apparent from the primary structure revealed in these studies. It must be concluded these arise from features of the secondary and tertiary structure. Of course, it may be assumed there are chelating groups introduced by the protein moiety into the extraplanar coordination positions which give rise to the various electrochemical po-

TABLE I

SOME PHYSICOCHEMICAL PROPERTIES OF PHOTOSYNTHETIC HEMATIN CYTOCHROMES[a]

Compound	Isoelectric Point (pH)	Molecular Weight	Number Heme Groups	E_0 (value) (pH 7.0)	Auto-oxidation	References
c-type cytochromes						
C. thiosulfatophilum — 554	Basic	~16,000	1	~+0.16	Slow	Gibson and Larsen (1955)
Chromatium — 552	5.4 ± 0.1	97,000	3	+0.01	Rapid	Bartsch and Kamen (1960)
R. rubrum — 550	~7.0	~16,000	1	+0.365	None	Vernon and Kamen (1954)
R. capsulatus — 550	<7.0	—	—	+0.32	None	Kamen and Vernon (1955)
R. palustris — 552	<7.0	—	—	+0.31	None	Kamen and Vernon (1955)
P. sativum — 555 (cytochrome f)	4.7	110,000	2	+0.38	None	Hill (1954)
E. gracilis — 552 (cytochrome f)	>6.0	—	—	+0.36	None	Nishimura (1959)
RHP						
Chromatium	5.5 ± 0.1	36,000	2	−0.005	Rapid	Bartsch and Kamen (1960)
R. rubrum	5.0 ± 0.1	28,000	1	−0.008	Rapid	Bartsch and Kamen (1958)

[a] Data have been selected from various references shown.

tentials observed. There is some knowledge about substitution groups which may account for these effects, as Falk and Nyholm (1957) have pointed out. Thus, replacement of a basic group such as the histidine imidazole residue, by a carboxyl or amino group can be expected to lower the electrochemical potential several hundred millivolts. The early potentiometric studies of Michaelis and Friedheim (1931) on complexes of iron suggest that pyrophosphate radicals could exert similar effects. These matters all await further research.

A fascinating new development in hematin biochemistry should be mentioned here in passing. The availability of bacterial forms of hematin compounds raises the possibility of using immunochemical procedures for studying these protein structures. Thus the bacterial heme proteins make good antigens when used to invoke specific antisera in mammalian systems. Newton and Levine (1959; Newton, 1958), in our laboratories, have demonstrated that whole cells, chromatophores, cell fragments, and soluble fractions can induce strong antigenic reactions when used to challenge rabbit sera. Orlando and Bartsch (private communication, 1960) extended these findings to the purified hematin compounds and obtained potent specific antisera which can be used to establish structural modifications in these compounds during isolation procedures. Furthermore, these sera have been used to demonstrate that the various hematin compounds (RHP and cytochrome c of *R. rubrum* and *Chromatium*) are immunologically different, even when isolated from the same photoactive structure. This work is being extended in the hope it will aid in structural analysis of the various hematin compounds isolated from photosynthetic structures.

Photochemical mechanisms involving participation of hematin compounds are possible. Iron, either as the free metal, or in any of its chelates (including heme) is an effective quencher of both single and triplet states of porphyrins, chlorophylls, and other conjugated macrocyclic systems; magnesium is not. It is conceivable that interaction between tetrapyrrolic chelates of iron and magnesium, resulting in deexcitation of the latter, initiates charge transfer reactions, leaving the former in an oxidized (or reduced state). There is a growing realization that the ring system of metal tetrapyrroles participates in oxidation-reduction reactions, as well as the central metal atom. The work of Gibson, Ingram, George and others has proven that this is so in the case of netmyoglobin acting as a peroxidase or catalase (for a recent review see King and Winfield, 1959).

In view of the highly conjugated ring system, it should not be difficult to remove (or add) electrons to the peripheral carbons creating transitory semiquinoidal forms of varying stability. An excited chlorophyll molecule involved in de-excitation by heme could leave the latter in an oxidized state with relation to the central iron and to the peripheral ring system. This sort of interaction is quite likely, but no experimental evidence has as yet been obtained to indicate its actuality. Experiments on model systems using flash spectrometry to study yields of triplet state chlorophyll, fluorescence yields, etc., in the presence and absence of iron chelates should provide the salient data.

In this connection, it should be mentioned that no thoroughgoing analysis for the presence of metals, other than iron, has ever been made in the variant hematin compounds associated with chloroplasts and chromatophores. It was mentioned above that it is important to determine whether these hematin compounds are qualitatively different from other hematin compounds in having a unique photochemical function. The presence of metals other than iron (e.g., copper, zinc, etc.) could be crucial in this respect.

Supposing that charge transfer reactions occur, a mechanism could be provided for direct energy storage in the following manner. Initial photon absorption would yield excitons which would migrate among chlorophyll molecules by inductive resonance until a specific site was reached where a particular chlorophyll molecule was attached to components of a transport chain including pyridine nucleotides, flavins, quinones, and hematin compounds. De-excitation of chlorophyll at this point could effect charge transfer from the hematin compound to chlorophyll, which in turn could reduce the pyridine nucleotides. The energy stored would be the product of the charge transferred by the electrochemical potential difference between the reduced pyridine nucleotide and the oxidized hematin compound.

It should be emphasized that the effective electrochemical potential of hematin compounds cannot be judged solely on the basis of the usual equilibrium

$$Fe^{++} = Fe^{3+} + \varepsilon^-$$

The possibility of a ferryl (Fe^{4+}) state creates another redox equilibrium

$$Fe^{3+} + H_2O = (FeO)^{++} + 2H^+ + \varepsilon^-$$

George (1959) has shown that in metmyoglobin, acting as a catalase, the potential of the intermediate ("Complex I") formed with various oxidizing agents is sufficiently positive (~ 1.0 volt) to oxidize peroxide. As is well known, metmyoglobin shows a typical reducing potential ($\sim +0.1$ volt) when assayed as the system,

$$Fe^{++} \rightarrow Fe^{3+} + \varepsilon^-$$

Similar conclusions can apply to all other hematin compounds. Thus, although the hematin compounds present in photosynthetic tissues have in no case sufficiently positive potentials, when measured as $Fe^{++} \rightarrow Fe^{3+} + \varepsilon^-$, to function in oxygen liberation or in peroxidation (Table I), they may well do so if transformed by photochemical activation to the ferri-ferryl system. This possibility is alternative to the suggestion of Hill that chlorophyll functions as an oxidase, in that it postulates a hematin compound as the ultimate photo-oxidase created by photochemical coupling with excited chlorophyll.

As remarked previously, the magnitude of the conjugated system of double bonds in ferri-porphyrins leads to the expectation that it should be oxidizable readily enough to enter into spontaneous charge transfer reactions with an activated porphyrin system such as is created by the excitation of chlorophyll. The resultant chemical species can be a semiquinone produced by removal of an electron either from carbon or nitrogen atoms of the tetrapyrrolic structure (King and Winfield, 1959). The reduced form of chlorophyll obtained as the other product could be a type of semichlorinogen, that is, chlorophyll with a partially reduced bridge carbon.

The further course of reaction in a system containing such a juxtaposition of a powerful reducing and oxidizing system could be dictated by competition between direct back-oxidation and reaction with other components present. Enzymic intervention could favor utilization of either the semichlorinogen in a reduction reaction or the semiquinoidal hematin oxidant in an oxidation reaction. One such enzyme, a TPN photoreductase, is known to exist from the work of San Pietro and Lang (1958). It is conceivable that the presence of this enzyme favors transfer of electrons from the chlorophyll semichlorinogen to TPN.

An interesting possible consequence of the reaction scheme pictured is the production of a ferryl hematin complex which can function either as a catalase or peroxidase. Which of the two functions is expressed might be dictated by the protein and by the organization of the heme groups. On this basis, the difference between photosynthetic green plant tissues (chloroplasts) and photosynthetic bacterial systems (chromatophores) would be attributable mainly to the nature of the protein attached to the functional heme groups participating in the photochemical charge transfer reactions. In the plant oxygen-evolving systems, the protein would confer catalase activity on the complex formed by photo-oxidation of the heme, while in the bacterial systems, the heme protein would express peroxidase activity after photo-oxidation. *A single reaction —photo-oxidation of heme groups by charge transfer to excited chlorophyll—would provide a unitary basis for all photosynthetic systems.*

There is no doubt that oxidation of hematin compounds in chloroplasts and chromatophores takes place as a dominating feature of photochemical activation of chlorophyll. As a single example from the many available, consider the findings of Olsen and Chance (1958, 1959), who observed the rapid changes in optical density of *Chromatium* suspensions when these were illuminated with photoactive infrared light under a variety of conditions (e.g., with and without air, oxidation inhibitors, etc.). When the cell suspensions were in transition from dark to light, a rapid loss of absorption at 422 mµ relative to the isobestic reference point (475 mµ) occurred, which was at least diphasic. On reverting to darkness, the suspensions recovered optical density in a triphasic reaction. The very rapid light effect (first phase) had a measured quantum yield of no less than one electron shifted per absorption of 2 quanta. This yield was of the same order as the over-all yield for the photosynthetic assimilation of CO_2 by *Chromatium*, showing that this reaction was not a secondary process. The absorption spectrum of the moiety responsible for the light-induced change of optical density at 422 mµ was that of a *c*-type cytochrome, similar to that isolated from *Chromatium* in our laboratory (Newton and Kamen, 1956).

Data on shifts in oxidation state of heme proteins induced by illumination obtained earlier in *R. rubrum* (Duysens, 1954; Chance and Smith, 1955) have been extended more recently by Smith and Baltscheffsky (1959) and indicate that these photochemical shifts are intimately associated with the photophosphorylation system of chromatophores. Of most interest is the recent finding of Nishimura and Chance that the rate of the primary heme protein photo-oxidation in *Chromatium* is independent of temperature down to —170° C, whereas the dark reduction is halted completely when the temperature is lowered to —20° C (Nishimura and Chance, 1959; private communication).

Duysens and Amesz (1957) and Olsen (1958) have found changes in fluorescence emission in photosynthetic systems which are initiated by light absorption and which can be interpreted as reduction of bound pyridine nucleotides. These phenomena parallel photo-oxidation of hematin compounds, but may occur at a somewhat lower rate, according to the most recent observations in Chance's laboratory (Chance and Olson, 1959). *In vitro* observations on the separation of reducing and oxidizing systems have been reported by Vernon (1959), using *R. rubrum* chromatophore suspensions.

Obviously much more work must be done in attempting to specify the time sequence of these light-induced reactions and to identify the reagents with the actual entities isolated and purified from the photoactive system.

The temptation to speculate further must be resisted until experimental evidence can be obtained on the nature of photochemically induced charge transfer reactions, if any, between porphyrin chelates of magnesium and iron. When these are available, it may be possible to rationalize reaction mechanisms which specify the manner of energy storage and consequent activation of substrates in the biosynthetic phase of photosynthesis.

ACKNOWLEDGMENTS

I take pleasure in expressing the appreciation of my colleagues and myself for the generous support given us in connection with researches from our laboratories cited. This aid has come from the C. F. Kettering Foundation, the National Institutes of Health, and the National Science Foundation.

REFERENCES

Arnon, D. I., Whatley, F. R., and Allen, M. B. (1954). *J. Am. Chem. Soc.* **76**, 6324.
Bartsch, R. G., and Kamen, M. D. (1958). *J. Biol. Chem.* **230**, 41.
Bartsch, R. G., and Kamen, M. D. (1960). *J. Biol. Chem.* **235**, 825.
Chance, B., and Olson, J. M. (1960). *Arch. Biochem. Biophys.* (in press).
Chance, B., and Smith, L. (1955). *Nature* **175**, 803.
Davenport, H. E., and Hill, R. (1952). *Proc. Roy. Soc.* **B139**, 327.
Duysens, L. N. M. (1954). *Nature* **173**, 692.
Duysens, L. N. M., and Amesz, J. (1957). *Biochim. et Biophys. Acta* **24**, 19.
Egami, F., Itahashi, M., Sato, R., and Mori, T. (1953). *J. Biochem. (Tokyo)* **40**, 527.
Elsden, S. R., Kamen, M. D., and Vernon, L. P. (1953). *J. Am. Chem. Soc.* **75**, 6347.
Falk, J. E., and Nyholm, R. S. (1957). "Current Trends in Heterocyclic Chemistry" (A. Albert, eds.), p. 130. Butterworths, London.
Frei, W., Reidmüller, L., and Almasy, F. (1934). *Biochem. Z.* **274**, 253.
Frenkel, A. (1954). *J. Am. Chem. Soc.* **76**, 5568.
Fujita, A., and Kodama, T. (1934). *Biochem. Z.* **273**, 186.
George, P. (1959). Private communication.
Gibson, J., and Larsen, H. (1955). *Biochem. J.* **49**, xxviii.
Hill, R. (1939). *Proc. Roy. Soc.* **B127**, 192.
Hill, R. (1951). *Symposia Soc. Exptl. Biol.* **5**, 222.
Hill, R. (1954). *Nature* **174**, 501.
Hill, R. (1958). *Proc. Intern. Congr. Biochem. 3rd Congr. Brussels 1955* p. 225.
Hill, R., and Scarisbrick, R. (1951). *New Phytologist* **50**, 98.
Hulcher, F. M., and Vishniac, W. (1958). *Brookhaven Symposia in Biol. No.* **11**, 348.
Ishimoto, M., and Koyama, J. (1955). *Bull. Chem. Soc. Japan* **28**, 231.
Kamen, M. D. (1955). *Bacteriol. Revs.* **19**, 250.
Kamen, M. D. (1956). *In* "Enzymes: Units of Biological Structure and Function" (O. H. Gaebler, ed.), p. 483. Academic Press, New York.
Kamen, M. D., and Vernon, L. P. (1954). *J. Bacteriol.* **67**, 617.
Kamen, M. D., and Vernon, L. P. (1955). *Biochim. et Biophys. Acta* **17**, 10.

Katoh, S. (1959). *J. Biochem.* (*Tokyo*) **46**, 629.
Keilin, D. (1925). *Proc. Roy. Soc.* **B98**, 312.
King, N. K., and Winfield, M. E. (1959). *Australian J. Chem.* **12**, 47.
Michaelis, L., and Friedheim, E. A. H. (1931). *J. Biol. Chem.* **91**, 343.
Newton, J. W. (1958). *Brookhaven Symposium in Biol. No.* **11**, 289.
Newton, J. W., and Kamen, M. D. (1956). *Biochim. et Biophys. Acta* **21**, 71.
Newton, J. W., and Kamen, M. D. (1957). *Biochim. et Biophys. Acta* **25**, 462.
Newton, J. W., and Levine, L. (1959). *Arch. Biochem. Biophys.* **83**, 456.
Nishimura, M. (1959). *J. Biochem.* (*Tokyo*) **46**, 219.
Nishimura, M., and Chance, B. (1959). *Symposium on Hematin Enzymes, Canberra, Australia, 1959.* (Pergamon, New York, in press).
Nishimura, M., and Chance, B. (1960). Private communication.
Olson, J. M. (1958). *Brookhaven Symposia in Biol. No.* **11**, 316.
Olsen, J. M., and Chance B. (1958). *Biochim. et Biophys. Acta* **28**, 227; (1959). *Arch. Biochem. Biophys.* (in press).
Orlando, J., and Bartsch, R. G. (1960). Private communication.
Paleus, S., and Tuppy, H. (1959). *Symposium on Hematin Enzymes, Canberra, Australia 1959.* (Pergamon, New York, in press).
Postgate, J. R. (1955). *Biochim. et Biophys. Acta* **18**, 427.
San Pietro, A., and Lang, H. M. (1958). *J. Biol. Chem.* **231**, 211.
Scarisbrick, R., and Hill, R. (1943). *Biochem. J.* **37**, xxii.
Smith, L. (1954). *Bacteriol. Revs.* **18**, 106.
Smith, L., and Boltscheffsky, M. (1959). *J. Biol. Chem.* **234**, 1575.
Smith, L., and Chance, B. (1958). *Ann. Rev. Plant Physiol.* **9**, 449.
Tamiya, H., and Yamagutchi, S. (1933). *Acta Phytochim.* (*Japan*) **7**, 233.
Tuppy, H., and Paleus, S. (1955). *Acta Chem. Scand.* **9**, 353.
Vernon, L. P. (1954). *Arch. Biochem. Biophys.* **51**, 122.
Vernon, L. P. (1959). *J. Biol. Chem.* **234**, 1883.
Vernon, L. P., and Kamen, M. D. (1954). *J. Biol. Chem.* **211**, 643.
Yaoi, H., and Tamiya, H. (1928). *Proc. Imp. Acad.* (*Tokyo*) **4**, 436.

Discussion

FRENCH: How many quanta can you absorb as electrons in cytochrome at once?

KAMEN: Two, assuming that the valence change would be from ferrous to ferryl.

ARNON: I am a little bit puzzled by your statement about two quanta per electron shift. I can not quite visualize two quanta for one electron.

KAMEN: Actually this is a minimum estimate and it could very well be one quantum. Assumptions were made to minimize the yield as much as possible, and it may be just one quantum per electron. If we can get a system with one cytochrome and one chlorophyll and nothing else, we will see one quantum per electron. But in the plants, you saw the reaction in the light was biphasic and I think that the efficiency that is being measured is not only that of the initial reaction, but of the following reactions too.

SMITH: Would the protochlorophyll complex have to have the order of one cytochrome associated with one chlorophyll?

KAMEN: At least one. It has to be close to it, otherwise this mechanism would not work.

HENDRICKS: So you would have to be picking up a Soret band of the order of twice the magnitude of that of chlorophyll alone. You ought to see it pretty directly.

SMITH: May I have a comment on this? I have looked for the cytochrome band at about 560 mμ and we have no indication in the purified product of a band there. In the original leaf you do have a band at 560 mμ and we have attributed that to cytochrome in the original leaf material but there is no indication of it in the holochrome; and as Dr. Hendricks says, the Soret band ought to show up. Actually the ratio of heights is about right for the red and the blue peaks in the protochlorophyll so I do not believe there is very much cytochrome there—not molecule for molecule.

KAMEN: Well, that surprises me because your preparation is the one you use for making cytochrome.

HENDRICKS: What do you visualize as the function of the RHP pigment?

KAMEN: The function of RHP in *Chromatium* to my mind is not very clear. In the case of *rubrum*, it may act as an electron transfer agent or as an oxidase. But in *Chromatium*, as far as I can tell, now, it is probably contributing to that second phase, which is the thermal phase. I might say something else which is even more peculiar. Chance finds five heme moieties. That is, Chance and Olson get five different hemes from these two phases, two in the first phase and three in the second, the dark phase. Now in the cytochrome that we isolate there are three hemes and in the RHP there are two—that adds up to five also. Now you see we only get two proteins out of the *Chromatium*, no matter how hard we try, and they keep getting five heme species in their kinetics. So if you believe in kinetics then there must be five different hemes there, and if you think that our heme analyses are right then we have got the five hemes, but they are split up into two groups and the question then arises as to whether there is a separation of function between the RHP and cytochrome *c* or whether they are part of the same mechanism. That is, you have to have two hemes in conjunction doing something, instead of just one. That might modify this picture if you have to have two hemes, but so far the only thing that I can think of that bears on this case is the situation in the sulfate reducers where, as you will recall, there is only one cytochrome—only one cytochrome and it is cytochrome *c* with a molecular weight of 12,000 and two hemes on it. So even in the simplest case where all you are doing is putting electrons on to sulfate with no jump at all in potential, you have to move at least two. In fact, you have to move at least two to get the energy for ATP from hydrogen up to sulfate and the drop there is 0.2 volts and it is remarkable to my mind that there are two hemes there for this purpose. And there is another notion that may have something to do with this. If there is only one cytochrome *c*, there ought to be something else of hemoprotein nature. I do not know of any system in nature which does not have more than one heme in it, whether it has only one cytochrome or not.

BOGORAD: Has anyone tried to produce this ferryl complex with the *c* cytochrome?

KAMEN: No, but it can be done, and of course this model reaction I have shown you in which we have the ferricyanide going in nitric acid to the nitroso compound is a good example of a similar kind of structure. Whether ferryl iron exists as such in the porphyrin is difficult to say, but, using the normal valence table, you can write some nice oxidase reactions leading to formation of ferryl ion. If you believe that any hematin compound can be transformed into a catalase or peroxidase by the photochemical reaction involved here, as it goes up to the ferryl state it can split anything, such as water or peroxide. It can oxidize peroxide when it is up that high. Then any hematin compound that you get out and isolate is a candidate for catalase action. Now in the case of bacteria, you might say the reason that they are bacteria and not plants is that in the photochemical step, the hematin compound that is formed is tied to a protein which only confers peroxidase activity, and in the case of the plant it is tied to a protein that confers catalase activity, so that the bacteria are on the road to becoming plants if only they can learn how to make peroxidase into catalase. This is a photochemical reaction. Now it does not exist before the photochemistry sets in and it is only transient.

Recently Shibata in Japan has split catalase into smaller fragments, in which case it becomes peroxidase and then, on reconstituting it, it becomes catalase again. And this suggests that maybe bacteria have not learned how to agglomerate this stuff yet. But it is possible to write mechanisms with this kind of ferric-ferryl system which will split oxygen out of peroxide, on one heme. It takes, however, the transfer of four electrons and the movement of two hydroxyl ions on to it. And so all you have to do is provide hydroxyl ions and a source of electrons and you can make oxygen quite readily without ever going through the intermediary of water itself. Well, I would like to emphasize that at this point there are no data; I took the opportunity to speculate here because I feel that the field needs some sort of fresh viewpoint, and that we have been sitting for a long time with no particular entry into this. I think that the inorganic chemists have got a very good approach now based on what happens with abnormal valence states in chelates and if we examine this we may have some hope that perhaps very shortly there will be some enzyme reactions that show up. Anyway it is photochemistry we have to deal with. I should end by telling you that you have to think logarithmically. The photosynthetic regime extends over 18 orders of magnitude in time in seconds, starting at 10^{-15} seconds and going down to 10^3 seconds which is the time in which a cell divides. The whole business of cell division is crammed into the last 3 orders of magnitude; biochemistry begins at 10^{-1} seconds, 2 orders of magnitude further on. But all the stuff before that is what we have to know about. And the photochemistry starts maybe at 10^{-9}, maybe at 10^{-5}, but it is between 4 and 6 orders of magnitude before the biochemistry starts, and so when we measure CO_2 fixation and photosynthesis and so on, we are doing archeology. Our interest now is in that photochemical phase which is between 10^{-9} and 10^{-1} seconds.

21

The Wavelength Dependence of Photosynthesis and the Role of Accessory Pigments

FRANCIS T. HAXO

Scripps Institution of Oceanography, University of California, La Jolla, California

Introduction

Interest in the function of pigments accompanying chlorophyll in diversely-colored algae arose largely from efforts of botanists at the turn of the last century to interpret the vertical distribution of marine algae in reference to the available submarine light field. The well known experiments of Engelmann (1882a, 1883, 1884) provided evidence that light absorbed by the masking pigments of brown algae, diatoms, red algae, and blue-green algae was available for photosynthesis, in addition to chlorophyll-absorbed light energy. In his view, these pigments represented evolutionary adaptations by which optical windows in the visible absorption spectrum of chlorophyll were effectively closed. For example, the presence of phycoerythrin in deep-growing red algae was interpreted as an adaptation permitting more effective absorption and utilization of the prevailing blue and green light. This interpretation met with an opposing view. Oltmanns and others argued that the observed vertical distribution of algae did not involve chromatic adaptation but rather a response to the prevailing light intensity gradient. In the modern view, the color and the intensity of light, as well as many other environmental factors, are recognized as interacting in complex ways to control the vertical distribution of algae.

Nevertheless, the essential premise upon which Engelmann's theory of complementary color adaptation was based proved correct. Rigorous investigations by more refined techniques over the past 20 years have confirmed his conclusion that fucoxanthin of diatoms (Dutton and Manning, 1941; Wassink and Kersten, 1946; Tanada, 1951) and of brown algae (Haxo and Blinks, 1950), phycocyanin of Cyanophyta (Emerson and Lewis, 1942), and phycoerythrin of the Rhodophyta (Haxo and Blinks, 1950) absorb light effectively for photosynthesis. Other studies have revealed something about the mechanism of accessory pigment participation in photosynthesis. Investigations by Dutton *et al.* (1943)

and by Wassink and Kersten (1944) on diatoms, by French and Young (1952) on red algae, and by Duysens (1952) on red and blue-green algae have shown that quanta absorbed by the dominant accessory pigments of these algae are able to excite chlorophyll fluorescence. Thus, a consistent and widely accepted picture seemed to emerge—the so-called energy transfer hypothesis—that accessory pigments function in photosynthesis only insofar as they are able to transfer the absorbed energy to chlorophyll a.

Indications of an additional role of accessory pigments have come from the recent studies conducted by Emerson and his co-workers* on the long-wavelength limit of photosynthesis in green algae and in red algae. These led to the important discovery that light absorbed by accessory pigments (for example chlorophyll b, in the case of *Chlorella*, and phycoerythrin, in the case of *Porphyridium*) improves the efficiency with which far-red light is used in photosynthesis. Thus, it would appear now that accessory pigments do something more than extend the absorption range of chlorophyll a; rather, they may be essential for its efficient operation.

The present paper will survey what is currently known concerning the wavelength dependence of photosynthesis in plants (other than the bacteria) representing the major types of pigment combinations. For additional discussions of various aspects of this problem, reference is made to the following reviewers: Rabinowitch (1945, 1951, 1956); Blinks (1954, 1955); French and Young (1956); Duysens (1956); Emerson (1958); and Haxo (1960).

A few comments should be made concerning the experimental methods employed to determine the photosynthetic activity of a given pigment and the limitations in interpreting these results. Two approaches have been used. Emerson and his collaborators have favored determinations of quantum yield as a function of wavelength, measured manometrically on partially or completely absorbing cell suspensions. Depressions in yield are indicative of inactive absorption by a given pigment or absorption by a less efficient pigment. The advantages of this method are: (*1*) errors in cell absorption measurements are minimized, particularly when totally absorbing suspensions are used; (*2*) quantitative information is obtained on pigment activity.

As a second approach, photosynthetic action spectra have been determined on partially absorbing thalli, using the polarographic technique for dissolved oxygen. The activity curves for photosynthetic oxygen

---

* See Emerson *et al.* (1956, 1957) and Emerson and Chalmers (1958).

production when plotted against thallus absorption provide, by the degree of correspondence and divergence of the superimposed curves, information on the relative activities of the individual pigments. The original design of the polarographic reaction vessel, which we owe largely to L. R. Blinks and C. M. Lewis (cf. Haxo and Blinks, 1950) has recently been modified by the present reviewer (Haxo, unpublished) to accommodate thin cell suspensions over a recessed, horizontal electrode, making possible comparative studies on a wide variety of unicellular algae. The advantages of the polarographic technique, as used by these workers, are the rapidity with which action spectra may be determined and the small size of the plant sample required. Its disadvantages are: (1) information is provided on the relative rates of photosynthesis only; (2) rates of respiration cannot be measured conveniently.

Perhaps the most serious limitation in interpreting the photosynthetic activity of pigments follows from the difficulties in measuring the absorption spectra of individual plastid pigments in the living cells. Estimates of fractional absorption by component pigments have of necessity been reconstructed from measurements on extracts. This method has not proved entirely satisfactory because the extent to which pigment absorption *in vivo* is distorted by wavelength displacement and geometrical flattening can be estimated only approximately. Direct *in vivo* measurements of the absorption spectra of chlorophylls and other pigments (French, 1958) will doubtless prove helpful in this regard.

Plants Containing Chlorophyll a, Chlorophyll b, and Conventional Carotenoids (Chlorophyceae, Euglenophyceae, and Higher Plants)

We will first of all consider action spectra for typical grass-green colored plants. The Chlorophyta and the euglenids resemble higher plants most closely in pigment composition, containing chlorophyll *a* and chlorophyll *b*, β-carotene, and one or more xanthophylls. Spectral assimilation curves for the typical green-colored representatives of both algal groups, as determined by Engelmann, indicated that in gross features photosynthetic effectiveness follows the combined absorption curves of the chlorophylls.

This situation is seen clearly from the manometric studies of Emerson and Lewis (1943) on *Chlorella* (Fig. 1). Two additional features were noted. A decline in photosynthetic efficiency was observed beyond the chlorophyll *a* peak in the red, an effect which was originally unexplained other than to suggest that the energy of far-red quanta may be too low to activate the chlorophyll molecule for participation in subsequent dark reactions. We will return later to Emerson's recent interpretation of the long-wavelength decline in photosynthesis in *Chlorella* and other algae. A second feature is the divergence in the absorption and activity curves between 440 and 560 mμ, where carotenoids absorb. Duysens (1952) has estimated that 40–50% of the light absorbed by carotenoids is available for photosynthesis, although the exact extent to which these pigments are active remains uncertain (Emerson, 1958). The ubiquitous occurrence of carotenoids in the photosynthetic apparatus of plants suggests that they may play an important role other than that of trapping

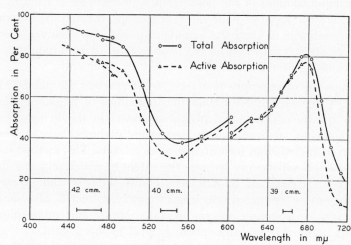

FIG. 1. Cell absorption and photosynthetic activity in *Chlorella* (Emerson and Lewis, 1943).

light for photosynthesis. This may be that of conferring protection against chlorophyll-catalyzed photo-oxidations, as suggested by the studies on the physiological effects of carotenoid deficiencies in photosynthetic bacteria (Sistrom et al., 1956; Cohen-Bazire and Stanier, 1958; Stanier, 1958) and in *Chlorella* (Claes, 1954; Kandler and Schötz, 1956).

Using polarographic techniques, photosynthetic action spectra similar to those from manometric studies were obtained by Haxo and Blinks (1950) for the blade-type marine alga *Ulva* and more recently by Haxo (unpublished) for *Chlorella*. The latter results are presented in Fig. 2 and are of some interest since the photosynthetic action spectra obtained by the two methods are in substantial agreement.

In both Chlorophyta and Euglenophyta, representatives occur in which chlorophyll is masked by an abundance of extra-plastid carotenoids, most frequently either β-carotene or astaxanthin (Goodwin and Jamikorn, 1954). Engelmann (1882b) recognized that masking pigments of the so-called hematochrome-type, as found in *Haematococcus* and *Trentepohlia* (*Chroolepus*) reduce the amount of light available for photosynthesis. This situation has been clearly substantiated by quantitative studies of Yocum and Blinks (1954) for the red phases of *Haematococcus, Trentepohlia,* and *Dunaliella*. In *Trentepohlia* the masking effect of extra-plastid carotenoids is sufficiently great to suppress photosynthesis in blue light almost completely.

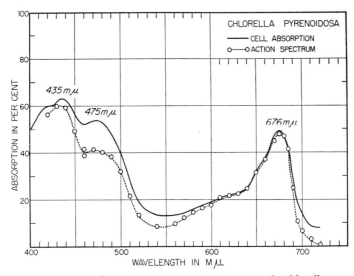

Fig. 2. Absorption and photosynthetic action spectra of *Chlorella pyrenoidosa* determined polarographically (Haxo, unpublished).

Only very limited data are available on the wavelength dependence of photosynthesis in higher plants. The action spectrum determined by Hoover (1937) for wheat plants in general resembles that of green algae. Differences may be attributed to the greater thickness and complexity of the tissue. This is further suggested by the fact that the action spectrum for the Hill reaction in isolated spinach chloroplasts is very similar to the action spectra for photosynthesis in *Chlorella* and in *Ulva* (Chen, 1952). We may refer here also to unpublished studies by M. Hommersand in the author's laboratory on the thin-leafed aquatic phanerogam *Elodea densa*, the detailed photosynthetic spectrum of which is almost an exact duplicate of the reported for *Ulva*.

Algae in Which Chlorophyll a Is Accompanied by Chlorophyll c and a Photosynthetically Efficient Carotenoid, Either Fucoxanthin or Peridinin (Phaeophyceae, Bacillariophyceae, and Dinophyceae)

In plant life in the ocean doubtless the two most abundant carotenoids are the polyoxy xanthophylls, fucoxanthin and peridinin. The former is characteristic of the diatoms and brown algae, the latter of dinoflagellates. Both pigments contribute strongly to cell absorption in the blue-green to green portion of the spectrum, and both occur within the plastid in company with chlorophylls a and c and other carotenoids.

Evidence for fucoxanthin-sensitized photosynthesis in diatoms has

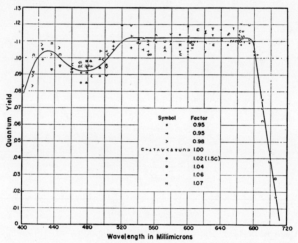

Fig. 3. Quantum yield of photosynthesis as a function of wavelength for *Navicula minima* (Tanada, 1951).

come from the investigations of Dutton and Manning (1941), Wassink and Kersten (1946), and Tanada (1951). Tanada's study of quantum yield as a function of wavelength in *Navicula minima* provides the most detailed assessment of this situation (Fig. 3). Since the quantum yield was found to be practically constant (about 0.11) from 680 down to 530 mμ, where about 85% of the absorbed light could be attributed to fucoxanthin, Tanada concluded that this carotenoid was about as effective as the chlorophylls in sensitizing photosynthesis. Other carotenoids appeared to be inactive in this respect.

Studies by a number of workers using broad band-pass filters, in particular those of Montfort (1940) and Levring (1947), have all pointed to a high photosynthetic activity for fucoxanthin in brown algae as well. The more detailed study by Haxo and Blinks (1950) in the thin-bladed marine alga *Coilodesme* suggests this also, since the curves for photosynthetic activity and thallus absorption deviate only slightly in the region of strong fucoxanthin absorption (Fig. 4).

The dinoflagellates are of particular interest since they, together with the diatoms, constitute wide-spread and dominant components of marine phytoplankton populations. The recent study by Haxo (1959) provides evidence for believing that the function of peridinin in the photosynthesis of dinoflagellates is analogous to that of fucoxanthin in

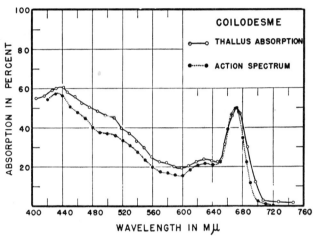

FIG. 4. Absorption and photosynthetic action spectra of the brown alga *Coilodesme californica* (Haxo and Blinks, 1950).

diatoms and brown algae. In *Gonyaulax polyedra* (Fig. 5), cell absorption and photosynthetic activity curves were found to be reasonably parallel in the region 520–560 mµ, where about 80% of the absorbed light could be attributed to peridinin. Other carotenoids appeared less active, as shown by the divergence between ca. 500 and 460 mµ. The marked decline in photosynthetic activity below 435 mµ was ascribed to inactive absorption by a pale yellow pigment which contributes most strongly to cell absorption in the near-ultraviolet.

Algae in Which Phycobilins Accompany Chlorophyll a Alone or Possibly with Chlorophyll d (Principally Rhodophyta and Cyanophyta)

The red and blue-green algae have long attracted interest because of their abundant content of the red and blue chromoproteins, the phycoerythrins and the phycocyanins. In these algae, the presence of three phycoerythrins (R-, C-, and B-) and three phycocyanins (R-, C-, and allo-) has been established (cf. Haxo and ÓhEocha, 1960). All of these pigments with the exception of allophycocyanin, which has not been adequately studied from this viewpoint, have been shown to absorb light actively for photosynthesis.

The photosynthetic action spectrum reported by Haxo and Blinks (1950) for *Porphyra nereocystis* (Fig. 6) is typical of the results obtained by these workers for marine red algae containing predominantly R-phycoerythrin. The marked parallelism observable in the middle of the spectrum between photosynthetic activity and phycoerythrin ab-

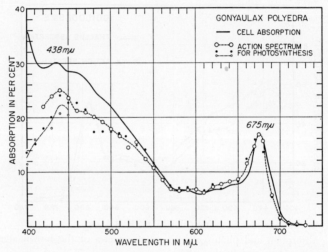

FIG. 5. Absorption and photosynthetic action spectra for the dinoflagellate *Gonyaulax polyedra* (Haxo, unpublished).

sorption provides clear evidence of an accessory function of this pigment in confirmation of the conclusion reached by most earlier investigators of this problem.

The marked depression in activity at both ends of the spectrum where absorption by chlorophyll *a* is high indicated a much lower effectiveness for light absorbed directly by chlorophyll. This situation was observed in all of the red algae studied by these workers and, with the additional studies of Yocum and Blinks (1954) and of Duysens (1952), led to the concept that a portion of the chlorophyll *a* in red algae exists in a physiologically inactive condition. This will be discussed subsequently in the light of Emerson's findings on the long-wavelength dependence of photosynthesis.

In general, these studies showed a low effectiveness for the carotenoids of most red algae, for example, *Porphyra* spp. and *Porphyridium cruentum*. However, a more detailed study of the deep-growing red algae may reveal a more active role for these pigments. This is suggested by the action spectra of *Myriogramme* and *Schizymenia* (Haxo and Blinks, 1950), which show a persistent parallelism between absorption and photosynthetic activity in the region 480–500 mμ, where carotenoids may be expected to contribute significantly to thallus absorption.

The photosynthetic effectiveness of B-phycoerythrin which was later isolated in the pure form from *Porphyra naiadum* by Airth and Blinks (1956) was also demonstrated by Haxo and Blinks (1950). B-Phyco-

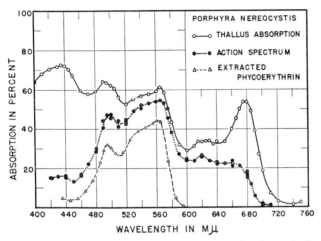

Fig. 6. Absorption and photosynthetic action spectra for the red alga *Porphyra nereocystis* (Haxo and Blinks, 1950).

erythrin is very likely identical to the phycoerythrin present in the unicellular red alga *Porphyridium cruentum*, which has also been shown to be effective in photosynthesis (Duysens, 1952; Haxo and Norris, 1953). Duysens' studies of photosynthesis in *Porphyridium* are of particular interest since they were combined with quantitative estimates of absorption by component pigments (Fig. 7). Also included are points for the effectiveness spectrum for chlorophyll *a* excitation as determined by French and Young (1952). The latter agrees well with that of photosynthesis and of phycoerythrin absorption and indicates some of the evidence for the hypothesis that energy is transferred from accessory pigments to chlorophyll *a*.

A high photosynthetic activity for C-phycoerythrin (*in vivo* maximum at 565–570 mμ of blue-green algae was shown by an action spectrum analysis of the bright red filaments of *Phormidium ectocarpi* [(Fig.

8); Haxo and Norris, 1953] and this situation has been confirmed by these workers in unpublished studies of two other species of *Phormidium* (*P. fragile* and *P. persicinum*).

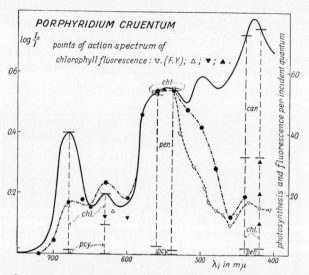

Fig. 7. Absorption spectrum (●—●), action spectrum for photosynthesis (●–·–●), and action spectrum for chlorophyll fluorescence (△ - - - - △) of the red alga *Porphyridium cruentum* (Duysens, 1952). The detailed fluorescence action spectrum is from French and Young (1952).

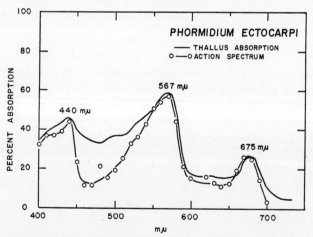

Fig. 8. Absorption and photosynthetic action spectra for the blue-green alga *Phormidium ectocarpi* (Haxo and Norris, 1953). The pronounced absorption peak at 567 mμ is attributable to C-phycoerythrin.

Rigorous assessment of the photosynthetic effectiveness of C-phycocyanin was first made by Emerson and Lewis (1942) in the case of the unicellular blue-green *Chroococcus*. The quantum yield of photosynthesis was found to be consistently high (about 0.08) throughout the region of strong phycocyanin absorption and to be equal to that of light absorbed directly by chlorophyll. Carotenoids, on the other hand, were found to be virtually inactive. The data of these workers for action spectra of partially absorbing suspensions are shown in Fig. 9.

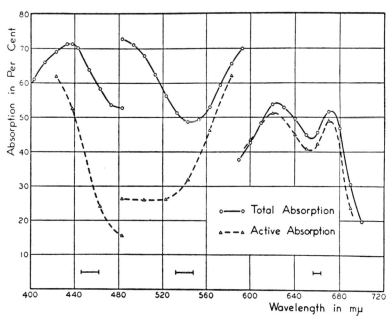

Fig. 9. Cell absorption and action spectra for photosynthesis in *Chroococcus*. Cell suspensions of different densities were used in the red, green, and blue regions of the spectrum (Emerson and Lewis, 1942).

Other phycocyanin-rich Cyanophyta, e.g., *Oscillatoria, Lyngbya, Anabaena*, and *Synechococcus*, resemble *Chroococcus* in containing a highly active phycocyanin but differ in showing appreciable (about one-half) inactive absorption by chlorophyll *a* (Haxo and Blinks, 1950; Duysens, 1952). A similar situation has been observed for C-phycocyanin in red algae, as shown in Fig. 10 for *Porphyridium aerugineum* (Haxo, unpublished), which lacks phycoerythrin.

Haxo and Allen (unpublished) have recently determined the photosynthetic action spectrum of the so-called "acid *Chlorella*," *Cyanidium caldarium*, which, although showing morphological affinities to the green algae, resembles the blue-green algae in containing C-phycocyanin and chlorophyll *a* alone (Allen, 1959). Their data are shown in Fig. 11. The results are entirely comparable to those reported for most red and blue-green algae in that light absorbed directly by chlorophyll is inefficiently utilized.

No algae are known to contain R-phycocyanin as an exclusive phycobilin. However, this pigment occurs in abundance in the "green" thalli of the red alga *Porphyra perforata* and the corresponding broadening of

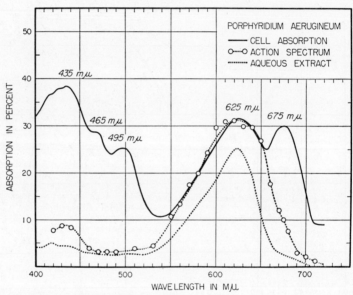

Fig. 10. Cell absorption and photosynthetic action spectra of the red alga *Porphyridium aerugineum*, which contains C-phycocyanin as its principal accessory pigment (Haxo, unpublished).

the photosynthetic activity curve toward the orange-red part of the spectrum confirms its photosynthetic effectiveness (Haxo and Blinks, 1950).

Algae in Which Chlorophyll *a* Is Accompanied by Chlorophyll *c* and by Phycobilins (Cryptophyceae)

The diversity in plastid pigmentation among photosynthetic members of the Cryptophyceae has long been noted. Indeed, this may be quite as great as that encountered in the Rhodophyta and the Cyanophyta. That this group of algal flagellates might contain water soluble pigments of the phycobilin type was only alluded to in the earlier literature (Geitler, 1924). Very recently investigations by Allen *et al.* (1959), by

Ó hEocha and Rafferty (1959), and by Haxo and Fork (1959) have all indicated that pigments of the phycoerythrin and phycocyanin type occur in the plastids of cryptomonads. Of additional interest is the fact that chlorophyll c appears to be a consistent component of the plastid pigments, in company with chlorophyll a and phycobilins. The presence of this accessory chlorophyll was suggested by the *in vivo* derivative spectra of *Cryptomonas ovata* as determined by French and Elliott

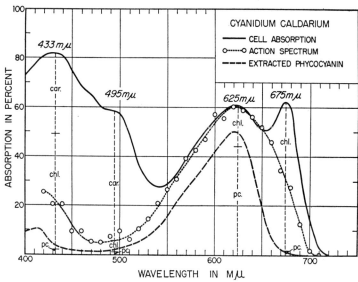

Fig. 11. Action spectrum for photosynthesis of *Cyanidium caldarium* compared with absorption spectra of intact cells and extracted C-phycocyanin. The relative contributions to cell absorption by chlorophyll a (chl.), phycocyanin (pc.) and by carotenoids (car.) are indicated by bars along the broken vertical lines (Haxo and Allen, unpublished).

(1958) and was established by isolation of the pure pigments from *Cryptomonas ovata, Rhodomonas lens,* and *Hemiselmis virescens* (Haxo and Fork, 1959). Thus, by contrast to the red and blue-green algae, the accessory pigment systems of cryptomonads are unusual in their content of both phycobilins and a second chlorophyll.

The photosynthetic action spectra of *Rhodomonas* sp. (Haxo and Fork, 1959), a red-colored marine flagellate, and *Hemiselmis* (Haxo and Belser, unpublished), a blue flagellate, are therefore of some interest. Figures 12 and 13 both reveal that the phycobilins of these cryptomonads absorb light actively in photosynthesis and that light absorbed directly by chlorophyll *a* is by and large more active than in the red and blue-green algae, a situation which may be due to the presence of chlorophyll c.

The Emerson Effect (Photosynthetic Enhancement)

Having thus surveyed the wavelength dependence of photosynthesis in plants representing the major pigment combinations, we may return to an unusual feature revealed by these studies. This is the paradoxical situation observed in red algae and in most blue-green algae that light

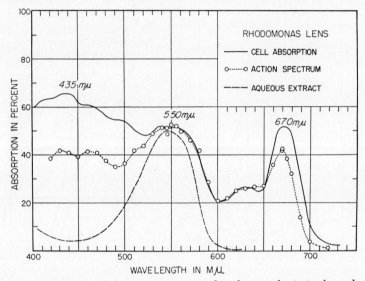

FIG. 12. Comparison of the action spectrum for photosynthesis in the red cryptomonad *Rhodomonas lens* with the absorption spectra of intact cells and extracted phycoerythrin. Chlorophyll a, chlorophyll c, and carotenoids also contribute to cell absorption (Haxo and Fork, 1959).

absorbed directly by chlorophyll a at the ends of the spectrum is utilized in photosynthesis less efficiently than light absorbed by accessory phycobilins (Haxo and Blinks, 1950; Duysens, 1952; Yocum and Blinks, 1954). With the demonstration by French and Young (1952) and by Duysens (1952) that an efficient energy transfer takes place between phycoerythrin and chlorophyll a, this situation seemed to receive adequate explanation from the postulate (Duysens, 1952; Yocum and Blinks, 1954) that two kinds of chlorophyll occur in red algae, one photosynthetically active and fluorescent, the other photosynthetically inactive and nonfluorescent, and, according to this postulate, phycoerythrin-absorbed light would be selectively transferred to active chlorophyll a.

The recent studies of Emerson and his co-workers (1956-1958) on the effects of temperature and supplementary light on the long-wavelength limit of photosynthesis in variously colored algae reveal a new role for accessory pigments and have an important bearing on the problem of "inactive chlorophyll." In *Chlorella pyrenoidosa* and in *Porphyridium cruentum*, the far-red limit of photosynthesis could be extended to longer wavelengths by lowering the temperature from 20°C to 5°C. The data of Emerson *et al.* (1957) for *Chlorella* are shown in

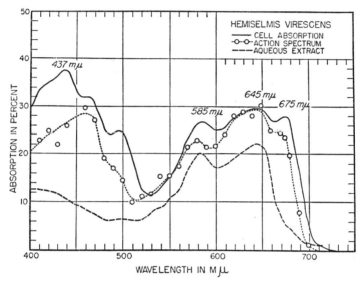

Fig. 13. Absorption and action spectra of the blue cryptomonad *Hemiselmis virescens*. The pigment of the aqueous extract is an aberrant phycocyanin (Haxo and Belser, unpublished).

Fig. 14. Even more striking results were obtained in *Porphyridium*, the range of full photosynthetic efficiency being extended from 650 to 670 mμ, which fell well within the red absorption band of chlorophyll *a*. This finding would seem to argue against the supposition that chlorophyll of red algae (and blue-green algae) has a uniquely lower efficiency than that of green algae, and this is supported by the recent quantum yield studies of Brody (1958) on *Porphyridium* in a region of the spectrum (644 mμ) where phycobilins and chlorophyll contribute about equally to cell absorption.

Supplementary light served to extend the high quantum yields in both red and green algae to even longer wavelengths than was possible with lowered temperatures. For *Porphyridium*, full photosynthetic effectiveness was extended to about 700 mμ (Brody, 1958) and to about the same wavelength for *Chlorella* (Fig. 14). Full efficiency could thus be obtained throughout almost the entire red absorption band of chlorophyll *a* in both groups of algae. Studies on the effectiveness of various wavelengths in improving the yield of photosynthesis in a band of far-red light (690 mµ and longer) in these and other algae suggested that enhancement follows the absorption spectrum of the characteristic accessory pigment(s) present (Emerson *et al.*, 1957; Emerson and Chal-

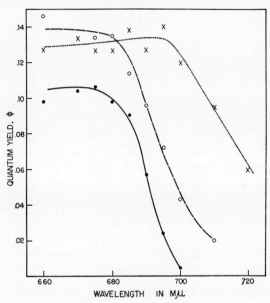

Fig. 14. Effect of temperature and supplementary light on quantum yield in *Chlorella*. The solid dots and continuous line are for 20°C; the open circles and dashed line for 5°C. The dotted curves and crosses are for measurements at 20°C with supplementary light (Emerson *et al.*, 1957).

mers, 1958; Emerson, 1958a). In *Chlorella*, peaks of effectiveness were found at 480 and 655 mµ, giving a curve resembling chlorophyll *b* absorption; whereas in *Porphyridium* the action spectrum showed a single peak at the 546 mµ mercury line, in accordance with absorption by phycoerythrin (Fig. 15). For the diatom *Navicula*, peak effectiveness was at about 540 mµ and a second smaller maximum occurred at about 645 mµ, suggesting activity for both fucoxanthin and chlorophyll *c*. For the blue-green alga *Anacystis*, maximum effectiveness was at about 600 mµ, implicating C-phycocyanin.

That accessory pigments function in the enhancement of photosynthesis in far-red light has been confirmed in unpublished studies by

Myers and French for *Chlorella* and by Blinks for red algae. Similar action spectrum studies by David Fork in the author's laboratory are also consistent with this view and are of further interest because they include some of the marine algae studied by Haxo and Blinks. In *Ulva*, as in *Chlorella*, the action spectrum for enhancement resembles the absorption spectrum of chlorophyll *b*. The only brown alga studied, *Endarachne Binghamiae*, showed clear evidence for effectiveness of both fucoxanthin and chlorophyll *c*. In the red alga *Porphyra perforata* (Fig.

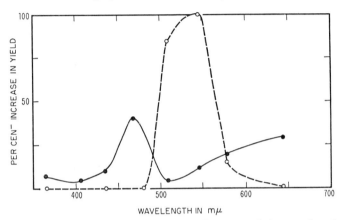

Fig. 15. Action spectra for the effect of supplementary light in enhancing photosynthesis in far-red light (690 mμ). KEY: — ● —, *Chlorella;* — o —, *Porphyridium cruentum*. Plotted from the data of Emerson (1958a).

16, upper curve) photosynthetic enhancement was most effective in the middle of the spectrum where phycoerythrin contributes strongly to thallus absorption. Similar results were obtained with several other red algae examined.

In order to explain the effect of supplementary light, Emerson suggested that maintenance of maximum photosynthetic efficiency in the far-red requires excitation of a pigment having an energy level higher than the first excited state of chlorophyll *a*. Thus, efficient photosynthesis would require simultaneous excitation of more than one photosynthetically active pigment. This function would be fulfilled by the accessory pigments—i.e., chlorophyll *b* in green algae, chlorophyll *c* in diatoms and brown algae, and phycobilins in red and blue-green algae. In this connection, Franck (1958) has theorized that a cooperation between the first excited singlet state of chlorophyll *a* and its metastable triplet state is a requirement for efficient photosynthesis. Rabinowitch (1959) has recently suggested that interpretation of the enhancement effect could be sought either in different photochemical functions of the several cell pigments, or in the existence of two or more forms of excited chlorophyll *a*. One of these would result from direct absorption in the far-red part of the spectrum, while the other could be obtained either by direct absorption of higher frequency quanta by chlorophyll *a* or by resonance transfer of the energy of the quanta from the excited accessory pigments.

Although the studies of Emerson and co-workers stressed chlorophyll *a* inefficiency as a long-wavelength phenomenon, the possibility was also considered that all the light absorbed by chlorophyll *a*—at long wavelengths and also at short wavelengths—must be supplemented by light absorbed by some accessory pigment to sustain maximum yields. In support of this, preliminary studies by Emerson and Chalmers (1958) with *Tribonema*, which lacked accessory pigments, failed to show enhancement in the far-red and showed a low quantum yield throughout the spectrum. Nevertheless, from the experimental evidence at hand it was unclear whether the low efficiency of chlorophyll *a* is a property of its absorption of low energy quanta in the far-red (a long-wavelength decline) or whether it is an intrinsic feature of chlorophyll *a*–sensitized photosynthesis and could thus be expected in any region of the spectrum in which chlorophyll *a* excitation takes place without appreciable excitation of accessory pigments.

In a direct test of the latter possibility, David Fork has recently demonstrated photosynthetic enhancement in the blue end of the spectrum in several species of marine red algae.* These offer the advantage over green algae of having low absorption by accessory pigments in the blue, as well as the red absorption band of chlorophyll *a*. Irradiation of *Porphyra perforata*, *Cryptopleura crispa*, and *Cryptonemia* sp. with the 436 and 546 mμ mercury lines, separately and paired, revealed an enhancement of photosynthesis at 436 mμ quite comparable to that observed when the 546 mμ line was paired with far-red light of wavelengths greater than 690 mμ. The action spectra for photosynthetic enhancement in the blue follow the absorption spectrum of phycoerythrin, in particular, and are quite similar to those obtained for enhancement in the far red. The results for *Porphyra perforata* are shown in Fig. 16.

Since it is possible in red algae to improve the effectiveness of utilization of quanta absorbed by chlorophyll *a* at both ends of the spectrum, it is unlikely that the Emerson Effect is dependent upon provision of higher energy quanta *per se*. The situation would seem to be restricted to the alternative proposed by Emerson that efficient photosynthesis is

---

* *Note added in proof:* This effect has been confirmed by Blinks [(1959). Personal communication; (1960). *Proc. Natl. Acad. Sci. U.S.* **46**, 327-333.]

possible only by the simultaneous excitation of chlorophyll *a* and an accessory pigment. The apparent restriction of inefficient photosynthesis in green algae to long wavelengths would, accordingly, be a reflection of the fact that this is the one region of the spectrum where the activity of chlorophyll *a* can be assessed in the absence of chlorophyll *b* excitation.

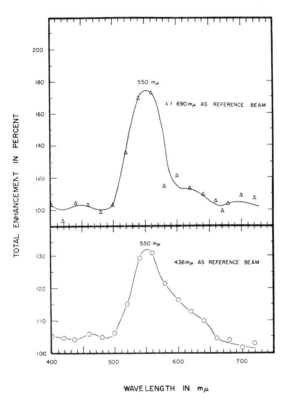

Fig. 16. Action spectra for photosynthetic enhancement by supplementary light in the red alga *Porphyra perforata*. In the upper curve, far-red light (wavelengths greater than 690 mμ) served as the reference beam. In the lower curve, the reference beam was the 436 mμ mercury line (Fork, unpublished).

This would also account for the low effectiveness in photosynthesis of chlorophyll-absorbed light at *both ends* of the spectrum in red and blue-green algae. It remains for future experimentation and consideration of other lines of evidence to reveal whether the low efficiency of chlorophyll *a*–absorbed light can best be explained in terms of a homogeneous chlorophyll *a* of intrinsically low effectiveness, or whether the concept of active and inactive chlorophyll *a* moieties should be retained.

References

Airth, R. L., and Blinks, L. R. (1956). *Biol. Bull.* **111**, 321-327.
Allen, M. B. (1959). *Arch. Mikrobiol.* **32**, 270-277.
Allen, M. B., Dougherty, E. C., and McLaughlin, J. J. A. (1959). *Nature* **184**, 1047-1049.
Blinks, L. R. (1954). *In* "Autotrophic Microorganisms" (B. A. Fry and J. L. Peel, eds.), p. 224. Cambridge Univ. Press, London and New York.
Blinks, L. R. (1955). *Ann. Rev. Plant Physiol.* **5**, 93-114.
Brody, M. (1958). Doctoral Thesis, University of Illinois, Urbana, Illinois.
Chen, S. L. (1952). *Plant Physiol.* **27**, 35-48.
Claes, H. (1954). *Z. Naturforsch.* **9b**, 461-469.
Cohen-Bazire, G., and Stanier, R. Y. (1958). *Nature* **181**, 250-254.
Dutton, H. J., and Manning, W. M. (1941). *Am. J. Botany* **28**, 516-526.
Dutton, H. J., Manning, W. M., and Duggar, B. M. (1943). *J. Phys. Chem.* **47**, 308-313.
Duysens, L. N. M. (1952). Doctoral Thesis. University of Utrecht, Holland.
Duysens, L. N. M. (1956). *Ann. Rev. Plant Physiol.* **7**, 25-50.
Duysens, L. N. M. (1959). *In* "The Photochemical Apparatus: Its Structure and Function." *Brookhaven Symposia in Biol.* **11**, 10-25.
Emerson, R. (1958). *Ann. Rev. Plant Physiol.* **9**, 1-24.
Emerson, R. (1958a). *Science* **127**, 1059-1060.
Emerson, R., and Chalmers, R. (1958). *Phycol. Soc. Am. News Bull.* **11**, 51-56.
Emerson, R., and Lewis, C. M. (1942). *J. Gen. Physiol.* **25**, 579-595.
Emerson, R., and Lewis, C. M. (1943). *Am. J. Botany* **30**, 165-178.
Emerson, R., Chalmers, R., Cederstrand, C., and Brody, M. (1956). *Science* **123**, 673.
Emerson, R., Chalmers, R., and Cederstrand, C. (1957). *Proc. Natl. Acad. Sci. U.S.* **43**, 133-143.
Engelmann, T. W. (1882a). *Botan. Z.* **40**, 419-422.
Engelmann, T. W. (1882b). *Botan. Z.* **40**, 663-669.
Engelmann, T. W. (1883). *Botan. Z.* **41**, 1-29.
Engelmann, T. W. (1884). *Botan. Z.* **42**, 81-105.
Franck, J. (1958). *Proc. Natl. Acad. Sci. U.S.* **44**, 941-948.
French, C. S. (1958). *In* "Photobiology," Proc. 19th Ann. Biol. Colloquium, p. 52-64. Oregon State College, Corvallis, Oregon.
French, C. S., and Elliott, R. F. (1958). *Carnegie Inst. Wash. Year Book* **57**, 278-286.
French, C. S., and Young, V. K. (1952). *J. Gen. Physiol.* **35**, 873-890.
French, C. S., and Young, V. K. (1956). *In* "Radiation Biology" (A. Hollaender, ed.), Vol. 3, pp. 343-391. McGraw-Hill, New York.
Geitler, L. (1924). *Rev. Algol.* **1**, 357-375.
Goodwin, T. W., and Jamikorn, M. (1954). *Biochem. J.* **57**, 376-381.
Haxo, F. T. (1959). *Proc. Intern. Botan. Congr., 9th Congr., Montreal, 1959* pp. 154-155.
Haxo, F. T. (1960). *In* "Handbuch der Pflanzenhpysiologie" (W. Ruhland, ed.), Vol. 5, pp. 349-363. Springer, Berlin.
Haxo, F. T., and Blinks, L. R. (1950). *J. Gen. Physiol.* **33**, 389-422.
Haxo, F. T., and Fork, D. C. (1959). *Nature* **184**, 1051-1052.
Haxo, F. T., and Norris, P. S. (1953). *Biol. Bull.* **105**, 374.

Haxo, F. T., and Ó hEocha, C. (1960). *In* "Handbuch der Pflanzenphysiologie" (W. Ruhland, ed.), Vol. 5, pp. 479-510. Springer, Berlin.
Hoover, W. H. (1937). *Smithsonian Inst. Publ. Misc. Collections* **95**, No. 21.
Kandler, O., and Schötz, F. (1956). *Z. Naturforsch.* **11b**, 708-718.
Levring, T. (1947). *Göteborgs Kgl. Vetenskaps-Vitterhets-Samhäll. Handl.* **5**, 1-89.
Montfort, C. (1940). *Z. physiol. Chem., Hoppe-Seyler's* **186**, 57-93.
Ó hEocha, C., and Rafferty, M. (1959). *Nature* **184**, 1049-1051.
Rabinowitch, E. I. (1945). "Photosynthesis and Related Processes," Vol. I. Interscience, New York.
Rabinowitch, E. I. (1951). "Photosynthesis and Related Processes," Vol. II, Part 1. Interscience, New York.
Rabinowitch, E. I. (1956). "Photosynthesis and Related Processes," Vol. II, Part 2. Interscience, New York.
Rabinowitch, E. I. (1959). *Plant Physiol.* **34**, 179-184.
Sistrom, W. R., Griffith, M., and Stanier, R. Y. (1956). *J. Cellular Comp. Physiol.* **48**, 473-515.
Stanier, R. Y. (1958). *In* "The Photochemical Apparatus. Its Structure and Function." *Brookhaven Symp. Biol.* **11**, 43-53.
Tanada, T. (1951). *Am. J. Botany* **38**, 270-283.
Wassink, E. C., and Kersten, J. A. H. (1944). *Enzymologia* **11**, 282-312.
Wassink, E. C., and Kersten, J. A. H. (1946). *Enzymologia* **12**, 3-32.
Yocum, C., and Blinks, L. R. (1954). *J. Gen. Physiol.* **38**, 1-16.

Discussion

Fox: If we entertain for a moment the assumption that photosynthetic plants had their beginnings in aqueous shallow regions, what is the evolutionary position of the bilichromoproteins? We do not see them in higher plants, do we?

Haxo: Well, I think most people would start their evolutionary scheme with chlorophyll *a* and, accepting Emerson's viewpoint, this would, in the absence of accessory pigments, have only moderate efficiency. In time, as accessory pigments came into the picture, they not only closed optical windows but they actually enhanced the activity of chlorophyll *a*.

French: I think there is a good deal of question as to how this enhancement effect comes about. I do not think we have much of an idea as to why we get an enhancement effect. One idea is that it might be in the pigment system, that is is part of the photochemistry that activating one pigment makes another one more able to work effectively. Another possibility is that the things come together at the chemical level some time after the photochemical part of it, that different products are made by chlorophyll *b* photosynthesis and chlorophyll *a* photosynthesis, and that these two things somehow get together and give more effective photosynthesis. But this must be in the oxygen-evolving part, because these are oxygen evolution measurements. Now there is one experiment of Jack Myers' that bears on this a little bit. He found that it was possible to give the two beams not only simultaneously to get enhancement, but to give one after the other up to 15 second intervals: 15 seconds of one, 15 seconds of the other. With a few seconds of one and a few seconds of the other you got almost as good an effect as you did with the two given together. Now that makes it look as though it were not highly immediate photochemistry with time constants that were very small, but the time constants of this thing were in the order of seconds. Now another thing that came out from your slides very nicely, I think, is that the enhancement effect requires activation of two pigments, one of which is chlorophyll b—let us just talk about *Chlorella* for a moment—one of which is chlorophyll b and possibly also the 670 form of chlorophyll a, and we have a slide that shows the same sort of an action with a hump at 670 mμ, so that apparently chlorophyll b and chlorophyll a 670 may be tied together. Perhaps we have energy transfer within that complex but that complex does one job and the long wavelength form of chlorophyll a, the 695 stuff, does another.

WEBER: In the same vein as Dr. French, it would be very interesting to have action spectra on the photosynthetic bacteria. I wonder how one would go about this, since they do not have such an easy indication of their activity. The second pigment may be connected with the oxygen-evolution mechanism which is photochemically activated because you always find these two pigments when you get oxygen evolution, while in the photosynthetic bacteria you find only one pigment and you do not have oxygen evolution.

HAXO: I am sure this would be a very interesting problem; however, I do not have any suggestions as to how to proceed with rapid measurements of action spectra in photosynthetic bacteria.

22

Automatic Recording of Photosynthesis Action Spectra Used to Measure the Emerson Enhancement Effect

C. STACY FRENCH, JACK MYERS,[*] AND GUY C. MCLEOD

Department of Plant Biology, Carnegie Institution of Washington, Stanford, California

To study the role of different pigments in photosynthesis, the absorption spectra of the living materials are compared with the action spectra for the photosynthetic reaction. Absorption spectra may now be plotted by commercial instruments in minutes with an accuracy approaching 0.1%. In contrast with absorption spectroscopy, quantitative photosynthetic action spectra are ordinarily obtained with point-by-point methods in which a precision of ± 1% is indeed good work.

One great advantage of a continuously recorded curve over point-by-point measurements is the far greater ease of detecting small irregularities attributable to the influence of pigments present in small amounts. We have therefore attempted to make an automatically recording spectrophotometer for the rapid plotting of photosynthetic action spectra. The possibility of rapidly measuring action spectra would allow the effects of various physiological conditions to be compared, and a continuously recorded curve would allow slight irregularities in action spectra to be assigned to specific pigments with confidence.

In order to make a continuous record of photosynthetic action spectra it is necessary to get an electrical signal proportional to the momentary rate of photosynthesis. This requirement is admirably fulfilled by the platinum electrode described by Blinks and Skow (1938) and used so successfully by Haxo and Blinks (1950) for measuring action spectra of photosynthesis with a point-by-point procedure.

We are indebted to both Drs. Haxo and Blinks for discussions and demonstrations on the use of platinum electrodes and to Dr. Haxo for the loan of a modified electrode [similar to that used by Haxo and Fork (1959)]. The particular value of a bare platinum electrode with a single layer of the photosynthetic cells on its surface is that the oxygen diffusion current gives a direct measurement of photosynthetic rate at any moment

---

[*] *Present address:* Departments of Botany and Zoology, University of Texas, Austin, Texas.

rather than measuring the concentration of oxygen in the bulk of a solution as do polarographic methods in stirred solutions.

The principle of automatic recording of photosynthetic action is as follows. The electrode current controls the light intensity by means of a servo system maintaining the rate of photosynthesis constant as the spectrum is traversed. This reduces the recording problem to one of continuously measuring the intensity of light incident on the cells as adjusted by the electrode output. Light from a grating monochromator, 5-mμ half-band width, is focused on the cells covering the electrode. An optical wedge driven by a servo motor under the control of the electrode regulates the light intensity. A fraction of this regulated light is deflected by a glass plate to a PbS cell. The output, IS, from the photocell is corrected by a photoelectric curve follower making the corrected signal, I, independent of the wavelength sensitivity of the photocell. The voltage from this corrector has a value independent of wavelength which is proportional to the number of incident quanta per unit time falling on the cell suspension. A servo-driven dividing mechanism produces from this corrected voltage a DC output, $1/I$, proportional to the reciprocal of the corrected voltage. This voltage, $1/I$, is plotted on a Brown recorder against wavelength to give the action spectrum for photosynthesis. The functional interrelations of the various parts are shown in the block diagram of Fig. 1.

Since the device operates by controlling the apparent rate of photosynthesis, uncorrected for changes in respiration in the dark, it is essential that the respiratory rate during a single sweep remain constant. It is therefore advisable to sweep limited parts of the spectrum between checks of the respiratory rate and to avoid rapidly sweeping across parts of the spectrum which might induce the chromatic transients discovered by Blinks (1954, 1957, 1959) and discussed by Myers and French (1960).

The possibility of rapidly repeating measurements of action spectra under various conditions, such as the presence of continuously applied background light of various colors, has made it feasible to investigate the action spectrum for the enhancement effect (Emerson, 1958) in a new way. Figure 2 shows the red part of the action spectrum of photosynthesis in *Chlorella* as measured with long-wavelength-red background light of 700 mμ absorbed by chlorophyll *a* only, with no background light, and with background light of 650 mμ absorbed partially by chlorophyll *b* and partially by chlorophyll *a*. Subtraction of these curves gives the effectiveness spectrum for each of the two pigment groups which when simultaneously activated produce the Emerson enhancement phenomenon.

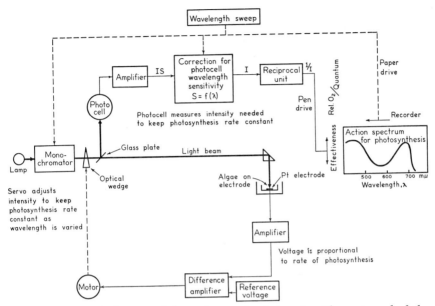

FIG. 1. A block diagram of the action spectrophotometer. The reciprocal of the relative number of quanta per second of the light incident on the cells required to maintain photosynthesis at a constant rate is continuously plotted against wavelength.

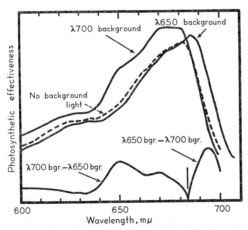

FIG. 2. Action spectra for photosynthesis in *Chlorella* showing the Emerson enhancement effect. Background light of 700 mμ absorbed by chlorophyll a (probably mainly by C_a 695) increases the effectiveness of photosynthesis by light absorbed by chlorophyll b,650, and by C_a 670, as shown by the difference curve below. Conversely, background light of 650 mμ stimulates photosynthesis by C_a 695 for which a part of the activation spectrum is shown by the longer wavelength branch of the difference curve. Rates of photosynthesis in relative units were 74 for the swept spectral lights and 132 for the background lights.

A similar experiment for a red alga is shown in Fig. 3. These two experiments confirm the conclusions of Emerson and Chalmers (1958) and of Blinks* that the function of chlorophyll b in green algae is analogous to that of the phycobilin accessory pigments of red algae, and that there are two separate photochemical steps in photosynthesis, each catalyzed by its own specific pigment system. Furthermore, the shoulder at 670 mµ on the difference curve of Fig. 2 suggests that in *Chlorella* the 670 form of chlorophyll a may also have the same function as does chlorophyll b.

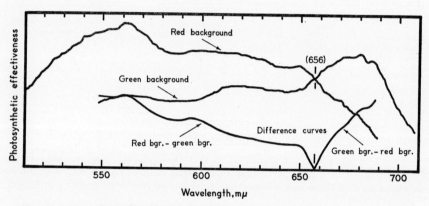

Fig. 3. Action spectra for photosynthesis in a red marine alga, *Porphyra perforata*. Red background light absorbed by chlorophyll a and green absorbed both by phycoerythrin and by carotenoids have very different influences on the recorded spectra due to the Emerson enhancement effect. The two branches of the difference curve are interpreted as being the effectiveness spectra for the enhancement of each of the two cooperating pigment systems.

The relative rates of photosynthesis by the measuring beam was 27 and from the background beams were 60 and 62. The background light came from a tungsten lamp and was filtered through 3 cm water and a "Calflex" infrared reflecting filter with 50% cut off at 695 mµ. For red light a Corning #2403 filter was added giving a band centering about 680 mµ and for green a #4084 filter giving a broad band peaking at about 520 mµ.

As Dr. Blinks has pointed out to us, the light passing through the cell layer is reflected by the bright platinum surface and again partially absorbed by the cells. The reflectance of platinum varies somewhat with wavelength and the fraction of light transmitted by the first pass through the cell layer also varies with cell absorption and therefore with wavelength. We did not make the measurements of light absorption necessary to correct this effect (cf. Haxo and Blinks, 1950). However, the resulting

---

* In this volume, see Paper No. 23.

distortion of the action spectra should be similar for any one batch of cells so the difference spectra shown for the Emerson effect should not include this error.

The automatic recording action spectrophotometer has so far been successful only with nonmotile algae, although experiments with *Euglena* and *Ochromonas* are being undertaken. Whether it will be practical for motile algae is still a question for the future. We do expect, however, that its use will contribute appreciably to understanding of the differences, if any, in the function of the various forms of chlorophyll a as they exist in different plants.

REFERENCES

Blinks, L. R. (1954). Autotrophic Microorganisms. *Symposium Soc. Gen. Microbiol.* pp. 224-246.
Blinks, L. R. (1957). "Research in Photosynthesis," pp. 444-449. Interscience, New York.
Blinks, L. R. (1959). *Plant Physiol.* **34**, 200-203.
Blinks, L. R., and Skow, R. K. (1938). *Proc. Natl. Acad. Sci. U.S.* **24**, 420-427.
Emerson, R. (1958). *Science* **127**, 1059-1060.
Emerson, R., and Chalmers, R. F. (1958). *Phycol. Soc. Am. News Bull.* **11**, 51-56.
Haxo, F. T., and Blinks, L. R. (1950). *J. Gen. Physiol.* **33**, 389-422.
Haxo, F. T., and Fork, D. C. (1959). *Nature* **184**, 1051-1052.
Myers, J., and French, C. S. (1960). *J. Gen. Physiol.* **43**, 723-736.

DISCUSSION

BLINKS: What about transients which might appear? With the slow sweep I suppose you do not get them, but I think perhaps in some algae it might become apparent.

FRENCH: This record stops at 600 mμ and we get into terrible trouble if we go to around 480 mμ. Chromatic transients are certainly a major difficulty in operating the thing.

VISHNIAC: It would be of great interest to use this on photosynthetic bacteria and on subcellular systems in which no oxygen evolution takes place. Is it correct to assume that perhaps pH changes could be similarly measured, or would they be too slow?

BLINKS: No, I do not think it would be much slower than this. The earliest experiments that we did with it 20 years ago showed very rapid responses. Again with transients one has to worry about CO_2 gushes and whatever acid gushes there are. In other words, you must know the time course.

23

Chromatic Transients in the Photosynthesis of Green, Brown, and Red Algae

L. R. BLINKS

Hopkins Marine Station of Stanford University, Pacific Grove, California

Chromatic transients have been described in the photosynthesis of red and green algae (Blinks, 1957, 1959). These consist of temporary alterations of photosynthetic rate when the wavelength of incident light is changed, even though the intensities have been adjusted to give equal steady state rates. A characteristic time course is indicated in Fig. 1a;

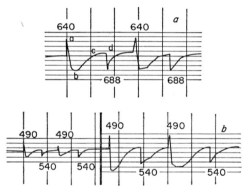

FIG. 1. Chromatic transients in *Ulva lobata*, generated by alternate exposures to light of: (a) 640 and 688 mµ; and (b) 490 and 540 mµ. A cusp ("a") occurs on going from 688 to 640 mµ; then there is a depression ("b")—here below the steady state, with slower recovery ("c"). On returning to 688 mµ there is another depression ("d") and recovery.

the rate at 688 mµ is first shown, then on changing the wavelength to 640 mµ there is an abrupt increase of oxygen evolution (cusp, "a") followed shortly by a decrease either to the original level, or below it (depression, "b"). From the latter there is recovery ("c") to the steady state rate. On return to 688 mµ, there is a depression ("d"), with recovery to the original rate. As indicated, these transients occupy several minutes, and may amount, at the peak values, to some 10 or 15% of the steady state rate. They may readily be followed by recording instruments such as the Speedomax, connected to a polarographic circuit, with the tissue in direct contact with a platinum electrode (Haxo and Blinks, 1950), polarized to reduce O_2.

These transients were originally explored by changing the wavelength of the illuminating monochromator (cf. Haxo and Blinks, 1950); they were limited to certain pairs of wavelengths which, because of lamp characteristics and proper regions of the action spectrum, happened to give equal, or nearly equal, rates. (The transients can still be seen, however, even though the steady states are not exactly equal.) Such pairs of wavelengths included those absorbed largely by chlorophylls a and b (as in Fig. 1a) and by mixed chlorophylls versus carotenoids (Fig. 1b) in the green alga *Ulva* (Blinks, 1959). In red algae (Blinks, 1957) such pairs could be found between chlorophyll a and the phycobilins (phycocyanin or phycoerythrin). It was concluded therefore that the transients were due to alternate absorption by the "fundamental" chlorophyll a, and by accessory pigments such as chlorophyll b, carotenoids, or phycobilins.

It seemed desirable to supplement these few fixed (and somewhat fortuitous) points with more complete sequences so as to obtain action spectra for the chromatic transients. This necessitated two (or more) sources of light: one, the monochromator, continuously variable; the other a fixed reference wavelength, absorbed largely by one pigment (e.g., chlorophyll a or phycoerythrin). The reference beam was obtained by focusing the image of a straight incandescent filament (or in some cases of a mercury arc) upon the tissue, with a proper interference filter inserted between the focusing lenses. Since most of the previous evidence indicated that alternations between chlorophyll a and accessory pigment absorptions were responsible, most of the action spectra were taken with long wavelengths of red light (centered at 702 mµ) as the reference beam; in a few cases other regions, such as 645 mµ (chlorophyll b), 560 mµ (phycoerythrin) and 490 mµ (carotenoids) were employed as the reference. Light from the monochromator could be alternated with these, at proper intensity, by means of electrical switching, or (to avoid lamp "on" effects) with synchronized camera shutters or a sliding mechanical shutter which interrupted one beam while exposing to the other. A few experiments were made with a rotating polarizer shutter (which permitted "mixing" as well as alternation.)

Larger transients often resulted from these alternations of filtered and monochromatic light, probably because of better restriction to light absorbed by "chlorophyll 690," at the wavelength of 702 mµ. Transients of 20 to 25% were sometimes found.

Although it is not clear how best to express the size of the transient, it was decided to employ the magnitude of the initial excursion "a" (as percentage of the steady photosynthetic rate). This does not differ too greatly from the depression "d" on returning to chlorophyll a absorption, and the average of "a" and "d" could probably be used without appreciably altering the action spectra.

Action Spectra

The simplest action spectrum so far found for chromatic transients is shown in Fig. 2. This was obtained with a very red species of the

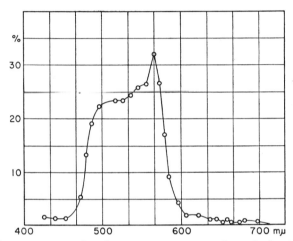

FIG. 2. Action spectrum for chromatic transients in the red alga *Porphyra thuretii*. The ordinates are the magnitudes of the cusp "a", expressed as percent of the steady photosynthetic rate. The peak at 560 mμ corresponds to the absorption maximum of phycoerythrin.

Bangiales, *Porphyra thuretii*, growing in fairly deep water. When the magnitude of the transients is plotted against wavelength (with the reference for alternation at 702 mμ) it is seen that the greatest values are found in the middle of the spectrum, in green light, with a peak at about 560 mμ. This corresponds very closely to the absorption maximum for phycoerythrin, and the curve is in fairly good agreement with the photosynthetic action spectrum for a related form, *P. nereocystis* (Haxo and Blinks, 1950). It differs from the latter in showing even less activity in the near red regions of the spectrum—due presumably to the presence of only a single chlorophyll (*a*), already activated at 702 mμ. (Chlorophyll *d*, if it exists here at all, seems in very low concentration—as in most red algae.) One can also conclude that there is very little phycocyanin present, as shown by the very slight activity in the region 615–650 mμ. Finally [just as in photosynthetic action spectra for most red algae (Haxo and Blinks, 1950)] there is very low activity in the blue and violet end of the spectrum. Apparently the carotenoids of red algae do not act as very efficient accessory pigments in the generation of chromatic transients any more than they do in photosynthesis. Conversely, however, if a very intense reference beam is employed at 440 mμ (the maximum for chlorophyll *a* in the violet), good transients are generated against light absorbed by phycoerythrin (in fact the action spectrum so derived looks rather like that of Fig. 2). In other words, it makes little difference whether chlorophyll *a* is activated in the *long* or the *short*

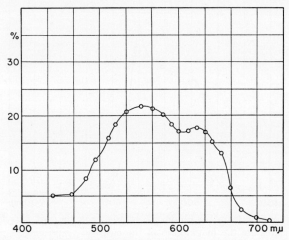

Fig. 3. Action spectrum of chromatic transients in *Porphyra perforata*. There is considerable activity in the region 610–650 mμ, due to a greater content of phycocyanin than in *P. thuretii*.

wavelength regions of its absorption; good transients are obtained on alternating either red or blue light with green.

A somewhat more complicated action spectrum is found with another species of *Porphyra*, *P. perforata*, which grows at high tide levels. This alga appears not red, but rather a slate grey, due to the large amounts of phycocyanin and allophycocyanin present (along with fair amounts of phycoerythrin). This is reflected in the action spectrum (Fig. 3) for transients, which are now large in the region 600–650 mμ (almost as large as in the green, 500–560 mμ). There is a slight depression in the region of 600 mμ due to the poorer absorption by phycobilins here. Again, there is little activity in the blue and violet (though, as with *P. thuretii*, 440 mµ as a reference beam generates transients almost as well against phycobilins as does 702 mµ).

Green algae, as might be expected from their very different pigment complex, show an action spectrum quite different from the red algae. Figure 4 represents *Ulva* as an example. The maxima are at the regions 480–490 mµ and 640–650 mµ. These no doubt represent carotenoid and chlorophyll *b* absorption, respectively. (The fairly high activity in the region 650–680 mµ may represent the "accessory" participation of several forms of chlorophyll *a*, found by French and Myers.*) There are quite

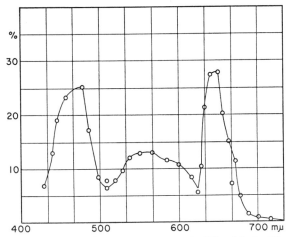

FIG. 4. Action spectrum of chromatic transients in *Ulva lobata*. The peak at 640–650 mµ represents chlorophyll *b*, while that at 450–480 mµ is probably due to carotenoids.

striking minima as well, lying close to the maxima (viz., 620–630 and 500–520 mµ). These are presumably regions of absorption largely by chlorophyll *a*. Reasonably high activity is found between these two regions, with perhaps a hint of two peaks or humps. This doubtless represents predominant absorption by chlorophyll *b*. There is some activity in the far blue and violet (appreciably more than in the red algae): this is doubtless due to carotenoid absorption (possibly also to some chlorophyll *b* participation).

Rather similar action spectra have been found in the related green algae *Enteromorpha* and *Monostroma*, and French and Myers (1960) have reported almost identical curves for the unicellular fresh-water alga, *Chlorella* (though the form of the transients differs). It is to be expected

---

* See Paper No. 22 by C. S. French and co-workers, in this volume.

that green algae rich in carotene (such as *Trentepohlia* or *Dunaliella*) may give different results in the short end of the spectra, just as they do in photosynthetic action spectra.

When we turn to the brown algae, the situation is not as clear. Most of the marine brown algae are not well adapted to transient measurement, being too thick. A few, like *Coilodesme*, though several cells thick, have their pigmented cells only at one surface of the thallus. *Punctaria*, new to such studies, is fairly thin, but sometimes displays only very slight transients. This variability is not understood, and seems unrelated to the steady rate. Figure 5 represents a fairly successful series of meas-

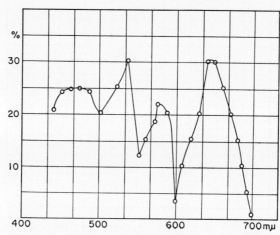

Fig. 5. Action spectrum of chromatic transients in *Punctaria occidentalis*. The peak at 650 mμ probably represents chlorophyll *c*—as perhaps does that at 570–580 mμ. The maximum at 540 is doubtless due to fucoxanthin, while the broad region 450–580 may be due to other carotenoids—possibly also to chlorophyll *c*.

urements with *Punctaria* in a thallus which did show good sized transients. It is seen that there are several maxima. That at 650 mμ is no doubt due to chlorophyll *c*. (Whether various forms of chlorophyll *a* occur in brown algae seems not to have been reported.) The very sharp depression at 600 mμ resembles that at 630 mμ in the green algae, but is even more striking. It may represent predominant absorption by chlorophyll *a*. Chlorophyll *c* probably is responsible for the increased activity in the region 570–590 mμ, and certainly fucoxanthin for the high peak in the region of 520–540 mμ. There is very little depression at 500 mμ, doubtless due to the good overlap of fucoxanthin absorption with that of other carotenoids. The latter, of course, again along with chlorophyll *c*, must account for the high transient activity in the blue and violet regions.

The large number of active accessory pigments in brown algae (which makes them so efficient in photosynthesis) is no doubt the cause of the complicated action spectrum for the chromatic transients.

Cause of the Transients

Identification of the pigments responsible for the chromatic transients seems fairly straightforward: apparently almost any *active* accessory pigment (which leaves out the carotenoids of red algae) can generate transients when it absorbs light after absorption by chlorophyll *a* alone; conversely, the latter can give rise to "return" transients when it absorbs light following absorption by other pigments. This indicates that the accessory pigments promote, at least for several minutes, a somewhat different photochemical process from that due to chlorophyll *a*.

What might this be? It must be admitted that we have no clear idea. It was earlier suggested (Blinks, 1957) that some sort of solarization might be involved, by which the accessory pigment activated photosynthesis very rapidly (cusp "a") but soon produced damage via photo-oxidation or other photosensitization (perhaps related to the high fluorescence of phycobilins). This led to the depression "b". Since the same time course is obtained in wavelengths absorbed by other pigments such as chlorophyll *b* or *c* and carotenoids, this is probably questionable. Fluorescence measurements made during the transient would be interesting, however. If the depression "b" were due to photo-oxidation, it seems hardly likely that it would persist for several minutes into the chlorophyll depression ("d"); more reasonable seems a respiratory increase of unspecified type. Emerson and Lewis found such increase of respiration by manometric methods, lasting for some minutes into the dark period, following illumination with blue light. It may be assumed that it occurred as well during illumination. The speed of the electrode method suggests that this respiration starts slightly *after* the onset of illumination—perhaps ½ minute later. This might represent the time taken for the production of certain more easily respired intermediates (newly phosphorylated compounds?) or for their diffusion from the plastid to respiratory centers (mitochondria). The latter would be especially evident in cells (like those of *Enteromorpha tubulosa*) where the plastid is closer to the electrode than is the hyaline cytoplasm.

Whatever the reason for the enhanced respiration, one may construct a formal scheme such as is shown in Fig. 6. If respiration is indeed increased by phycobilins or other accessory pigments, its persistence and decay during the subsequent light absorption by chlorophyll indicates that the respirable substance is now being consumed (or is not being produced as fast by the activity of chlorophyll). This, in fact, seems the essence of the transients—that one pigment is doing a somewhat different thing from another. This could also be the cause of the Emerson effect—the enhancement, over and above strict additivity, of photosynthesis produced by two superimposed wavelengths of light.

It might be suggested that the Emerson effect persists slightly after darkening, so that following illumination by red light (702 mμ), the

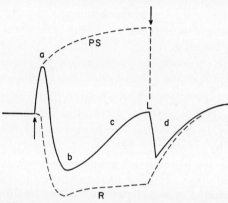

Fig. 6. Scheme indicating a possible explanation of the chromatic transients. The photosynthetic rate at a given wavelength (e.g., 702 mμ) is first indicated. On changing the wavelength (e.g., to 650 mμ) at the first arrow, the photosynthetic rate (PS) is enhanced (cusp "a") but, as shown by the dotted line, respiration (R) soon increases, causing the depression "b". However, photosynthesis continues to rise (dashed line, causing recovery "c"). On reverting to the original wavelength (e.g., 702 mμ) the photosynthetic rate immediately falls, while the respiration remains increased, giving a depression of net oxygen evolution. The latter eventually returns to its original steady rate in a few minutes, as respiration decreases.

stage is set for an extra effect of green light (560 mμ). This could cause the cusp "a". Why, however, there should be a depression "d" in the opposite sequence is difficult to see.

Actually some degree of formal explanation for the transients is afforded by the dark-light transients (e.g., Fig. 7), in which somewhat different time courses are often (though not invariably) found at different wavelengths. Thus in green algae at 688 mμ there may be a slow increase of O_2 production, rising over 2 or 3 minutes, while at 640 mμ there is often a more abrupt rise, a cusp, recession, and later rise. The latter resembles the chromatic transient at 640 mμ. Since it occurs without immediately preceding chlorophyll absorption, it seems that no preconditioning, or persistence of the Emerson effect, is necessary. That the cusp in such a dark-light transient is not as sharp as in alternations, may be only because it is swamped out in the very abrupt rise always found during establishment of the enhanced diffusion gradient of O_2 in the light. Occasionally there is a much sharper cusp when "accessory" light is *added* to red (Emerson effect).

Recent work (Blinks, 1960) with *Enteromorpha tubulosa* indicates that the transients are indeed due to a different time course of respira-

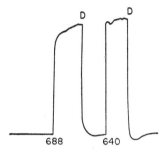

FIG. 7. Dark-light transients at the indicated wavelengths (688 and 640 mµ). There is a slower rise to the steady rate in the first case, while in the second there is often found a faster rise and a cusp. D indicates darkening; there is frequently an enhanced respiration (shown by a temporary depression below the steady dark level), following exposure to wavelengths absorbed by accessory pigments.

tion at different wavelengths. If photosynthesis is interrupted at very short intervals following illumination at 650 mµ, the succeeding respiration is found to be at first unaffected (during the cusp "a"); then becomes markedly enhanced (during the depression "b"); and finally returns almost to normal (recovery "c"). Conversely, illumination at 702 mµ gives rise to a very *early* enhancement of respiration, followed by recovery. This would account for the depression "d" during the first moments of absorption by chlorophyll *a*, as well as for the slow rise of oxygen evolution on going from dark to 702 mµ.

REFERENCES

Blinks, L. R. (1957). In "Research in Photosynthesis" (H. Gaffron, ed.), pp. 444-449. Interscience, New York.
Blinks, L. R. (1959). *Plant Physiol.* **34**, 200-203.
Blinks, L. R. (1960). *Science* **131**, 1316.
Haxo, F., and Blinks, L. R. (1950). *J. Gen. Physiol.* **33**, 389-422.
Myers, J., and French, C. S. (1960). *J. Gen. Physiol.* **43**, 723-736.

DISCUSSION

HENDRICKS: All the patterns of transients you show seem to resolve into three sorts of components—a rise, a quick drop, and then a subsequent rise. It suggests one that keeps rising and meets another one that is started further down.

BLINKS: This is what I have been indicating with the respiration diagram. But I think geometry comes into this. We cannot regard these as homogeneous systems. It may well be that the cytoplasm is doing something which is different and has a different time delay, perhaps because it is a little further from the electrode than the plastid. If, say, respiration were to come slightly after the initial evolution of oxygen, it might be merely due to the 2 or 3 μ more diffusion distance between where most of the mitochondria are and where the plastid is against the electrode. The time course can become more complicated as you use thinner cells. You can get a species of green alga with a plastid all on one side. You can put that now against the electrode and illuminate from that side. The response is then extremely fast. One gets sometimes a little sharp spine. This is the dark-light transient. One can pick up very rapid arrival of something, possibly from the nearest plastid surface, since if the tissue is turned around this is much delayed. So I think geometry cannot be forgotten.

WEBER: Is it correct that these do not involve any background light?

BLINKS: No, there can be, but that tends to suppress them. This probably would be worth doing. A little bit has been done with it, but mostly it just makes them less conspicuous.

24

Chemical Participation of Chlorophyll in Photosynthesis

WOLF VISHNIAC

Department of Biology, Brookhaven National Laboratory, Upton, New York[*]

When Willstätter, at the turn of the century, began his monumental studies of the pigments of green leaves, he assumed that the chlorophylls were closely connected with what were then unique biological events—the fixation of carbon dioxide and the evolution of oxygen. Throughout his work one can find as one important theme the view that carbon dioxide fixation may be mediated by the chlorophyll molecule itself. It was only during his study on carbon dioxide assimilation that he abandoned the hope of finding such a reaction. Since then the chlorophylls appeared to have taken on in the minds of most investigators the exclusive role of photosensitizing pigments, i.e., the physical collectors of the radiant energy which they absorb, although there have been occasional attempts to implicate chlorophyll itself in one or another chemical reaction. As one example we may cite a paper by Emerson, Stauffer, and Umbreit (1944) in which the authors proposed among other things that in the light an energy-rich phosphate was formed on the chlorophyll molecule. More recently it was proposed on a purely speculative basis by Korkes (1956) that light induced a change in chlorophyll which converted magnesium to a univalent state, thereby making it a strong reducing agent. There have also been a number of spectrophotometric observations of absorption changes during the illumination of intact photosynthesizing cells, some of which have been tentatively attributed to chlorophyll itself (Strehler and Lynch, 1957). While some of these changes appear to be correlated consistently with early photosynthetic events, it appears not yet possible to interpret these optical changes in terms of specific reactions. In this paper, three topics will be discussed, each one of which deals with a different possible role of the chlorophyll

---

[*] On leave of absence from the Department of Microbiology, Yale University. Experimental work mentioned herein was carried out in part at the Brookhaven National Laboratory under the auspices of the U. S. Atomic Energy Commission, and in part at Yale University supported by grants from the National Science Foundation, the National Institutes of Health, and the Yale University Medical School Fluid Research Fund.

molecule in photosynthesis: (a) the direct carboxylation of chlorophyll as recently proposed by Warburg; (b) the incorporation of tritium from tritium-labeled water during photosynthesis; and (c) the apparent close coupling between chlorophyll and carotenoids in a purple bacterium.

Some years ago, Warburg and his associates (1954) undertook a study of a possible oxygen precursor in *Chlorella*. On the assumption that in the dark during respiration some material might be formed which upon illumination might give rise to oxygen, Warburg studied the rate of oxygen evolution during a dark-light transition. He found that immediately upon the illumination of a cell suspension which had previously been respiring in the dark, there was a rapid rate of oxygen evolution which then decreased to a smaller linear rate. Warburg assumed that the linear rate represented a steady state photosynthesis, and extrapolated this linear rate to zero time. The intercept on the ordinate gave Warburg an amount of oxygen which he took to be equivalent to the amount of precursor formed in the dark, an amount which he called the "oxygen capacity." In several experiments under a variety of conditions it appeared that on a molar basis the oxygen capacity of *Chlorella* was equal to the total chlorophyll content of the cells used. This phenomenon could be repeated by returning the cells to darkness, allowing them to respire, and reilluminating them. Further studies showed that the recovery of the oxygen capacity required not only the presence of oxygen in the dark, but also the presence of carbon dioxide (Warburg and Krippahl, 1956a). This suggested to Warburg a CO_2-fixing reaction closely linked to oxygen evolution.

Several chemical agents, notably fluoride, affected cells in the dark and gave rise to the evolution of fixed amounts of carbon dioxide (Warburg and Krippahl, 1956b). Under anaerobic conditions, cells which had previously been respiring in the dark would upon the addition of fluoride liberate a fixed amount of carbon dioxide, which again was equal on a molar basis to the chlorophyll content of the cells and fluoride in this experiment. There existed apparently a 1:1:1 relationship between the oxygen capacity, the total chlorophyll content, and the fluoride-induced CO_2 liberation in a suspension of algal cells. The effect of fluoride was reversible; that is, fluoride could be washed off the cells, and after respiration in the dark in the presence of CO_2, the effect of fluoride could again be demonstrated. These observations led Warburg to devise a scheme (Warburg and Krippahl, 1956b) in which CO_2 was fixed and reduced on, and O_2 evolved from, the chlorophyll molecule itself. Briefly, Warburg envisaged the opening of the isocyclic ring with the formation of an aldehyde group, which in the dark during respiration was transferred to a suitable acceptor, and replaced with carbon dioxide to form a new carboxyl group. This carboxyl group would be split off by fluoride in the dark, or in the light the carboxyl group would be reduced to the aldehyde level with the concomitant formation of a peroxide on the adjacent carbon atom. The elimination of oxygen from the peroxide would then reconstitute the original chlorophyll molecule. This particular scheme, although not expressly retracted by Warburg, probably has been abandoned by him, but at the time it seemed important, regardless of the interpretation, to attempt to demonstrate the alleged stoichiometry between oxygen, CO_2, and chlorophyll.

Dr. Fuller and the present author accordingly tested a number of algae for the existence of the oxygen capacity and the fluoride-dependent CO_2 evolution. In addition to *Chlorella*, we also used *Scenedesmus* and *Euglena*. The cultures were grown both at the low pH which Warburg specified and at pH 6.8 without affecting the results. Furthermore, since Warburg et al. (1956) had discussed the role of fluctuating illumination, we used at times a clock-driven variable transformer to vary the light intensity during a 24-hour period in such a way that low light intensities occurred at the beginning and at the end of the 24 hours, with a peak in light intensity 12 hours after the beginning of the cultivation period. The experiments require the use of very dense algal suspensions, since the observation of 5 or 10 μmoles of CO_2 necessitates the presence of a sufficient number of algae to contain 5 or 10 μmoles of chlorophyll. As a result the suspensions are approximately 40% packed cells.

We were able to confirm without any difficulty the numerical relationships which Warburg had reported. Table I shows that in three instances, using all three organisms, the amount of CO_2 evolved in the dark by the addition of fluoride corresponded closely to the total amount of chlorophyll present, regardless whether continuous or fluctuating illumination had been used during cultivation. In one instance, a 48-hour

TABLE I

FLUORIDE-DEPENDENT CO_2 EVOLUTION AND OXYGEN CAPACITY

Alga	Age (hours)	CO_2 (μmoles)	Chlorophyll (μmoles)	Illumination
Chlorella	24	4.46	4.68	Fluctuating
Scenedesmus	24	11.2	11.4	Fluctuating
Euglena	24	23.8	22.7	Continuous
Euglena	48	10.4	35.6	Continuous
Scenedesmus	24	11.6[a]	11.6	Fluctuating

[a] Micromoles of O_2.

culture of *Euglena* did not conform to the 1:1 ratio, possibly because after 24 hours the chlorophyll synthesis continues beyond the physiological needs of the organisms. The last experiment in the table shows that the value of the oxygen capacity in a 24-hour culture of *Scenedesmus* also agreed closely with total amount of chlorophyll.

It was next decided to investigate whether the carbon dioxide which was released by fluoride originated in the chlorophyll molecule or not. In the experiment described in Fig. 1, a suspension of *Euglena* containing 22.7 µmoles of chlorophyll was treated with fluoride, giving rise to 23.8 µmoles of carbon dioxide. The resulting cell suspension, here called "decarboxylated" *Euglena*, was washed free of fluoride in the dark and in the absence of carbon dioxide. The cells were then resuspended and allowed to respire in the dark in presence of oxygen and $C^{14}O_2$. The "recovered" *Euglena* contained approximately 5 million cpm of C^{14}. Ten per cent of the suspension was again treated with fluoride and the liberated carbon dioxide was trapped in baryta. In the resulting barium carbonate, 400,000 cpm of C^{14} were found. It can be concluded from this observation that carbon dioxide released by fluoride originates in a location on which carbon dioxide is also fixed, since the carbon dioxide released in this experiment was of the same specific activity as the original carbon dioxide. The major portion of the cell suspension was killed by boiling and no radioactivity or carbon dioxide was lost during this process. The extraction with acetone and subsequent purification of chlorophylls revealed no radioactivity in the chlorophyll, while the major amount of radioactivity, approximately four million cpm, could be obtained in an aqueous extract. Chromatography of the aqueous extract located the radioactivity almost exclusively in glutamic acid, specifically in the α-carboxyl as determined by treatment with ninhydrin. It appears therefore that a reversible carboxylation of γ-aminobutyric acid may take place in *Euglena*, and that the resultant glutamic acid is the source of the fluoride-labile carbon dioxide.

Since these experiments have been performed, Warburg himself has indicated that he believes glutamic acid to be decarboxylated by the fluoride treatment (Warburg *et al.*, 1956) and there appears to be little reason for relating this activity to the carbon metabolism of photosynthesis. There remains only the curious stoichiometric relationship between chlorophyll and the fluoride-dependent CO_2 evolution, as well as the oxygen capacity. A number of speculative schemes might be proposed to account for the results observed, but there is insufficient experimental evidence at present to attach much significance to any one of them.

24. CHEMICAL ROLE OF CHLOROPHYLL

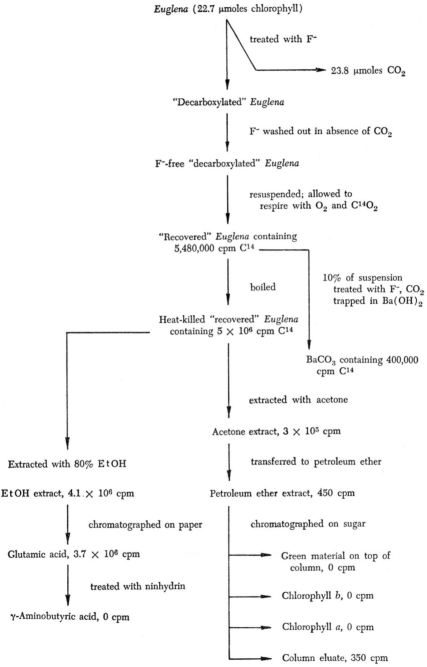

FIG. 1. Fractionation of *Euglena* labeled with $C^{14}O_2$ after fluoride treatment.

A recurrent suggestion in the interpretation of photosynthetic events is that chlorophyll itself undergoes reduction and reoxidation, and thus initiates the electron flow which results in the reduction of carbon dioxide to carbohydrates. The experimental work of Krasnovsky (1948) and the proposals of Franck (1957) both point in that direction. Dr. Irwin Rose and the present author have attempted to use tritium as a tracer to investigate the possible oxidation-reduction reactions which chlorophyll may undergo (Vishniac and Rose, 1958). The feasibility of such studies was suggested by several lines of investigation. The work of Vennesland and her collaborators has demonstrated the stereospecificity of pyridine nucleotide reaction in hydrogen transport (Levy and Vennesland, 1957). Thus, if several dehydrogenases are coupled sequentially, then tritium or deuterium originating in one substrate may be found in the pyridine nucleotide after reduction and reoxidation as a result of the opposing stereospecificities of the dehydrogenases. Similarly, it might be possible to recover in chlorophyll tritium originating in labeled water if hydrogen in its transfer from water to final acceptors passes through chlorophyll as the result of either nonspecific or opposing stereospecific reactions. It had furthermore been shown by Weigl and Livingston (1952) that in neutral organic solvents chlorophyll did not exchange any of its hydrogens with the deuterium of labeled water.

In our experiments, we used intact cells of *Chlorella*, chloroplasts, fragments of chloroplasts from spinach and pea leaves, and acetone powders of chloroplast fragments. The various preparations were illuminated for periods ranging from 30 seconds to 5 minutes in a buffer containing tritium-labeled water. Our preliminary experiments (Table II) indicate that chlorophyll a was consistently more radioactive after illumination than chlorophyll isolated from dark controls. Subsequent experiments with improvements in technique reduced the dark incorporation to near 0. Concerning the acetone powder which was used: if

TABLE II
INCORPORATION OF TRITIUM IN CHLOROPHYLL a

Material	Chlorophyll a (cpm/μmole)	
	Light	Dark
Chlorella	17,200	4,550
	10,600	2,610
Chloroplast fragments	16,300	3,870
Acetone powder	762	83
	952	144
	1,100	0

a suspension of chloroplast fragments is dried with acetone at about —5° C., most of the pigments are extracted. However, a small amount of chlorophyll is left behind and cannot be extracted with organic solvents unless the proteins of the particle are denatured. Such denaturation can be effected by heating or by a variety of chemical treatments including the use of urea. We have routinely used urea to remove the chlorophyll firmly bound to the particles at the conclusion of the experiment. Control experiments showed that the treatment of such particles with urea in the presence of tritium-labeled water did not lead to the introduction of label into chlorophyll. The isolated radioactive pigment was chromatographed, its specific activity was determined, and in some instances converted to pheophytin and rechromatographed. The pheophytin was shown to retain the radioactivity of the chlorophyll from which it was made, indicating that the radioactivity was associated with the chlorophyll molecule and not with a contaminating impurity.

While mild acid treatment of the labeled chlorophyll did not remove the radioactivity from the molecule, treatment with alkaline reagents led to a rapid loss of tritium. These observations suggest that the radioactivity is located on the isocyclic ring, presumably the tritium has replaced the protium on carbon-10, since treatment with alkali is known to cause allomerization, that is, the cleavage and oxidation of that ring. The extent of the labeling was always low. Assuming the incorporation of one tritium per molecule of chlorophyll, we found only between 0.5 and 5% of the chlorophyll molecules labeled. The labeling seemed to be confined to chorophyll *a*, little or no radioactivity was found in purified chlorophyll *b*. There was considerable discrepancy between the total radioactivity in a pigment solution prior to chromatography and the radioactivity recovered in chlorophyll *a*. A systematic investigation of the chromatograms, both on paper and on sugar columns, revealed that the major portion of radioactivity remained at the origin. By increasing the ketone content of the solvent system used, either acetone in petroleum ether on sugar columns or methyl isobutylketone in decane on paper, additional pigment spots could be moved away from the origin. Figure 2 shows a paper chromatogram with such an additional pigment spot, and a tracing obtained from an automatic radioactivity scanner through which the chromatogram was fed. It can be seen that some radioactivity is associated with chlorophyll *a* but that most of the radioactivity is located in the unknown pigment. The unknown pigment possesses an absorption spectrum of the chlorophyll type but its absorption peaks are located further in the red. Assuming similar molecular weights and extinction coefficients at the absorption peak, the unknown pigment in this particular experiment had 25 times the specific activity of chlorophyll *a*. We are at present engaged in studying the nature and role of this new pigment.

The final topic concerns the close functional and spatial association of some carotenoids with bacterial chlorophyll in the chromatophores of the photosynthetic bacterium *Chromatium* as observed by Bergeron and Fuller (1959).* The absorption spectrum of these particles, similar to that of the whole bacterium, shows features characteristic of the purple

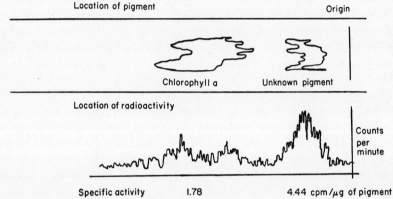

FIG. 2. Chromatogram of pigments extracted from chloroplast acetone powder after illumination in tritium water.

bacteria in general; that is, in addition to the absorption peaks which can be ascribed to bacteriochlorophyll and the carotenoids and which can be obtained in the extracted pigments, there is a considerable absorption in the infrared, particularly around 890 and 850 mµ, which does not seem to correspond to any known pigment. When the particles were suspended in 0.5 M sucrose it was found that photophosphorylation could be obtained when either the carotenoids were illuminated (at 490 to 500 mµ) or when the light was received by chlorophyll (810 mµ). However, transfer of the particles to distilled water, while not impairing the rate of photophosphorylation when the long wavelengths of illumination was used, reduced photophosphorylation at the shorter wavelengths at which the carotenoids received light. Concomitant with the inability of the carotenoids to act as accessory photosynthetic pigments was a partial disappearance of the infrared absorption peaks in the spectrum of the particle. These observations were interpreted by the authors to mean that carotenoids can transfer energy to bacteriochlorophyll only when

---

* See also Anderson and co-workers (1960).

they are in close physical proximity, and that such proximity also gives rise to the absorption peaks in the infrared. The entry of water into the particles in the hypotonic medium resulted in a swelling which apparently dissociated the carotenoid layer from the chlorophyll layer, leading both to a loss of the infrared peaks and to a loss of the ability of the carotenoids to act as accessory photosynthetic pigments.

A prudent working hypothesis concerning its role must therefore assign to chlorophyll more than the ability to absorb and transmit radiant energy. The recognition of the chemical role of chlorophyll may thus in some measure vindicate the spirit of Willstätter.

REFERENCES

Anderson, I. C., Bergeron, J. A., and Fuller, R. C. (1960). *Arch. Biochem. Biophys.* (in press).
Bergeron, J. A., and Fuller, R. C. (1959). *Nature* **184**, 1340.
Emerson, R. L., Stauffer, J. F., and Umbreit, W. W. (1944). *Am. J. Botany* **31**, 107.
Franck, J. (1957). *In* "Research in Photosynthesis" (H. Gaffron, ed.), p. 142. Interscience, New York.
Korkes, S. (1956). *Ann. Rev. Biochem.* **25**, 685-734.
Krasnovsky, A. A. (1948). *Doklady Akad. Nauk S. S. S. R.* **60**, 421.
Levy, H. R., and Vennesland, B. (1957). *J. Biol. Chem.* **228**, 85.
Strehler, B. L., and Lynch, V. H. (1957). *In* "Research in Photosynthesis" (H. Gaffron, ed.), pp. 89-99. Interscience, New York.
Vishniac, W., and Rose, I. A. (1958). *Nature* **182**, 1089.
Warburg, O., and Krippahl, G. (1956a). *Z. Naturforsch.* **11b**, 52.
Warburg, O., and Krippahl, G. (1956b). *Z. Naturforsch.* **11b**, 179.
Warburg, O., Krippahl, G. Schröder, W., and Buchholz, W. (1954). *Z. Naturforsch.* **9b**, 769.
Warburg, O., Schröder, W., and Gattung, H.-W. (1956). *Z. Naturforsch.* **11b**, 654.
Warburg, O., Klotzsch, H., and Krippahl, G. (1957). *Z. Naturforsch.* **12b**, 266.
Weigl, J. W., and Livingston, R. (1952). *J. Am. Chem. Soc.* **74**, 4160.

DISCUSSION

ARNON: Having been in Warburg's laboratory as a visitor during the period when these experiments with CO_2-chlorophyll ratios were going on, it may be incumbent upon me to make a comment on this subject. That is, as Dr. Vishniac has pointed out, Warburg has withdrawn the idea, which he originally believed in very strongly, that there is a constant ratio between CO_2 and chlorophyll. His withdrawal was rather unobtrusive; two short notes in *Z. Naturforsch.* where he pointed out that what he thought was a chlorophyll-CO_2 compound turned out to be a decarboxylation of glutamic acid.

Now then next he became very excited over the idea that the carboxylation reaction might be the carboxylation of γ-aminobutyric acid. So then he tried γ-aminobutyric acid as a CO_2 acceptor and found that that did not work. So he abandoned that idea too. The third point was that he did not have a 1:1 ratio of glutamic acid to chlorophyll. Now what is left then, after all this is said and done? He said that he was convinced that glutamic acid has some very important relation to CO_2 assimilation and that the capacity to decarboxylate glutamic acid by the fluoride reaction is also in some important way related.

WEBER: Some of you may have seen a letter in *Nature* about a year ago by Dr. Teale in which he prepared a model system which consisted of micelles in which he incorporated chlorophyll and fucoxanthin and he was able to show, by studying the action spectrum of the fluorescence of chlorophyll, that light absorbed by the fucoxanthin was transferred quantitatively to the chlorophyll. With up to a ratio of two molecules of fucoxanthin to one molecule of chlorophyll the transfer of energy is practically 100%. If you have more than two molecules, then to all intents and purposes the excess over two molecules of fucoxanthin just cannot transfer. And if one introduces into the micelles small amounts of alcohol or propylene glycol or some agent that one would expect to penetrate into the micelle, they seem to break down the complex between the fucoxanthin and the chlorophyll and energy transfer stops completely. So the evidence seems to be that, at least in the case of fucoxanthin and chlorophyll, you get a complex by some hydrogen bonding between the partners. When this is broken, you cannot transfer energy any longer. You need this intimate association because fucoxanthin, and, in general, all carotenoids, do not fluoresce at all. Their only chance of transfer, because of their short lifetime of excited state, is when they are forming a practically molecular complex.

VAN NIEL: I would like to mention that, in connection with the work of Duysens, it is a bit surprising to find that carotenoid-absorbed light is just as effective for photophosphorylation by purple bacteria chromatophores as light absorbed by the chlorophyll. The Duysens experiments show that the transfer from carotenoid to chlorophyll is not 100% effective, and in the experiments on phototaxis in purple bacteria as well as in the experiments on growth rates, the activities of light absorbed chiefly by carotenoids and that absorbed by the chlorophylls differ by a factor of about 10.

VISHNIAC: I did not want to give the impression that the rates were necessarily the same, but they certainly differ by much less than a factor of 10.

ARNON: If conditions were such that in the particular experiments much higher rates of photophosphorylation were obtainable, the drop might have occurred but would have gone undetected.

25

Studies with Photosynthetic Bacteria: Anaerobic Oxidation of Aromatic Compounds

STANLEY SCHER* AND MICHAEL H. PROCTOR†

Hopkins Marine Station of Stanford University, Pacific Grove, California
and
Department of Bacteriology, University of Wisconsin, Madison, Wisconsin

Bacterial photosynthesis presumably differs from green plant photosynthesis in being strictly dependent upon accessory hydrogen donors. Compounds which serve as hydrogen donors for anaerobic photosynthesis in light can also be oxidized aerobically in darkness. This might suggest that illumination yields oxygen; however, oxygen evolution by photosynthetic bacteria has never been detected, even when luminous bacteria are used as indicators (Molisch, 1907). Measurements of redox potentials provide evidence that oxidizing substances are formed when suspensions of photosynthetic bacteria are exposed to light (Roelofson, 1935; Wassink, 1947).

Van Niel (1941, 1949) has argued that in all photosyntheses the primary photochemical reaction is concerned with the generation from H_2O of a reducing and oxidizing compound. This does not imply that the water molecule is split into a hydrogen atom and a hydroxyl radical. Kamen (1956, 1957) has hinted that the "splitting of water" as such may not be involved in the initial light reaction, but can occur instead as a result of the transfer and storage of energy during back-oxidations, which have been postulated by Hill (1939). Oxygen as hydrogen acceptor is not a requisite to the Hill postulate. The oxidizing component generated in light can be a substance other than oxygen, yet functionally equivalent to oxygen for the metabolism of photosynthetic bacteria.

Arnon (1959) has suggested an hypothesis in which oxygen is formed during electron transfer subsequent to the photochemical act. The photochemical reaction common to green plants and photosynthetic bacteria is

---

* Aided by a postdoctoral fellowship from the American Cancer Society to the senior author. *Present address:* Laboratory of Comparative Biology, Kaiser Foundation Research Institute, Richmond, California.

† *Present address:* Department of Biochemistry, Cambridge University, Cambridge, England.

visualized as ejection of an electron from chlorophyll with formation of an oxidant and a reductant which recombine by passage through a chain of catalysts including pyridine nucleotides, flavins (or quinones), and cytochromes—high-energy phosphate bonds being formed during this process (see also Baltscheffsky, 1960). According to Arnon's hypothesis, there is no need to postulate that the function of the external hydrogen donor in photosynthetic bacteria is to reduce a precursor of oxygen. The hydrogen donors that have so far been studied in Arnon's experiments are those that can reduce pyridine nucleotides in dark reactions, thus supplying the reducing power for the conversion of carbon dioxide to cell material. With such hydrogen donors, the quantum requirement for bacterial photosynthesis would be expected to be independent of the chemical nature of the hydrogen donor, as was previously found for hydrogen, thiosulfate, and tetrathionate in purple and green sulfur bacteria (Wassink et al., 1942; Larsen et al., 1952; Larsen, 1953, 1954). It would be of interest to determine whether the quantum requirement is increased with hydrogen donors, such as succinate, that cannot reduce pyridine nucleotides in dark reactions.

If light and oxygen are equivalent, then photosynthetic bacteria should be able to carry out, under strictly anaerobic conditions in the light, oxidations that normally require molecular oxygen. To test this hypothesis, we can determine whether photosynthetic bacteria can utilize hydrogen donors that require molecular oxygen or its equivalent for their oxidation.

Aromatic Compounds as Hydrogen Donors

From the extensive studies on the biological oxidation of aromatic compounds such as benzoic acid, protocatechuic acid, and catechol, we know that molecular oxygen is an obligatory oxidant for these substrates (Stanier, 1950). If such compounds can serve as hydrogen donors for bacterial photosynthesis, then we may be able to determine the nature and fate of the oxidizing substance formed in light.

With this in mind, we prepared elective cultures in which benzoate, or other aromatic compounds, were provided as organic hydrogen donors for nonsulfur photosynthetic bacteria. From these elective cultures, several strains of purple bacteria have now been obtained in pure culture. On the basis of microscopic observations, as well as nutritional experiments in chemically defined media, the organisms have been provisionally classified as strains of *Rhodopseudomonas*. They grow well on acetate, lactate, succinate, and α-ketoglutarate, but not with formate or citrate. Since the growth factor requirement of these strains is satisfied by *p*-aminobenzoic acid in microgram quantities, they are most probably representatives of the *R. palustris* group (Hutner, 1946).

We have also tested representative species of the Athiorhodaceae from the culture collection at Hopkins Marine Station to determine their ability to utilize benzoate or other aromatic compounds. Most strains of *Rhodopseudomonas palustris, R. capsulatus, R. spheroides, R. gelatinosa,* and some strains of *Rhodospirillum rubrum* grow well anaerobically in the light with benzoate as substrate.

With benzoate as hydrogen donor, growth is proportional to the concentration of benzoate. At concentrations above 60 mg % a considerable lag is observed before growth is evident. Optimum growth with benzoate as substrate is obtained in slightly alkaline media.

By analogy with *Pseudomonas,* the enzymes for the oxidation of aromatic compounds should be inducible (Stanier, 1947). In manometric experiments, benzoate-grown cells oxidize benzoate, protocatechuic acid, and catechol at the same rate without a lag. Lactate-grown cells attack these substrates only after a lag. An induction period is also involved in the oxidation of monohydroxy-substituted benzoic acids, even by benzoate-grown cells. It thus seems logical to conclude that monohydroxy-substituted benzoic acids are not intermediate products in benzoate oxidation by *Rhodopseudomonas.*

Gaffron (1941) observed that phenylpropionic acid is attacked by *Rhodovibrio* (presumably a *Rhodopseudomonas*) in the light; the remaining benzoic acid can be recovered quantitatively from the medium. Since the ability to carry out anaerobic oxidation of benzoate appears to be wide-spread among the nonsulfur purple bacteria, it is possible that these results were obtained in short-term manometric experiments with cells not previously adapted to benzoate.

In preliminary experiments with Hughes press extracts of benzoate-grown cells, protocatechuic acid appears to be decarboxylated to catechol. When benzoate serves as substrate, β-keto acids can be detected by the Rothera reaction and by catalytic decarboxylation with aniline citrate. Until the enzymes which operate in the anaerobic oxidation of benzoate by *R. palustris* have been investigated in detail, we cannot say with certainty that the pathway of benzoate oxidation follows the scheme proposed by Stanier (1950) for *Pseudomonas.* Nevertheless, the apparent involvement of a dihydroxy-substituted benzoic acid as the first oxidation product and the immediate attack by benzoate-grown cells on protocatechuic acid and catechol argue in favor of an oxidation mechanism that has so far been observed only in cases where molecular oxygen can participate.

Stanier and his associates (Stanier, 1948, 1950; Sleeper and Stanier, 1950; Sleeper, 1951) and Parr *et al.* (1949) have shown that the primary attack on benzoic acid involves a reaction in which two hydroxyl groups are simultaneously introduced into the benzene nucleus. This type of process, a "peroxidation," does not occur under anaerobic conditions, even when normally effective hydrogen acceptors such as methylene blue or tetrazolium compounds are present; it requires molecular oxygen or peroxide. In the case of catechol, Hayaishi and co-workers (1955) consider a direct cleavage by means of a peroxide-type addition complex a likely mechanism.

Judging from our knowledge of oxygen transferases (Mason, 1957) in which molecular oxygen is incorporated into the substrate, one would expect that the oxidation of benzoate in nonsulfur photosynthetic bacteria would be mediated by an iron protein, probably an iron-porphyrin protein. If benzoate oxidation is similar to that of catechol, polyvalent iron may be involved in the fission of the aromatic carbon-carbon bond (Mason, 1957). When measured as

$$Fe^{2+} \rightarrow Fe^{3+} + e^-$$

hematin compounds present in photosynthetic tissues do not have sufficiently positive potentials to function in peroxidation, but they may well do so if transformed by photochemical activation to higher oxidation levels.* The oxidation of aromatic compounds by *Rhodopseudomonas* spp. under strictly anaerobic conditions in light suggests that the photochemical reaction of these organisms involves the formation of a high-potential oxidizing entity that can form peroxides from water. A tentative scheme for benzoate oxidation by *Rhodopseudomonas* is given in Eq. 1.

$$\text{C}_6\text{H}_5\text{COOH} \xrightarrow[\text{Fe}^{IV\text{-}V(?)}]{2[OH]} \text{C}_6\text{H}_3(\text{OH})_2\text{COOH} \xrightarrow{-CO_2} \text{C}_6\text{H}_4(\text{OH})_2 \xrightarrow[\text{Fe}]{2[OH]} \text{C}_6\text{H}_2(\text{OH})_2(\text{COOH})_2 \quad (1)$$

Possible Functions of the Hydrogen Donor in Photosynthesis

Since no oxygen-evolving system is present in photosynthetic bacteria, Gest and Kamen (1959) have suggested that one of the functions of the accessory hydrogen donor might be the prevention of destructive photooxidative reactions. Photosensitized oxidations have been observed in green plants only under extreme environmental conditions, such as high light intensities or high partial pressures of oxygen. The substrates for these reactions are either cellular reserve materials or externally sup-

---

* See Paper No. 20, by M. D. Kamen, in this volume.

plied substrates (Rabinowitch, 1945). Pigments are not the first substrates to undergo photo-oxidation in plants; they remain intact as long as other oxidizable materials are available to the cells. Only after the supply of substrates is exhausted are the photoreactive pigments attacked. Thus pigment destruction is not a primary effect of the oxidizing action of intense light (Sirvonal and Kandler, 1958).

If the addition of substrates can prevent damage caused by photodynamic action, then the Gest and Kamen hypothesis may hold for both green plant and bacterial photosyntheses. It should be possible to induce a requirement for an external hydrogen donor in green plants by increasing the oxygen pressure or intensity of illumination. Under these conditions the oxygen-evolving system would be inadequate to remove the excess oxidant and an accessory hydrogen donor would become obligatory.

When certain algae are adapted to anaerobic conditions, they exhibit a metabolism reminiscent of the photosynthetic bacteria. Nakamura (1937) has found that the diatom *Pinnularia* and the blue-green alga *Oscillatoria* can reduce carbon dioxide with hydrogen sulfide, and that the evolution of oxygen was replaced by the deposition of sulfur globules in the cells. Gaffron (1939, 1940) has also noted that molecular hydrogen, glucose, or other organic substances can act as hydrogen donors for photoreduction of carbon dioxide by *Scenedesmus*. The chrysophyte flagellate *Ochromonas malhamensis* fails to grow in the light unless a suitable organic substrate is present (Hutner et al., 1953). In the presence of hydrogen donors such as glucose or glycerol, *O. malhamensis* can grow by photoreduction and can carry out essentially bacterial photosynthesis (Vishniac and Reazin, 1957).

So far, we have discussed bacterial photosynthesis as a process comparable to green plant photosynthesis, involving the photochemical production of an oxidizing moiety, which, while not molecular oxygen proper, nevertheless is equivalent to O_2 for the oxidation of benzoic acid and other aromatic compounds requiring the participation of a high-potential oxidant. This is in agreement with the interpretation of bacterial photosynthesis as a photochemical action through which dissociation of water produces an oxidant (van Niel, 1941). It must be admitted however, that recent developments have suggested other possibilities; Arnon (1959) has postulated a "cyclic" photosynthetic phosphorylation reaction as the common denominator of both green plant and bacterial photosyntheses. In the case of green plants, there is an additional "noncyclic" phosphorylation reaction leading to the formation of both reduced pyridine nucleotide and adenosine triphosphate. The photosynthetic bacteria are considered to be incapable of this reaction, and to derive reduced pyridine nucleotide from the external hydrogen donor.*

But if the only function of light is the generation of phosphate bond energy [a suggestion which was made earlier by Emerson et al. (1944)], then the oxidation of benzoate should proceed with oxidants other than molecular oxygen or peroxide. Thus far, no knowledge is available of benzoate oxidation occurring under anaerobic conditions with oxidants such as nitrate, sulfate, etc., although Gaffron (1933) has shown that nitrate can serve as an oxidant for fatty acid oxidation by a strain of nonsulfur purple bacteria in the dark. Our experiments with nitrate, nitrite, and sulfate as hydrogen acceptors for the anaerobic oxidation of benzoate and other substrates by *Rhodopseudomonas* have been negative. Nitrate reduction with various aliphatic compounds has been observed, but not with aromatics such as benzoate (Allen and van Niel, 1952). Elective cultures for sulfate reducers with aromatic compounds have also yielded negative results so far. We are aware of only one report that benzoate is decomposed under anaerobic conditions in darkness (Buswell and Hatfield, 1936). These workers used crude cultures of methane-producing bacteria; thus far, no pure cultures of methane bacteria are known which use benzoate anaerobically.

Summary

The experiments to date with benzoate and other aromatic compounds indicate that bacterial photosynthesis involves the production of an oxidant equivalent to molecular oxygen produced as a consequence of a photochemical reaction. The close similarity between green plant photosynthesis and bacterial photosynthesis is also suggested as resulting from the need for, or use of, oxidizable substrates in both types of photosynthesis, i.e., by photosynthetic bacteria under normal conditions and by organisms carrying out green plant photosynthesis under special conditions.

Acknowledgment

It is a pleasure to acknowledge the guidance of Professor C. B. van Niel, who took an active interest in the work, and provided many helpful suggestions in the preparation of the manuscript.

---

* Certain reactions that occur in photosynthetic bacteria, e.g., the synthesis of succinate from propionate and carbon dioxide by *Chlorobium* (Larsen, 1951) and the synthesis of poly-β-hydroxybutyrate from acetate by *Rhodospirillum* (Stanier et al., 1959) do not require reduction of CO_2; for these the generation of adenosine triphosphate by "cyclic" photophosphorylation is sufficient.

References

Allen, M. B., and van Niel, C. B. (1952). *J. Bacteriol.* **64**, 397-412.
Arnon, D. I. (1959). *Nature* **184**, 10-21.
Baltscheffsky, H. (1960). *Biochim. et Biophys. Acta* **40**, 1-8.
Buswell, A. M., and Hatfield, W. D. (1936). *Illinois State Water Survey Bull.* **32**.
Emerson, R. L., Stauffer, J. F., and Umbreit, W. W. (1944). *Am. J. Botany* **31**, 107-120.
Gaffron, H. (1933). *Biochem. Z.* **260**, 1-17.
Gaffron, H. (1939). *Nature* **143**, 204-205.
Gaffron, H. (1940). *Am. J. Botany* **27**, 273-283.
Gaffron, H. (1941). Unpublished results referred to in Franck, J., and Gaffron, H. *Advances in Enzymol.* **1**, 200-262.
Gest, H., and Kamen, M. D. (1959). In "Handbuch der Pflanzenphysiologie" (W. Ruhland, ed.), Vol. 5, in press. Springer, Berlin.
Hayaishi, O., Katagiri, M., and Rothberg, S. (1955). *J. Am. Chem. Soc.* **77**, 5450-5451.
Hill, R. (1939). *Proc. Roy. Soc.* **B127**, 192-210.
Hutner, S. H. (1946). *J. Bacteriol.* **52**, 213-221.
Hutner, S. H., Provasoli, L., and Filfus, J. (1953). *Ann. N. Y. Acad. Sci.* **56**, 852-862.
Kamen, M. D. (1956). In "Enzymes: Units of Biological Structure and Function" (O. H. Gaebler, ed.), pp. 483-498. Academic Press, New York.
Kamen, M. D. (1957). In "Research in Photosynthesis" (H. Gaffron, ed.), pp. 149-163. Interscience, New York.
Larsen, H. (1951). *J. Biol. Chem.* **193**, 167-173.
Larsen, H. (1953). *Kgl. Norske. Vidensk. Selskab. Skr.* **1**, 1-199.
Larsen, H. (1954). In "Autotrophic Microorganisms" (B. A. Fry and J. L. Peel, eds.), pp. 186-201. Cambridge Univ. Press, London and New York.
Larsen, H., Yocum, C., and van Niel, C. B. (1952). *J. Gen. Physiol.* **36**, 161-171.
Mason, H. S. (1957). *Advances in Enzymol.* **19**, 79-233.
Molisch, H. (1907). "Die Purpurbakterien nach neuen Untersuchungen." Fischer, Jena.
Nakamura, H. (1937). *Acta Phytochim. Japan* **9**, 189-229.
Parr, W. H., Evans, R. A., and Evans, W. C. (1949). *Biochem. J.* **45**, xxix-xxx.
Rabinowitch, E. I. (1945). "Photosynthesis and Related Processes," Vol. 1, Chapter 18. Interscience, New York.
Roelofson, P. A. (1935). "On the Photosynthesis of the Thiorhodaceae." Dissertation, Utrecht, Holland.
Sirvonal, C., and Kandler, O. (1958). *Biochim. et Biophys. Acta* **29**, 359-368.
Sleeper, B. P. (1951). *J. Bacteriol.* **62**, 657-662.
Sleeper, B. P., and Stanier, R. Y. (1950). *J. Bacteriol.* **59**, 117-127.
Stanier, R. Y. (1947). *J. Bacteriol.* **54**, 339-348.
Stanier, R. Y. (1948). *J. Bacteriol.* **55**, 477-494.
Stanier, R. Y. (1950). *Bacteriol. Revs.* **14**, 179-191.
Stanier, R. Y., Doudoroff, M., Kunisawa, R., and Contopoulou, R. (1959). *Proc. Natl. Acad. Sci. U. S.* **45**, 1246-1260.
van Niel, C. B. (1941). *Advances in Enzymol.* **1**, 263-328.
van Niel, C. B. (1949). In "Photosynthesis in Plants" (J. Franck and W. E. Loomis, eds.), pp. 437-495. Iowa State College Press, Ames, Iowa.

Vishniac, W., and Reazin, G. H., Jr. (1957). *In* "Research in Photosynthesis" (H. Gaffron, ed.), pp. 239-242. Interscience, New York.
Wassink, E. C. (1947). *Antonie van Leeuwenhoek* **12**, 281-293.
Wassink, E. C., Katz, E., and Dorrestein, R. (1942). *Enzymologia* **10**, 285-354.

26

Fluorescence Parameters and Photosynthesis

G. WEBER

Department of Biochemistry, University of Sheffield, Sheffield, England

The main pigments involved in the absorption of radiant energy by the photosynthetic systems show conspicuous fluorescence when in solution in relatively pure condition, but very much less when *in situ* in the organism. Since the fluorescent emission results in a loss of the energy originally absorbed by the system a general negative correlation between fluorescence and energy utilization may be expected. The parameters characteristic of the emitted radiation may also give important information concerning properties of the system closely related to its capacity of energy utilization. The parameters that will be examined in detail in the following are:

1. The absolute quantum yield of the fluorescence;
2. The fluorescence spectrum;
3. The fluorescence excitation spectrum;
4. The lifetime of the excited state;
5. The polarization of the fluorescence.

When the fluorescence is incoherent dipole radiation (case of all known organic materials) these parameters contain all the relevant information about the emitted fluorescence. In this paper the information obtainable will be discussed in relation to available data.

Absolute Quantum Yield

Chlorophyll *a* in solution shows a maximum yield in ethyl ether solution of 0.33. This value seems now well established by two independent methods of measurement: (*a*) comparison with dipolar scattering (Weber and Teale, 1957); and (*b*) total emission measurements by the integrating sphere (Latimer, 1956). Similarly chlorophyll *b* has yielded a value of 0.12. In *Chlorella* the quantum yield, measured by similar methods, is 0.025 ± 0.003. Diatoms show quantum yields in the neighborhood of 0.01 (Teale, 1959). The fluorescence yield can best be discussed by reference to the scheme of Fig. 1. G is the ground state; S_1, S_2, S_3 .... are excited singlet electronic states reached in absorption of light. Fluorescence emission takes place from the lowest vibrational levels of S_1 (fluorescence state). On absorption of light the molecule may reach the state S_2 or S_3, instead of S_1, but in any case the excess energy over that of the fluorescent state is lost to the medium in a time of the order of 10^{-11} second (Weber and Teale, 1958). Only the energy of the molecule in the fluorescent state can be utilized in photosynthesis. This utilization is not the only process by which energy can be dissi-

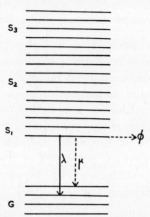

FIG. 1. Energy levels involved in partially competitive mechanism between chlorophyll a fluorescence and photosynthesis. KEY: G, ground state; S_1, S_2, S_3, singlet electronic excited states. Radiative transitions in full line. Radiationless transitions in broken lines.

pated. In general the fluorescence level may be depopulated by three types of processes: (a) radiationless transitions to the ground state, proceeding at rate μ; (b) emission of fluorescence (rate λ); (c) transfer of the energy to an unspecified system capable of converting it into chemical free energy (this "rate of photosynthesis" will be denoted by ϕ). At this point it is preferable to consider an over-all rate of radiationless transition μ without attempting to decide whether more than one process is involved.

With these definitions of the rates we can now define the absolute quantum yields of the three processes:

q_λ = Fluorescence yield = quanta emitted/quanta absorbed = $\lambda/(\lambda + \phi + \mu)$

q_ϕ = Photosynthesis yield = quanta utilized in photosynthesis/quanta absorbed = $\phi/(\lambda + \phi + \mu)$

q_μ = Dark deactivation yield = quanta lost by radiationless processes/quanta absorbed = $\mu/(\lambda + \phi + \mu)$

When $\mu = 0$ we have the case of *total competition* between fluorescence and photosynthesis. Inhibition of photosynthesis at the level of utilization of the energy from the excited state of chlorophyll ($\phi = 0$) must result in a complementary increase in the quantum yield of fluorescence, reaching the value 1 when "photosynthesis" is totally inhibited. Increases in fluorescence yield of the order of 3 with over-all inhibition of photosynthesis have been observed by several authors (Wassink and Katz, 1939; Shiau and Franck, 1947), but the greatest increases, observed after use of inhibitors of the substituted urea type (Teale, private communication), are only of the order of 5, thus raising the absolute quantum yield of fluorescence to a maximum of 0.1. Clearly we cannot have a case of total competition between fluorescence and photosynthesis and other radiationless losses must also be included.

In the case of partial competition between fluorescence and photosynthesis, calling Q_λ the maximum fluorescence yield, observed when "photosynthesis" is inhibited ($\phi = 0$), we have from the preceding definitions:

$$Q_\lambda = \lambda/(\lambda + \mu)$$

and in general,

$$\phi/\lambda = (1/q_\lambda) - (1/Q_\lambda) \tag{1}$$

so that

$$q_\phi = 1 - (q_\lambda/Q_\lambda) \tag{2}$$

Thus the observed increase in fluorescence yield by a factor of 5 is consistent with a yield of photosynthesis as high as 80%.

The possibility must be considered that the energy utilization in photosynthesis takes place from a state different from the fluorescent state. Figure 2 shows this *"noncompetitive"* case. T is here a metastable

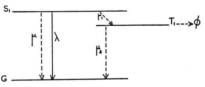

Fig. 2. Electronic energy levels involved in noncompetitive mechanism between chlorophyll *a* fluorescence and photosynthesis. Same conventions as in Fig. 1. S_1, fluorescence state; T_1, low-lying metastable state; G, ground state.

state of energy lower than that of the singlet state. The existence of this state is now well substantiated for chlorophyll *a* (Livingstone, 1955) and for chlorophyll *b* (Linschitz and Sarkanen, 1958; Claesson, 1959) in solutions, although attempts to demonstrate its existence *in vivo* have not been successful (Rosenberg et al., 1957). Formally this case is identical with that of partial competition of Fig. 1, the transition μ, between S and T taking here the place of ϕ. The only difference is in the existence of a process of dark deactivation μ_2 which is noncompetitive with the fluorescence. The experimental data so far available do not explicitly demand the existence of this process, nor are they sufficient to exclude it.

Two different processes are seen to quench the fluorescence. Energy utilization with rate ϕ and dark deactivation with rate μ. If it is admitted, in agreement with some direct (Emerson and Arnold, 1932) and much indirect evidence (e.g., Wessels and van der Veen, 1956; Spikes, 1956) that energy utilization takes place at a number of centers very small compared with the total number of chlorophyll molecules, ϕ becomes different for the different molecules and has the meaning of an average rate of transfer of energy to a molecule in which there is an overwhelming probability of it being utilized in photosynthesis. The actual *quenching* of the fluorescence can then take place in any chlorophyll molecule with rate μ, or in a restricted number of privileged chlorophyll molecules with a rate ϕ very much greater than λ or μ. The over-all rate of energy utilization ϕ is then determined by the average rate of transfer from an ordinary chlorophyll molecule to a privileged one. The first type of quenching (rate μ) will be called "distributed quenching," the second (rate $\bar{\nu}$) will be called "terminal quenching."

Fluorescence Excitation Spectrum

The fluorescence excitation spectrum and the fractional absorption spectrum of the chlorophylls and pheophytins in solution are identical (Weber and Teale, 1958). Thus in solution there is no indication of the existence of tautomeric forms of chlorophyll with different quantum yields.

The fluorescence excitation spectrum of *Chlorella* and other photosynthetic organisms is complex, indicating the existence of energy transfer from several accessory pigments and from chlorophyll *b* to chlorophyll *a*, the pigment absorbing at the longer wavelengths (Duysens, 1952). The transfer of electronic energy from the carotenoids to chlorophyll *a* can be demonstrated in model systems of micelles containing both pigments in high local concentration, yet maintaining a small over-all optical density, if a dilute micelle suspension is used (Teale, 1958).

The decrease in the over-all quantum yield of photosynthesis on illumination with light of wavelength longer than the absorption maximum of chlorophyll has been reported by several authors (e.g., Emerson and Lewis, 1943). In interpreting these observations it is necessary to keep in mind the errors in the absorption measurements introduced by the scattering of the exciting light. Particularly important in this respect is the recently studied specific scattering (Latimer and Rabinowitch, 1956; Latimer, 1959) that accompanies the anomalous dispersion at the edges of an absorption band. It is doubtful whether the opal glass method of Shibata and co-workers (1954) introduces more than a partial correction to the scattering error.

Obviously the method to be relied upon in this case is the integrating sphere. Teale (1959) found a constant fluorescence quantum yield in *Chlorella* (within 5%) over the range of excitation 600–700 mµ. This constancy in the fluorescence yield was observed only in dilute suspensions in which the fluorescence was uniformly emitted from the whole illuminated volume, when the absorption was measured by the integrating sphere. In these studies the specific scattering was eliminated or much reduced by using in the fluorescence path a polarizing filter oriented so as to exclude the vertical component of the scattering and by the use of a suspending medium of high refractive index (concentrated sucrose, propylene glycol) for the *Chlorella*. The separation of the exciting from the fluorescent light requires a filter with a sharp cutoff at about 710 mµ, and thus precludes the observation of the quantum yield at longer wavelength excitation. The data are shown in Fig. 3, which includes the values observed in concentrated suspensions.

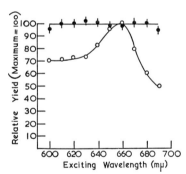

Fig. 3. Relative fluorescence efficiency of *Chlorella* for different exciting wavelengths (from Teale, 1959). *Open circles:* thick suspension with maximum optical density 2. *Dark circles:* thin suspension with maximum optical density 0.2.

The observations show that if more than one type of chlorophyll *a* molecule is present, this second type is either in small proportion or has the same fluorescence yield as the main chlorophyll population.

Lifetime of the Excited State of Fluorescence

The lifetime of the excited state of chlorophyll *a* and *b* in solution has been determined by Dimitrievsky and co-workers (1957) and by

Brody (1958). Both studies gave values of 5 mµseconds for chlorophyll *a* in various solvents. It is known that if the processes that decrease the absolute quantum yield from unity are competitive with the emission

$$\tau = q \cdot \tau_n \qquad (3)$$

where τ is the actual (observed) lifetime and τ_n the natural lifetime of the oscillator (also called by some authors the emissive lifetime). The natural lifetime can be calculated from the oscillator strength of the absorption band of lowest frequency (e.g., Förster, 1951). For chlorophyll *a* this is 15 mµseconds so that the observed lifetime agrees very well with the determined quantum yield of 0.33 already quoted. In *Chlorella*, Brody measured a lifetime of the excitation of 2 mµseconds and similar values were reported by Dimitrievsky for *Chlorella* and *Elodea*. Use of the simple equation (3) above would give $\tau \cong 0.3$ mµseconds while the observed value is 5 to 10 times longer. The discrepancy between the observed and calculated lifetimes may yield information about the photosynthetic unit. Equation 3 applies when the observed chlorophyll molecules are a homogeneous population. Homogeneous means in this context that at the moment of the excitation all molecules have the same *a priori* probability of emission, and heterogeneous that we can divide the excited molecules into groups having different *a priori* probabilities of emission and therefore different quantum yields. The observed quantum yield will be

$$\overline{q} = \sum_i q_i n_i / \sum_i n_i \qquad (4)$$

where n_i is the number of emitting units in group i having yield q_i. As stated the equation assumes that all the groups have the same fluorescent spectrum, but it is also valid for groups of molecules having different fluorescent spectra if n_i denotes the number of quanta emitted by group i under the given excitation conditions.

Since the observed lifetime of the excited state is weighted with respect to emitted intensity I,

$$\overline{\tau} = \sum \tau_i I_i / \sum I_i = \sum \tau_i q_i n_i / \sum q_i n_i \qquad (5)$$

To each homogeneous group Eq. 3 is applicable so that $\tau_i = q_i \tau_n$; τ_n being the same for all groups.

$$\overline{\tau}/\tau_n = \sum q_i^2 n_i / \sum q_i n_i = (1/\overline{q}) \left(\sum q_i^2 n_i / \sum n_i \right) \qquad (6)$$

Equation 6 shows that $\overline{\tau} \geqslant \tau_n \cdot \overline{q}$.

We must now discuss the possible causes of inhomogeneity in the chlorophyll population:

1. There are two types of chlorophyll *a* molecules, differing in their quantum yield of fluorescence. The minimum number of the second type would be found, in accordance with Eq. 4, when the difference in the quantum yields of the two types is a maximum. If the yield of species 1 is q_1 and the yield of species 2 is 0, then

$$\bar{q} = f_1 q_1 \quad (7)$$
$$q_1 = \tau/\tau_n$$

and from quoted values f_1, the fraction of molecules with the higher yield is 1/7. Since there must be *a priori* differences in the probability of emission of the two types of molecules, these differences must exist at a time before the actual excitation takes place, and would be expected to give rise to changes in the absorption spectrum. The quenching of fluorescence by complex formation (Weber, 1948; Epple and Förster, 1954) in which there are two types of molecules—free, with nearly the normal quantum yield, and bound in a nonfluorescent complex—will be the solution equivalent of this case. A change in the absorption spectrum may be found in these cases (Weber, 1949; Epple and Förster, 1954) although this may be very small and difficult to detect in practice. While there is at present no definite evidence for the existence of a second type of chlorophyll *a* molecules in the proportion required, it is not possible to exclude this possibility at the present time. However, without postulating the existence of two kinds of molecules, another type of heterogeneity may be found if the quenching of the fluorescence involves the transfer of the excitation energy to a particular molecule by resonance transfer, or indeed by any mechanism competitive with the emission:

2. In this second type, the heterogeneity is one of position with respect to the quenching or trapping center, responsible for the terminal quenching already defined. Consider, for example the linear array of Fig. 4; the dotted line represents a chlorophyll molecule where there is overwhelming probability of quenching ($\phi \gg \lambda$), and the full lines the ordinary chlorophyll molecules.

A molecule separated from T by n others will be quenched only if the excitation energy travels by transfer n steps all in the same direc-

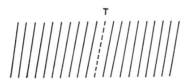

FIG. 4. Schematic arrangement of fluorescent molecules (solid lines) and quencher T.

tion, which requires an average of n^2 steps in a random walk. For each n value there is a different probability of quenching, and the exact calculation of $\bar{q}$ and $\bar{\tau}$ requires a knowledge of the spatial arrangement of the molecules in the chloroplast.

A case similar to this occurs in the quenching of the fluorescence in solution due to resonance transfer of the excitation to a nonfluorescent molecule. Approximate calculations of $\bar{q}$ and $\bar{\tau}$ for this system can be made, on the assumption that the quenching is due to transfer of the excitation to the nearest quenching neighbour.

If R is the characteristic distance for the transfer (that is, the distance between fluorescer and quencher at which the transition probability of transfer $\bar{v}$ equals the transition probability of emission λ), it is known (Perrin, 1932; Förster, 1947), that

$$\bar{v}(r)/\lambda = R^6/r^6 \tag{8}$$

where r is the actual distance between the two molecules.

The quantum yield $q(r)$ of molecules having a nearest quenching neighbor at distance r is

$$q(r) = \lambda/[\lambda + \bar{v}(r)] = r^6/[R^6 + r^6] \tag{9}$$

and the fraction of such molecules is (e.g., Chandrasekhar, 1943)

$$H(r) = 4\pi r^2 L \exp(-4\pi r^3 L/3),$$

with L equal to the number of quencher molecules in 1 ml. Thus, using Eqs. 4 and 6, we have

$$\bar{q} = \int_{2a}^{\infty} q(r)H(r)dr \Big/ \int_{2a}^{\infty} H(r)dr$$

$$\tau/\tau_n = \int_{2a}^{\infty} q^2(r)H(r)dr \Big/ \int_{2a}^{\infty} q(r)H(r)dr \tag{10}$$

$2a$ is the distance of maximal approach between the molecules. Figure 5 gives the values of relative yield and lifetime for this case as a function of quencher concentration. The concentration has been normalized by the use of the parameter $\beta = 4\pi R^3 L/3$. Thus $\beta = 1$ indicates that on the average 1 molecule of quencher is found within a spherical volume of radius R. Quantitatively the calculations are in agreement with the results of Galanin and Lewschin (1951) on the quenching of the fluorescence of 3-aminophthalimide by a variety of colored quenchers, and also with the studies on the quenching of the fluorescence of fluorescein by erythrosin. In the latter case Förster (1947) found $q/q_0 = \frac{1}{2}$ when the erythrosin concentration c was $7.10^{-4} M$, while Schmillen (1953) found $\tau/\tau_0 = \frac{1}{2}$ when $c = 1.5 \; 10^{-3} M$. The ratio of the two concentrations predicted by the graph is 2.3, the observed ratio 2.1.

The relation of the trapping center in the photosynthetic unit to the average chlorophyll molecule is analogous to the relation of the average fluorescent molecule to its nearest quenching neighbor. The difference between the two cases is that in the former the excitation is directly transferred from fluorescent molecule to quencher, while in the latter it has to travel from one to the other through a number of contiguous

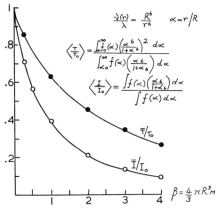

FIG. 5. Changes in relative quantum yield and relative lifetime of the excited state as a function of the quencher concentration (proportional to β) in quenching by transfer of the excited state.

fluorescent molecules which are excited transitorily. When the quenching center has been rendered inoperative, transfer of the excited state among the fluorescent molecules will increase the quenching no further. Experimentally it will be possible to detect this event by simultaneous measurements of $\bar{q}$ and $\bar{\tau}$. Since only the homogeneous, distributed quenching is then present the relation $\bar{\tau} = \bar{q}\tau_n$ will be obeyed. The use of Eq. 2 will then permit a determination of the maximum possible quantum yield of photosynthesis.

Polarization of the Fluorescence

The polarization of the fluorescence of molecules in solution reaches a maximum value in a viscous medium in which molecular rotational relaxation time is much longer than the lifetime of the excited state (Perrin, 1929; Weber, 1953). Depolarization of the fluorescence in dilute solution in nonviscous solvents occurs as a result of the brownian molecular rotations, in concentrated rigid solutions as a result of transfer of the excited state (Perrin, 1929). The first cause of depolarization is sensitive to changes in the temperature and viscosity of the medium, while the second is not. Following the observation of Arnold and Meek (1956) that the fluorescence from chloroplasts is poorly polarized ($p = 0.03$) as compared with that of chlorophyll a in castor oil ($p = 0.42$), the present author has attempted to distinguish between the two possible causes of depolarization just mentioned by measuring the polarization of the red chloroplast fluorescence as a function of temperature. Between 2°C and 38°C, the fluorescence from *Brassica* chloroplasts and from *Chlorella* does not change by more than 5%. If the original depolarization is assumed to be due to rotations in a medium of which the viscosity changes with temperature in an ordinary manner (2–3% per degree centigrade), a decrease in polarization of about —35% was expected. It can then be concluded that transfer of the excitation rather than molecular rotation is involved in the loss of polarization.

On simple geometrical considerations (Weber, 1954) Eq. 11 can be derived

$$(1/p) - (1/3) = [(1/p_\infty) - (1/3)] [1 + (3/2) \overline{\sin^2\theta}\, \overline{n}] \quad (11)$$

where p is the observed polarization, $\overline{\sin^2\theta}$ the average sine square of the molecular oscillators involved in any one transfer, and $\overline{n}$ the average number of transfers. p_∞ is the polarization observed in the absence of transfers ($n = 0$). It is clear that if $\overline{\sin^2\theta} = 0$, that is, in transfer between parallel oscillators, any number of transfers will leave the polarization unchanged and any calculation of $\overline{n}$ from p and p_∞ will require a knowledge of the orientation. When this is random, as in solutions, $\overline{\sin^2\theta}$ may be calculated explicitly (Weber, 1954) and, if c is the molar concentration; then

$$\begin{aligned}(1/p) - (1/3) &= [(1/p_\infty) - (1/3)] (1 + Ac) \\ A &= [(4\pi Nc)/(15 \times 10^3)] [R^6/(2\alpha)^3]\end{aligned} \quad (12)$$

The latter equation can be used to determine R from a plot of $1/p$ against c and is shown in Fig. 6 for solutions of chlorophyll a in liquid paraffin. It is seen that the linear law is well followed at the lower concentrations with slope corresponding to $R = 36$ Å. The minimum of p at the higher concentrations is undoubtedly due to self-quenching which decreases the lifetime of the excited state. Concentrated chlorophyll solutions show dimerization, the dimers having negligible quantum yield as compared with the monomer (Weber and Teale, 1958). Transfer of the excitation to a dimer through a chain of monomers will result in quenching, and we may expect a peculiar law of concentration quenching under this condition.

The theory of concentration depolarization already developed (Weber, 1954) can be adapted to this case. If it is assumed that only

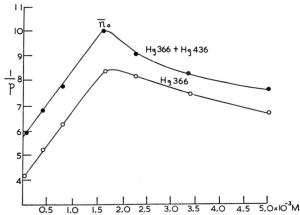

FIG. 6. Polarization of the fluorescence of chlorophyll a in liquid paraffin. *Solid circles:* excitation with Hg 366 and Hg 436. *Open circles:* excitation with Hg 366.

fluorescent monomers and nonfluorescent dimers exist in solution then the equilibrium constant K is related to α the degree of dissociation of the dimers and c the molar concentration of chlorophyll a by the equation

$$K = [2\alpha^2/(1-\alpha)]c \qquad (13)$$

With an argument already developed (Weber, 1954),

$$\bar{n} = \alpha Ac/[1 + Ac(1-\alpha)] \qquad (14)$$

where A is given by Eq. 12 and is the slope of the straight line in the plot of $1/p$ against c, when c tends to 0. The condition of $\bar{n}$ maximum obtained as usual from $d\bar{n}/dc = 0$ yields

$$\begin{aligned}\alpha_0 &= (Ac_0 - 1)/Ac_0 \\ K &= [2(Ac_0 - 1)^2/Ac_0]c_0\end{aligned} \qquad (15)$$

where α_0 and c_0 are the values of the degree of dissociation and concentration obtaining at the minimum value of p or maximum $\bar{n} = \bar{n}_0$). From the values in Fig. 4,

$$\begin{aligned}n_0 &= 1.35 \\ c_0 &= 1.7 \times 10^{-3}M \\ \alpha_0 &= 0.75 \\ K &= 7.7 \times 10^{-3}M\end{aligned}$$

The low dimerization constant is to be expected because of the low solubility of chlorophyll *a* in liquid paraffin.

The ideas of transfer of the excitation to the dimer through a chain of monomers may be used to develop a theory of the concentration quenching of the fluorescence of chlorophyll *a* solutions: The probability of *n* transfers among monomers when no dimers are present is

$$[\bar{v}/(\lambda + \bar{v})]^n$$

where v is an average rate of transfer. In general this probability is

$$[\bar{v}/(\lambda + \bar{v})]^n \alpha^{n+1} \tag{16}$$

since *n* transfers among monomers require $n + 1$ monomers.

The relative quantum yield of the solution F/F_0 equals

$$F/F_0 = \frac{\Sigma\,[\,\bar{v}/(\bar{v}+\lambda)\,]^n \alpha^{n+1}}{\Sigma\,[\,\bar{v}/(\bar{v}+\lambda)\,]^n} = \frac{\alpha}{1 + (\bar{v}/\lambda)(1-\alpha)} \tag{17}$$

and replacing $\bar{v}/\lambda$ and α by their respective values Ac and

$$\tfrac{1}{2}\sqrt{1 + (8c/K)}$$

we find,

$$F/F_0 = 2/[(1 + \sqrt{1 + 8(c/K)} + AK(c/K)(\sqrt{1 + (8c/K)} - 1)] \tag{18}$$

in which the concentrations are expressed in K units, and the additional parameter AK is introduced, A being the initial slope in the plot of $1/p$ against c. Thus the only adjustable parameter in the quenching measurements is the dimer dissociation constant K, since A is independently obtained from polarization measurements. Careful measurements of the concentration dependence of the fluorescence of solutions of chlorophyll *a* in ether have been made by Watson and Livingstone (1950). Using $A = 980$ (obtained from the polarization measurements), and assuming that the probability of transfer is to a first approximation independent of the nature of the solvent, and that $K = 0.22\,M$, the curve of Fig. 7 is found. The values of Watson and Livingstone are superimposed. They fit the quenching curve with a standard deviation of 1.5%. There seems to be little doubt then that transfer of the excitation through the monomers to the nonfluorescent dimers explain the experimental results of depolarization as well as quenching. The easy dimerization of chlorophyll in solutions demands an explanation of the absence (or, at least, the relative rarity) of dimers in the chloroplast. If the polarization data of the chloroplast referred to a solution, to which Eq. 11 is applicable, it would be concluded that 12–15 transfers take place. In an organized structure the number of transfers must be much larger since $\overline{\sin^2\theta}$ must be smaller than the corresponding random value.

Thus the proportion of dimers must be very small, even if it is assumed that trapping of the excitation in a dimer molecule does not lead to its loss to the photosynthetic process. The low proportion of dimers (Brody, 1958) demands some structural characteristic in the

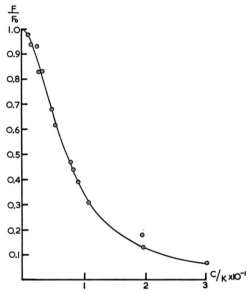

FIG. 7. Concentration quenching of chlorophyll *a* in ethyl ether from data of Watson and Livingstone (1950), compared with thoretical curve of Eq. 18.

chloroplast that makes the relation between neighboring molecules different from the relations in solution. Such a situation is not unique; while crystals of anthracene have a high fluorescence efficiency, self-quenching by molecular interactions is predominant in concentrated solutions.

Excitation Energy Transfer

The observations on polarization and yield of fluorescence, as well as direct physiological observations, make it necessary to assume the rapid transfer of the excited state to a reactive center, the number of such centers being of the order of 1/100 of the chlorophyll molecules. If transfer of the excitation among the chlorophyll *a* molecules proceeds by random walk until a reacting center is reached, 10^4 such transfers will be required on the average during the lifetime of the excited state if the center is to be reached with a high probability. Rabinowitch and Brody (1958) suggested that the minimal time for a transfer between chlorophyll molecules is of order 10^{-12} second and thus that only 100–200 transfers take place during the lifetime of the excitation. There appears to be no reason to use a value of 10^{-12} second as the transfer time between neighboring molecules, even if some nuclear vibrations are required for the transfer as the recent observations of Lyons and White (1958) appear to indicate. On the other hand, if only the experimental data are considered, the study of the sensitized fluorescence of mixed crystals of naphthalene and of anthracene by Wright (1955) has shown that some 10^4 transfers among anthracene molecules may take place during the 3 to 4 mµsec of the lifetime of the excitation. Sensitization of the fluorescence by virtually nonfluorescent species (quantum yield less than 10^{-3}) is known to occur in reduced diphosphopyridine nucleotide (Weber, 1957) and in mixtures of fucoxanthol and chlorophyll a (Teale, 1958). From these data it appears that at present there is no need to postulate a mechanism other than that of resonance energy transfer among the chlorophyll molecule as an explanation for the funneling of the excitation energy to a few acceptor centers.

Relation of the Delayed Light to Fluorescence

From the original studies of Strehler and Arnold (1951), as well as from more recent observations of Arnold and his co-workers (Arnold and Davidson, 1954; Arnold and Sherwood, 1956), the existence is well known of a luminescence of algae and chloroplasts having the same spectral distribution as the fluorescence but with a lifetime at least a million times longer. Although the quantum yield of this delayed light is not known, it is believed to be only a small fraction of that of the fluorescence.

Two main opinions have been held as regards the origin of the delayed light. According to the first (Strehler, 1957), it is in a general way a chemiluminescence resulting from a reversal of photosynthesis. Its low yield indicates the improbable reversibility of the mechanism of energy utilization. In the second view, the delayed light indicates the existence of a solid state (semiconductivity) mechanism in the chloroplasts. The luminescence follows the trapping of photoelectrons ejected from the chlorophyll molecules which are the radical holes. The return of the electrons to the traps results in the reappearance of the singlet electronic excited state of chlorophyll from which the ordinary fluorescence takes place. The importance of the phenomenon depends on the relation of the semiconductivity phenomenon to photosynthesis. The returning electrons may be thought of as those lost to the general semiconductivity process required in photosynthesis (Tollin et al., 1958) or the semiconductivity itself may be considered as a parasitic phenomenon accompanying photosynthesis but having no other effects than an insignificant decrease in its quantum yield (Brugger and Franck, 1958).

The difficulty of accepting a general semiconductor state as an obligatory step in the mechanism of energy utilization in photosynthesis resides in the very small quantum yield of photoconductivity of organic substances in films, which should be the nearest analog to the chloroplast situation. This poor semiconductivity must be due in a large measure to the strong coupling of the two electrons (ground and excited) with opposite spins in the molecule. This strong coupling must prevent the separation of the electron from the hole, and would be much decreased by reaction with a nearby *different* molecule, in general a reductant or electron donor D. Now the excited electron can be separated from the hole and eventually transferred to an electron acceptor A. A series of reactions can then be envisaged (Scheme 1).

(1) $Chl_{\uparrow\downarrow} + h\nu \longrightarrow Chl_{\uparrow}{}^{\uparrow}$

(2) $Chl_{\uparrow}{}^{\uparrow} \cdots\cdots\cdots Chl_{\uparrow}{}^{\uparrow} D$

(3) $Chl_{\uparrow}{}^{\uparrow} D \longrightarrow Chl_{\uparrow\downarrow}{}^{\uparrow} D^{+}$

(4) $Chl_{\uparrow\downarrow}{}^{\uparrow} \cdots\cdots\cdots Chl_{\uparrow\downarrow}{}^{\uparrow} A$

(5) $Chl_{\uparrow\downarrow}{}^{\uparrow} A \longrightarrow Chl_{\uparrow\downarrow} A^{-}$

(I)

The distinctive point of this scheme is that it requires resonance transfer (2) to bring the excited electron in contact with the electron donor D, which may be expected to exist in small amounts in comparison with the chlorophyll molecules. Stages (1), (2), and (3) occur during the lifetime of the excitation of the chlorophyll. Distributed quenching occurs principally during stage (2) and terminal quenching can be identified with stage (3). Step (4) is the migration of an electron among the chlorophyll molecules, or to some structure closely related in space with the chlorophyll molecules, and (5) is its reaction with an electron acceptor, either spatially fixed as D, or freely diffusible. The delayed light emission results from the reversal of step (3) alone, or (3), (4), and (5) successively. The delayed light may be expected to have a complex decay kinetics dictated not only by the probability of reversal of steps (3), (4), and (5) but also by the disappearance of D^+ (to D) and A^- (to A) by chemical reactions. The net result of all the steps is the creation of a redox pair

$$D + A + h\nu \rightarrow D^+ + A^-$$

with a free energy that may be calculated from the redox potential of the partners. If D is ferrocytochrome f and the A^-–A couple has a redox potential like the DPN–DPNH couple the over-all increase in free energy would be 0.85 volt or nearly 20 kg cal per photon.

Such a redox couple could furnish materially, and probably energetically, the conditions for ATP and TPNH production required for the fixation of carbon dioxide.

In the above scheme both semiconductivity and reversal of the chemical reactions play a part in the production of the delayed light, which is seen not as a spurious phenomenon unrelated to energy utilization, but as one directly connected with the primary photochemical process in photosynthesis.

REFERENCES

Arnold, W., and Davidson, J. B. (1954). *J. Gen. Physiol.* **57**, 667.
Arnold, W., and Meek, E. S. (1956). *Arch. Biochem. Biophys.* **60**, 82.
Arnold, W., and Sherwood, H. K. (1956). *Proc. Natl. Acad. Sci. U. S.* **43**, 105.
Brody, S. (1958). *Science* **128**, 838.
Brody, S., and Rabinowitch, E. (1957). *Science* **125**, 565.
Brugger, J. E., and Franck, J. (1958). *Arch. Biochem.* **75**, 465.
Chandrasekhar, S. (1943). *Revs. Modern Phys.* **15**, 86.
Claesson, S. (1959). *Nature* **183**, 661.
Dimitrievsky, O. D., Ermolaev, B. L., and Terenin, A. N. (1957). *Doklady Akad. Nauk S. S. S. R.* **114**, 751.
Duysens, L. N. M. (1952). Transfer of excitation energy in photosynthesis. Doctoral Thesis, Utrecht, Holland.
Emerson, R., and Arnold, W. (1932). *J. Gen. Physiol.* **15**, 391.
Emerson, R., and Lewis, C. M. (1943). *Am. J. Botany* **30**, 165.
Epple, R., and Förster, Th. (1954). *Z. Elektrochem.* **58**, 783.
Förster, Th. (1947). *Ann. Physik.* [6] **2**, 55.
Förster, Th. (1951). "Fluoreszenz organischer Verbindungen." Vandenhoeck & Ruprecht, Göttingen, Germany.
Galanin, A., and Lewschin, B. (1951). *Zhur. Eksptl. i Teoret. Fiz.* **21**, 120.
Latimer, P. (1956). *Science* **124**, 585.
Latimer, P. (1959). *Plant Physiol.* **34**, 193.
Latimer, P., and Rabinowitch, E. (1956). *J. Chem. Phys.* **24**, 488.
Linschitz, H., and Sarkanen, K. (1958). *J. Am. Chem. Soc.* **80**, 4826.
Livingstone, R. L. (1955). *J. Am. Chem. Soc.* **77**, 2179.

Lyons, L. E., and White, J. W. (1958). *J. Chem. Phys.* **29**, 447.
Perrin, F. (1929). *Ann. phys.* [10] **12**, 169.
Perrin, F. (1932). *Ann. phys.* [10] **17**, 283.
Rabinowitch, E., and Brody, S. S. (1958). *J. Chem. Phys.* **55**, 927.
Rosenberg, J. D., Takashima, S., and Lumry, R. (1957). *In* "Research in Photosynthesis," p. 85. Interscience, New York.
Schmillen, A. (1953). *Ann. Physik* [12] **135**, 294.
Shiau, Y. G., and Franck, J. (1947). *Arch. Biochem.* **14**, 253.
Shibata, K., Benson, A. A., and Calvin, M. (1954). *Biochim. et Biophys. Acta* **15**, 461.
Spikes, J. D. (1956). *Plant Physiol.* **31**, xxxii.
Strehler, B. L. (1957). "Research in Photosynthesis" (H. Gaffron, ed.), p. 118. Interscience, New York.
Strehler, B. L., and Arnold, W. (1951). *J. Gen. Physiol.* **34**, 809.
Teale, F. W. J. (1958). *Nature* **181**, 415.
Teale, F. W. J. (1959). Report to U. S. Army Research and Development in Europe (DA-91-508-EUC-281).
Tollin, G., Sogo, P. B., and Calvin, M. (1958). *J. chim. phys.* **55**, 919.
Wassink, E. C., and Katz, E. (1939). *Enzymologia* **6**, 145.
Watson, A., and Livingstone, R. L. (1950). *J. Chem. Phys.* **18**, 802.
Weber, G. (1948). *Trans. Faraday Soc.* **44**, 185.
Weber, G. (1949). *Biochem. J.* **47**, 114.
Weber, G. (1953). *Advances in Protein Chem.* **8**, 415.
Weber, G. (1954). *Trans. Faraday Soc.* **50**, 554.
Weber, G. (1957). *Nature* **180**, 1049.
Weber, G., and Teale, F. J. W. (1957). *Trans. Faraday Soc.* **53**, 646.
Weber, G., and Teale, F. W. J. (1958). *Trans. Faraday Soc.* **54**, 640.
Wessels, J. S. C., and van der Veen, R. (1956). *Biochim. et Biophys. Acta* **19**, 548.
Wright, G. T. (1955). *Proc. Phys. Soc. (London)* **B68**, 929.

Discussion

FRENCH: I was wondering if there is some way of running the experiment on delayed light production to see if the photoconductive effect had an influence on the decay of the delayed light. Is there any possibility of doing an experiment of that sort?

WEBER: Yes, it should be possible.

FRENCH: I am afraid that the difficulty is that the fluorescence from the light that you use to make it photoconduct would overlap with the delayed light spectrum, would it not?

WEBER: Yes, there is no information on the real amount of delayed light that comes out of these structures, because measurements have to be done at least more than a millisecond afterwards. The only way in which one could do the experiment is by using a method of measuring the lifetime of the excited state in such a way that one could distinguish a DC and an AC component in the fluorescent light, and the DC component would be the one of the delayed light and the AC component would be that of fluorescence. This is what I hope to be able to do.

Author Index

A

Abelson, P. H., 84, 85, *101*
Abramowicz, E., 114, *126*
Abramsky, T., 243, 245, 246, *256*
Ahrne, I., 259, 266, 267, *275*
Airth, R. L., 182, 183, 186, 187, 200, *201*, 347, *358*
Akabori, S., 130, *143*
Albers, V. M., 40, *50*
Albert, A., 54, *65*
Aldag, H. U., 56, 57, *66*
Aldrich, R. A., 249, *255*
Allard, H. A., 303, *319*
Allen, M. B., 4, 9, 39, 49, *50*, 129, 136, 137, 138, 141, *143*, 181, 184, *201*, 327, *334*, 350, *358*, 392, *393*
Almasz, F., 324, *334*
Altman, K. I., 262, *274*
Amesz, J., 333, *334*
Andersen, S. T., 88, 96, 98, *101*
Anderson, I. C., 60, 62, 63, *65*, 384, *385*
Anderson, J. M., 215, 216, 219, 220, *224*
Anderson, K. J. I., 163, *166*
Anderson, L. J., 88, 98, *102*
Anderson, R. C., 56, *66*
Arnold, W., 398, 404, 408, *410*, *411*
Arnon, D. I., 327, *334*, 387, 391, *393*
Aronoff, S., 212, *213*
Arth, G. E., 56, *66*
Arvanitaki, A., 284, *293*
Augustinsson, K. B., 54, *65*

B

Babushkin, L. N., 115, *125*
Bachmann, B., 237, *255*
Bader, G., 181, 193, *202*
Baker, B. L., 91, 94, 95, *101*, *102*
Ball, E. G., 288, *293*
Baltscheffsky, H., 388, *393*
Bannister, T. T., 182, *201*
Banse, K., 86, 87, 97, *102*
Barghoorn, E. S., 83, *102*, 131, *143*
Barlow, A. J. E., 58, *65*

Bartsch, R. G., 325, 326, 329, 330, *334*, *335*
Bassham, J. A., 215, *224*
Baudisch, O., 96, *102*
Bauer, L., 60, *65*
Bazhanova, N. V., 223, *224*
Beatty, R. A., 88, *102*
Belcher, J. H., 90, 98, *102*
Bellis, D. R., 40, *51*
Bendix, S. W., 108, 110, 117, *125*
Benitez, A., 35, 37, 40, *51*, 176, *179*, 259, 263, 264, 266, 267, *274*, *275*
Benson, A. A., 40, *51*, 182, *202*, 399, *411*
Berezovskii, V. M., 54, *65*
Bergeron, J. A., 207, *213*, 384, *385*
Bergmann, W., 139, *143*
Bermes, E. W., 220, *224*
Bezinger, E. N., 187, *202*
Bielig, H. J., 15, *29*
Bigelow, R. S., 131, 139, *143*
Bigwood, E. J., 189, *202*
Bishop, N. I., 80, *81*
Blanchard, G. C., 55, *66*
Blass, U., 215, *224*
Blinks, L. R., 41, *51*, 181, 182, 183, 186, 187, *201*, *202*, 339, 340, 341, 343, 344, 345, 346, 347, 349, 350, 352, 356, *358*, 359, 361, 362, 364, 365, 367, 368, 369, 370, 373, *375*, *375*
Blum, H. F., 108, *125*
Blumer, M., 92, 94, 97, *102*
Bogorad, L., 231, 232, 233, 236, 239, 240, 241, 242, 243, 244, 246, 247, 248, 249, 250, 254, *254*, *255*
Bohm, J. A., 113, *125*
Bohn, G. M., 295, *300*
Boltscheffsky, M., 333, *335*
Bonner, J., 210, *213*
Booij, H. L., 238, *254*, *256*
Boresch, K., 136, 141, *143*
Borthwick, H. A., 303, *319*, *320*
Bott, P. A., 18, *29*
Bourelly, J., 39, *50*

Bowness, J. M., 158, 159, *166*
Bracher, R., 109, 112, *125*
Bradbury, J. H., 190, *201*
Braun, A., 124, *125*
Braun, A. C., 208, *213*
Brawerman, G., 8, *9*
Breger, I. A., 84, *102*
Brin, G. P., 182, 186, *202*
Brockmann, H., 24, *28*
Brody, M., 340, 352, 353, *358*
Brody, S. S., 400, 407, 408, *410, 411*
Brown, J. S., 45, 48, *50*
Brown, P. K., 159, *165*
Brown, P. S., 159, *165*
Brugger, J. E., 409, *410*
Bruk, E. S., 88, *103*
Buchholz, W., 378, *385*
Buder, J., 111, *125*
Bünning, E., 109, 111, *125*
Bullock, E., 243, 246, 247, *254*
Burkholder, P. R., 112, *125*
Buswell, A. M., 392, *393*
Butler, W. L., 303, *321*

C

Calvin, M., 40, *51,* 60, *66,* 150, *165,* 182, *202,* 215, *224,* 399, 409, *411*
Cameron M. P., 54, *66*
Capenos, J., 157, 159, *166*
Carpenter, A. T., 241, 248, *254*
Cathey, H. M., *320*
Cederstrand, C., 49, *50,* 340, 352, 353, 354, *358*
Chaiet, L., 56, *66*
Chalmers, R. F., 49, *50,* 141, *143,* 340, 352, 353, 354, 356, *358,* 364, *365*
Chalazontis, N., 284, *293*
Chance, B., 39, *50,* 326, 333, *334, 335*
Chandrasekhar, S., 402, *410*
Chargaff, E., 8, *9*
Chattaway, F. W., 58, *65*
Chen, S. L., 343, *358*
Chertok, I., 98, *102*
Chichester, C. O., 208, 209, 210, *213, 214*
Cholnoky, L., 224, *224*
Christian, W., 74, *81,* 268, *275*
Chung, D., 187, *202*
Claes, H., 5, 8, 9, *9,* 212, *213,* 342, *358*
Clayton, R. K., 111, *125*

Cleasson, S., 397, *410*
Clendenning, K. A., 4, *9,* 108, *125*
Cohen-Bazire, G., 1, *9, 10,* 211, *213,* 342, *358*
Cohn, F., 107, *125*
Cole, H., 146, *165*
Colmano, G., 48, *51,* 152, *166,* 269, 270, *275*
Comar, C. L., 63, *65*
Comfort, A., 227, *254*
Contis, G., 149, 157, 159, *166*
Contopoulou, R., 392, *393*
Conway, W. G., 25, *28*
Cook, A. H., 53, *66*
Cook, R. M., 40, *51*
Cookson, G. H., 232, 243, 248, *255*
Coomber, J., 268, 270, *274*
Cooper, F. P., 178, *179*
Copeland, H. F., 130, *143*
Copeland, J. E., 39, *50*
Corcoran, E. F., 88, *102*
Coulson, C. B., 189, *201*
Cowperthwaite, J., 54, 59, 60, *66*
Crane, F. L., 80, *81*
Crane, S. C., 17, 19, *28*
Craston, D. F., 90, 97, *103*
Creitz, G. I., 86, *102*
Crowe, M. O'L., 54, 55, *65*

D

Dangeard, P. A., 109, 110, *125*
Davenport, H. E., 324, *334*
Davidson, J. B., 408, *410*
Della-Rosa, R. J., 262, *274*
Denekas, M. O., 94, *102*
Detwiler, S. R., 160, *165*
Deuel, H. J., Jr., 14, 15, *29, 30*
Deveze, P., 57, *66*
Dewey, D. L., 130, *144*
Dewey, V. C., 62, *65, 66*
Dhéré, C., 91, *102*
Dimelow, E. J., 28, *28*
Dimitrievsky, O. D., 399, *410*
Dorough, G. D., 215, *224*
Dorrestein, R., 388, *394*
Doudoroff, M., 392, *393*
Dougherty, E. C., 129, 130, 132, 136, 137, 138, 139, *143,* 181, 184, *201,* 350, *358*
Downs, R. J., 303, *320*

AUTHOR INDEX 415

Dresel, E. I. B., 233, 248, 249, 255
Drews, G., 110, 125
Duggar, B. M., 339, 358
Dunning, H. N., 84, 91, 92, 94, 95, 101, 102
Duranton, J., 260, 261, 262, 274
Durham, L. J., 266, 275
Durston, A. D., 108, 125
Dutton, H. J., 339, 344, 358
Duysens, L. N. M., 333, 334, 340, 342, 346, 348, 349, 352, 358, 398, 410

E

Eckles, C. H., 19, 29
Egami, F., 325, 334
Eggerer, H., 210, 213
Egle, K., 40, 51
Elbers, P. F., 3, 9, 149, 165
Elliott, R. F., 44, 45, 50, 351, 358
Elofson, R. M., 91, 94, 95, 102
Elsden, S. R., 325, 334
Emerson, R., 49, 50, 141, 143, 339, 340, 341, 342, 349, 352, 353, 354, 355, 356, 358, 362, 364, 365, 377, 385, 392, 393, 398, 410
Emery, K. O., 88, 90, 93, 97, 102
Engelmann, T. W., 107, 108, 109, 125, 181, 201, 339, 342, 358
Epple, 401, 410
Erdman, J. G., 92, 94, 102
Eriksson, I-B., see Eriksson-Quensel, I-B.,
Eriksson-Quensel, I-B., 181, 183, 186, 201, 203
Ermolaev, B. L., 399, 410
Evans, R. A., 390, 393
Evans, W. C., 390, 393
Eversole, R. A., 162, 165
Evstigneev, V. B., 182, 186, 202

F

Fair, W. R., 40, 51
Falk, J. E., 233, 248, 249, 255, 330, 334
Famintzin, A., 114, 125
Filfus, J., 391, 393
Fischer, H., 69, 72, 169, 170, 171, 172, 173, 175, 176, 177, 178, 178, 229, 255, 259, 274
Fisher, L. R., 15, 21, 22, 23, 28
Flint, L. H., 319
Föh, H., 279, 293

Förster, T., 271, 274, 400, 401, 402, 403, 410
Fogg, G. E., 39, 50, 90, 98, 102
Folkers, K., 56, 66
Fork, D. C., 3, 10, 36, 51, 134, 136, 140, 143, 181, 202, 351, 352, 358, 361, 365
Forrest, H. S., 56, 57, 58, 59, 65, 66
Fowdon, L., 187, 202
Fox, D. L., 12, 13, 14, 16, 17, 18, 19, 24, 26, 28, 28, 29, 88, 91, 96, 98, 102, 103, 108, 125, 296, 300
Fraenkel-Conrat, H., 190, 200, 201
Franck, J., 355, 358, 382, 385, 397, 409, 410, 411
Frank, G., 114, 125
Frank, S. R., 153, 154, 155, 165
Frei, W., 324, 334
French, C. S., 38, 40, 41, 42, 44, 45, 50, 80, 200, 201, 264, 273, 274, 275, 340, 347, 348, 351, 352, 358, 362, 365, 371, 375
Frenkel, A., 327, 334
Frey-Wyssling, A., 148, 149, 165, 165
Friedheim, E. A. H., 77, 79, 81, 330, 335
Fritsch, F. E., 130, 143
Fujita, A., 324, 334
Fujiwara, T., 130, 143, 187, 189, 190, 200, 201
Fuller, R. C., 60, 62, 63, 65, 384, 385
Funk, H. B., 54, 55, 59, 60, 66

G

Gaffron, H., 389, 391, 392, 393
Galanin, A., 402, 410
Galmiche, J. M., 260, 261, 262, 274
Garner, W. W., 303, 319
Gates, M., 54, 65
Gattung, H.-W., 379, 380, 385
Geitler, L., 39, 51, 109, 125, 350, 358
George, P., 331, 334
Gerould, J. H., 19, 29
Gest, H., 390, 393
Gibson, J., 325, 329, 334
Gibson, K. D., 233, 234, 237, 255
Giese, A. T., 259, 266, 267, 275
Gillbricht, M., 86, 97, 102
Gleason, L. S., 60, 66
Glebovskaya, E. A., 94, 102

AUTHOR INDEX

Glover, J., 14, 29
Godnev, T. N., 221, 224
Goedheer, J. C., 149, 165, 193, 201, 271, 272, 274
Gössel, I., 109, 111, 125
Goldie, E. H., 22, 28
Goldschmidt, M., 178, 178
Goldsmith, T. H., 21, 29, 158, 159, 165
Goodwin, T. W., 1, 3, 4, 5, 6, 7, 9, 12, 13, 14, 15, 20, 29, 69, 70, 72, 96, 111, 125, 131, 133, 134, 143, 169, 178, 205, 206, 208, 213, 342, 358
Gorham, E., 88, 102
Gothan, W., 84, 102
Grady, J. R., 88, 90, 93, 97, 102
Graham, J. R., 50, 51
Grangaud, R., 22, 29
Granick, S., 37, 38, 51, 63, 65, 154, 165, 231, 232, 233, 234, 235, 236, 237, 239, 242, 243, 244, 246, 249, 250, 251, 254, 255, 257, 258, 259, 274
Grassé, P.-P., 130, 143
Gravrilova, V. A., 182, 186, 202
Gray, C. H., 190, 191, 192, 198, 201
Gray, W. D., 58, 65
Greenblatt, C. L., 149, 166
Griffiths, L. A., 208, 213
Griffiths, M., 1, 9, 211, 213, 342, 359
Grinstein, M., 255, 255
Gross, J. A., 6, 9
Gullberg, J. E., 117
Gunderson, K., 88, 96, 98, 101
Gyorgyfy, C., 224, 224

H

Habermann, H. M., 73, 81
Halldal, P., 42, 44, 51, 108, 112, 125, 181, 182, 201
Hamackova, J., 84, 103
Hansen, W. E., 92, 94, 102
Hardin, G., 169, 179
Harris, J. I., 190, 200, 201
Harris, S. A., 56, 66
Hart, R. W., 271, 275
Hartman, M., 15, 29
Hartshorne, J. N., 111, 125
Harvey, H. W., 86, 102
Hatfield, D., 59, 65
Hatfield, W. D., 392, 393

Haug, A., 3, 10
Haupt, W., 113, 119, 125
Haurowitz, F., 181, 202
Haxo, F. T., 3, 4, 9, 10, 25, 26, 28, 28, 29, 36, 41, 51, 96, 103, 108, 125, 134, 136, 140, 143, 181, 182, 184, 185, 186, 189, 202, 210, 214, 339, 340, 341, 342, 344, 345, 346, 347, 348, 349, 350, 351, 352, 358, 361, 364, 365, 368, 369, 370, 375
Hayaishi, O., 390, 393
Heidingsfeld, I., 109, 125
Heikel, T., 251, 255
Heilbron, I. M., 3, 4, 10, 96, 102, 130, 131, 143
Heinrich, M. R., 62, 65, 66
Heinze, P. H., 303
Hendricks, S. B., 303, 319
Henning, U., 210, 213
Henry, B., 146, 165
Hewitt, L. F., 80, 81
Hill, R., 80, 81, 323, 324, 329, 334, 335, 387, 393
Hirose, H., 4, 10, 39, 51
Hitchon, B., 91, 94, 95, 102
Hjertén, S., 186, 203
Hodge, A. J., 150, 165
Hodgson, G. W., 91, 94, 95, 101, 102
Hogan, A. G., 13, 29
Hollenberg, G. J., 183, 202
Holt, A. S., 35, 51, 71, 169, 170, 171, 172, 174, 175, 177, 178, 178, 179
Holt, W. W., 40, 51
Hooker, H. D., 108, 126
Hoover, W. H., 343, 359
Hopkins, F. G., 53, 66
Hradil, G., 91, 102
Hubbard, R., 158, 164, 165, 166
Huisman, T. H. J., 189, 202
Hulcher, F. M., 325, 334
Hussong, H. E., 122, 126
Hutner, S. H., 6, 10, 54, 66, 389, 391, 393

I

Iodice, A. A., 233, 255
Ishimoto, M., 324, 334
Itahashi, M., 325, 334
Iwai, K., 61, 66

J

Jackson, A. H., 243, 246, 248, 255
Jacobs, E. E., 172, 174, 175, 178
Jaffe, H., 250, 255
Jahn, T. L., 6, 9
Jamikorn, M., 4, 5, 6, 9, 342, 358
Jencks, W. P., 28, 30
Jenkins, J. A., 207, 213
Jensen, E. A., 83, 103
Jensen, S. L., 211, 213
Johnson, A. W., 243, 246, 247, 254
Johnstone, D. B., 55, 66
Jones, E. A., 56, 58, 66
Jones, J. D., 84, 85, 102
Jones, R. F., 181, 187, 202

K

Kaczka, E. A., 56, 66
Kalenda, N. W., 92, 94, 102
Kamen, M. D., 230, 255, 325, 326, 327, 329, 333, 334, 335, 387, 390, 393
Kandler, O., 342, 359, 391, 393
Kaneko, Y., 56, 57, 66
Kaplan, I. R., 69, 70, 72, 173, 179
Karrer, P., 54, 66, 96, 98, 102
Katagiri, M., 390, 393
Katoh, S., 325, 335
Katsurai, T., 182, 186, 190, 203
Katz, E., 173, 179, 388, 394, 397, 411
Keilin, D., 324, 335
Kennedy, D., 284, 293
Kennedy, G. Y., 228, 255
Keresztesy, J. S., 56, 66
Kersten, J. A. H., 339, 340, 344, 359
Kessel, I., 210, 213
Kidder, G. W., 62, 65, 66
Kiessling, W., 153, 166
Kikuchi, G., 236, 237, 238, 255, 256
Kimmel, J. R., 187, 189, 190, 202
King, N. K., 330, 332, 335
Kirschfeld, S., 55, 61, 66
Kittredge, J. S., 96, 103
Kivkutsan, F. R., 187, 202
Klebs, G., 113, 126
Kleerekoper, H., 85, 102
Klöcker, A., 57, 66
Klotzsch, H., 385
Knorr, H. V., 40, 50
Kodama, T., 324, 334
Koenig, H., 96, 98, 102
Kohlbecker, R., 311
Kon, S. K., 15, 21, 22, 23, 28, 96, 102
Korkes, S., 377, 385
Korte, F., 56, 57, 66
Koski, V. M., 153, 165, 260, 263, 264, 269, 274, 275 (see also Young, V. M. K.)
Kosobutskaya, L. M., 42, 51, 267, 274
Koyama, J., 324, 334
Kozminski, Z., 86, 102
Krasovskaya, T. A., 215, 223, 224
Krasnovskii, A. A., see Krasnovsky, A. A.
Krasnovsky, A. A. (Krasnovskii, A. A.), 42, 51, 182, 186, 202, 260, 264, 267, 274, 382, 385
Kratz, W. A., 182, 202
Krinsky, N. I., 1, 10
Krippahl, G., 378, 385
Kritzler, H., 24, 29
Kuhn, R., 15, 29, 53, 66, 288, 293
Kumin, S., 234, 256
Kunisawa, R., 392, 393
Kupke, D. W., 259, 266, 267, 269, 275
Kylin, H., 96, 102, 110, 126, 181, 182, 202

L

Labbe, R. F., 233, 249, 251, 255, 257, 274
Lambrecht, R., 69, 72
Land, D. G., 132, 143
Lane, H. C., 60, 66
Lang, H. M., 332, 335
Larsen, B., 3, 10
Larsen, H., 69, 70, 71, 72, 173, 179, 325, 329, 334, 388, 392, 393
Lascelles, J., 236, 255
Latimer, P., 40, 51, 271, 274, 395, 399, 410
Laurens, H., 108, 126
Laver, W. G., 234, 237, 255
Lederer, E., 19, 29, 96, 97, 98, 102
Legge, J. W., 190, 191, 192, 193, 194, 196, 202, 229, 230, 238, 255
Lemberg, R., 181, 182, 190, 191, 192, 193, 194, 196, 200, 201, 202, 229, 230, 238, 255
Levin, H. L., 54, 66
Levin, O., 186, 203
Levine, L., 330, 335
Levring, T., 344, 359

Levy, A. L., 187, 190, 200, *201, 202*
Levy, H. R., 382, *385*
Lewis, C. M., 339, 341, 342, 349, *358*, 398, *410*
Lewis, N. B., 186, *203*
Lewschin, B., 402, *410*
Leyon, H., 149, *166*
Lijinsky, W., 208, *213*
Lind, E. F., 60, *66*
Linsbauer, K., 114, *126*
Linschitz, H., 397, *410*
Littman, E. R., 99, *102*
Litvin, F. F., 260, 264, *274*
Livington, R., 382, *385*
Livingstone, R. L., 397, 406, 407, *410, 411*
Lochmann, E. R., 55, 61, *66*
Lochte, H. L., 99, *102*
Lockwood, W. H., 251, *255*
Loeb, J., 109, *126*
Loeffler, J. E., 251, 255, 257, 259, 265, 266, 267, *274, 275*
Lubimenko, V. N., 40, *51*
Lukton, A., 208, 210, *213, 214*
Lumry, R., 397, *411*
Lunenschloss, A., *311*
Luntz, A., 108, *126*
Lwoff, A., 12, *29*
Lwoff, M., 12, *29*
Lyakhnovich, Y. P., 221, *224*
Lynch, V. H., 377, *385*
Lynen, F., 210, *213*
Lyons, L. E., 408, *411*
Lythgoe, B., 96, *102*
Lythgoe, R. S. ,163, *166*

M

McAlister, E. D., *319*
McDonald, H. J., 220, *224*
McDonald, S. F., 243, 246, 248, *255*
McIlrath, W. J., 254, *254*
McKendell, L. V., 220, *224*
Mackinney, G., 205, 207, 208, 209, 210, 212, *213, 214*
McLaughlin, J. J. A., 136, *143*, 181, 184, *201*, 350, *358*
McLean, J. D., 150, *165*
McNutt, W. S., 56, 57, 59, *65, 66*
McSwiney, R. R., 231, *255*

Maevskaya, A. N., 215, 223, *224*
Mainx, F., 112, *126*
Manning, W. M., 35, *51*, 169, 170, *179*, 339, 344, *358*
Manten, A., 111, *126*
Manunta, C., 19, *29*
Margarot, J., 57, *66*
Markham, E., 243, 246, 247, *254*
Marks, G. S., 250, *254, 255*
Marriott, F. H. C., 157, *166*
Mason, H. S., 390, *393*
Massonet, R., 22, *29*
Mast, S. O., 108, 109, 112, *126*
Masuda, T., 56, 57, *66*
Mattson, F. H., 14, *29*
Mauzerall, D., 235, 242, 243, 249, *255*
Maxwell, S. S., 109, *126*
Mazza, F. P., 77, *81*
Medem, F. G., 15, *29*
Meek, E. S., 404, *410*
Mehl, J. W., 14, 15, *29, 30*
Meinschein, W. G., 85, *102*
Mellon, A. D., 149, 153, 154, 155, 157, 159, 162, 163, *166*
Mercer, F. V., 150, *165*
Metzner, B., 60, *66*
Metzner, H., 60, *66*
Metzner, P., 69, 72, 173, *179*
Michaelis, L., 330, *335*
Mikhaĭlovnina, A. A., 96, 98, *103*
Miller, G. L., 163, *166*
Miller, W. H., 159, *166*
Millott, N., 13, 15, 16, *29, 30*, 280, 282, 283, 284, 286, 287, 288, 289, 290, 291, 292, *293*
Minnaert, K., 3, *9*, 149, *165*
Mittenzwei, H., 69, *72*
Möller, H., 182, 198, *202*
Mohr, H., *311*
Molisch, H., 175, *179*, 387, *393*
Montfort, C., 344, *359*
Moon, P., 117, *126*
Moore, C. V., 230, *255*
Moore, J. W., 92, 94, *102*
Moore, S., 189, *202*
Mori, T., 325, *334*
Morley, H. V., 33, 35, *51*, 169, 171, 178, *179*
Morton, R. A., 14, *29*

AUTHOR INDEX

Mosebach, G., 113, *126*
Mozingo, R., 56, *66*
Muir, H. M., 177, *179*, 230, *255*
Mulliken, R. A., 206, *213*
Muraveisky, S., 98, *102*
Myers, J., 50, *51*, 57, 58, 59, *65*, *66*, 182, *202*, 362, *365*, 371, *375*

N

Naftalin, L., 189, *202*
Nagy, E., 224, *224*
Nakagawa, S., 61, *66*
Nakamura, H., 391, *393*
Nakayama, T. O. M., 9, *9*, 208, 210, 211, 212, *213*, *214*
Nathan, H. A., 54, 55, 56, 58, 59, 60, 62, 63, *65*, *66*
Nathanson, N., 28, *30*
Neilands, J. B., 15, *29*
Neuberger, A., 177, *179*, 230, 233, 234, 237, *255*
Neve, R. A., 249, *255*
Newton, J. W., 325, 327, 330, 333, *335*
Nicholas, R. E. H., 231, *255*
Nicholson, D. C., 192, 198, *201*
Nishida, G., 251, *255*
Nishimura, M., 1, *10*, 206, *213*, 325, 327, 329, 333, *335*
Noack, K., 153, *166*
Norris, K. H., 303, *321*
Norris, P. S., 181, 182, 186, 189, *202*, 347, 348, *358*
North, W. J., 295, 296, 297, 300, *300*
Novelli, D. G., 88, 91, 98, *102*
Nüssler, W., 178, *178*
Nultsch, W., 109, *126*
Nyholm, R. S., 330, *334*

O

O'Carra, P., 197, 198, 199, 200, *202*
óhEocha, C., 136, *143*, 181, 182, 183, 184, 185, 186, 187, 188, 189, 190, 193, 194, 195, 196, 197, 198, 199, 200, *202*, 345, 351, *359*
Oku, S., 19, *29*
Olson, J. M., 333, *334*, *335*
O'Reachtaire, M., 183, *202*
Orlando, J., 330, *335*
Orr, W. L., 88, 90, 93, 97, *102*

P

Pakashina, E. V., 42, *51*
Palade, G. E., 149, 152, 153, 155, *166*
Paleus, S., 328, *335*
Palmer, L. S., 17, 19, *29*
Pánczél, M., 224, *224*
Pantin, C. F. A., 295, 296, *300*
Papenfuss, G. F., 110, *126*
Parker, M. W., 303, *319*
Parr, W. H., 390, *393*
Paul, K. G., 200, *202*
Peake, E., 91, 94, 95, *102*
Pease, M., 17, *29*
Perrin, F., 402, 403, 404, *411*
Peterson, E. A., 187, *202*
Petracek, F. J., 25, *29*
Pfeffer, M., 55, *66*
Philpott, D. E., 159, *165*
Phinney, H. K., 88, *103*
Pickels, E. G., 163, 164, *166*
Pieper, A., 110, *126*
Pinckard, J. H., 96, *103*
Pirenne, M. H., 157, *166*
Pires, G., 254, *254*
Piringer, A. A., *303*
Platt, J. R., 206, *213*
Pleshkov, B. P., 187, *202*
Postgate, J. R., 324, *335*
Price, L., 259, 265, *275*
Pringsheim, E. G., 110, *126*
Pringsheim, O., 110, *126*
Prosser, C. L., 284, *293*
Provasoli, L., 6, *10*, 391, *393*
Prunty, F. T. G., 231, *255*
Purcell, A. E., 210, *213*
Purrmann, R., 53, *66*

Q

Quilliam, J. P., 163, *166*

R

Rabinowitch, E. I., 38, 40, *51*, 97, *103*, 148, *166*, 192, *202*, 340, 355, *359*, 391, *393*, 399, 408, *410*
Radchenko, O. A., 84, 92, 94, 97, *103*
Radin, N. S., 230, *255*
Rafferty, M., 351, *359*
Raftery, M., 136, *143*, 181, 185, 187, 188, 189, 190, 196, *202*

Ragetli, H. W. J., 187, 203
Ramsey, V. G., 92, 94, 102
Rauen, H. M., 62, 66
Reazin, G. H., Jr., 391, 394
Reidmüller, L., 324, 334
Resuhr, B., 319
Richards, F. A., 86, 102, 103
Richert, D. A., 233, 236, 255, 256
Rick, C. M., 207, 213
Rickes, E. L., 56, 66
Riedmair, J., 175, 178
Rikhireva, G. T., 260, 264, 274
Riley, G. A., 83, 103
Rimington, C., 231, 232, 233, 238, 243, 248, 249, 251, 254, 255, 256
Rittenberg, D., 230, 255, 256
Robinson, R., 246, 256
Rockstein, M., 15, 29
Roelofson, P. A., 387, 393
Rogozinski, F., 17, 29
Rose, I. A., 382, 385
Rosenberg, J. D., 397, 411
Rotfarb, R. M., 221, 224
Rothberg, S., 390, 393
Roux, E., 260, 261, 262, 274
Rubenstein, M., 15, 29
Russel, C. S., 233, 235, 236, 243, 245, 246, 256

S

Sachs, P., 231, 256
Sager, R., 4, 6, 10, 39, 50, 149, 152, 166, 254, 256
St. George, R. C., 163, 166
Salomon, K., 262, 274
San Pietro, A., 332, 335
Saperstein, S., 208, 213
Sapozhnikov, D. I., 215, 223, 224
Sargent, M. C., 12, 29
Sarkanen, K., 397, 410
Sato, R., 325, 334
Savinov, B. G., 96, 98, 103
Scarisbrick, R., 324, 334, 335
Schatz, A., 6, 10
Scheer, I. J., 159, 166
Schendel, H. E., 187, 202
Schicke, H. F., 56, 57, 66
Schlegel, H. G., 3, 10
Schmid, R., 233, 234, 256

Schmidt-Nielsen, S., 18, 29
Schmillen, A., 403, 411
Schneiderhöhn, G., 109, 125
Schötz, F., 342, 359
Schram, E., 189, 202
Schroeder, H., 83, 103
Schröder, W., 378, 379, 380, 385
Schuette, H. A., 18, 29
Schütte, K. H., 187, 202
Schulman, M. P., 233, 236, 255, 256
Schultz, F., 84, 103
Schwarting, A. E., 73, 81
Schwertz, F. A., 149, 150, 166
Scott, J. J., 233, 241, 248, 254, 255
Scully, N. J., 303, 319
Senn, G., 114, 126
Seybold, A., 40, 51
Shapiro, S. A., 96, 98, 103
Shaw, K. B., 243, 246, 247, 254
Shemin, D., 230, 231, 233, 234, 235, 236, 237, 238, 243, 245, 246, 248, 255, 256
Shepherd, H. G., 220, 224
Sherwood, H. K., 408, 410
Sheshina, L. S., 84, 92, 94, 97, 103
Shiau, Y. G., 397, 411
Shibata, K., 40, 51, 182, 202, 265, 271, 274, 275, 399, 411
Shin, E., 109, 111, 126, 155, 158, 166
Shiota, A., 64, 66
Shlyk, A. A., 221, 224
Siedel, W., 192, 194, 198, 202
Siegel, S., 299, 300
Siegelman, H. W., 303, 320, 321
Silberman, H., 69, 70, 72, 173, 179
Simon, H., 60, 66
Sirvonal, C., 391, 393
Sisakyan, N. M., 187, 202
Sissens, M. E., 111, 125
Sistrom, W. R., 1, 9, 342, 359
Skopintsev, B. A., 88, 103
Skow, R. K., 361, 365
Sleeper, B. P., 390, 393
Smith, D. G., 189, 202
Smith, E. L., 163, 164, 166, 187, 189, 190, 202
Smith, G. M., 130, 134, 136, 143
Smith, J. H. C., 34, 37, 40, 51, 69, 72, 80, 81, 81, 153, 165, 166, 176, 179,

AUTHOR INDEX

200, *201*, 254, *256*, 259, 260, 261, 262, 263, 264, 266, 267, 268, 269, 270, 271, 272, 273, *274*
Smith, L., 326, 333, *334*, 335
Smith, P. V., 85, *103*
Sober, H. A., 187, *202*
Sogo, P. B., 409, *411*
Sørensen, N. A., 18, *29*
Southwick, B. L., 56, *66*
Spielberger, G., 170, 176, *178*
Spikes, J. D., 398, *411*
Stahl, E., 114, *126*
Stanier, R. Y., 1, *9*, *10*, 69, 72, 130, *143*, 211, *213*, 262, 263, *275*, 342, *358*, *359*, 388, 389, 390, 392, *393*
Starr, M. P., 208, *213*
Stauffer, J. F., 377, *385*, 392, *393*
Steemann-Nielsen, E., 83, *103*
Stern, A., 173, *179*
Stolfi, G., 81, *81*
Stoll, A., 172, 177, *179*
Strain, H. H., 1, 5, *10*, 35, 37, *51*, 131, 134, 135, 136, *143*, *144*, 154, *166*, 169, 170, *179*
Strasburger, E., 107, 109, 112, *126*
Strehler, B. L., 377, *385*, 408, *411*
Strother, G. K., 146, 152, 160, *166*
Sumner, F. B., 24, *29*
Sutherland, M. D., 14, *29*
Svedberg, T., 181, 182, 183, 186, 190, *203*
Swabey, Y. S., 90, *103*
Swift, H., 254, *254*
Swingle, S. M., 182, *203*
Szables, J., 224, *224*

T

Takamatsu, K., 1, *10*, 206, *213*
Takashima, S., 165, *166*, 397, *411*
Tamiya, H., 324, *335*
Tanada, T., 339, 344, *359*
Tarr, E., 28, *30*
Tazawa, M., 111, *125*
Teale, F. W. J., 395, 396, 398, 399, 404, 408, *411*
Terenin, A. N., 399, *410*
Thomas, J. B., 3, 9, 149, *165*, *166*
Thompson, G. A., 210, *213*

Thompson, S. Y., 15, 23, *28*
Thompson, T. G., 86, *103*
Thuret, G., 110, *126*
Tilden, J. E., 39, *51*
Tischer, J., 96, *103*
Tiselius, A., 182, 185, 186, *203*
Tollin, G., 409, *411*
Toole, E. H., 303, *320*
Toole, V. K., 303, *320*
Torto, F. G., 1, *10*
Towner, G. H., 40, *51*
Trask, P. D., 88, 96, *103*
Treibs, A., 91, 92, 94, 97, *103*
Treviranus, L. C., 107, *126*
Trumpy, B., 18, *29*
Trurnit, H. J., 48, *51*, 152, *166*, 269, 270, *275*
Tuppy, H., 328, *335*
Turano, A. M., 157, 159, *166*
Turian, G., 210, *214*
Tyler, S. A., 131, *144*

U

Uda, H., 19, *29*
Ullrich, H., 110, *126*
Umbreit, W. W., 377, *385*, 392, *393*
Updegraff, D. M., 88, 91, 98, *102*

V

Vallentyne, J. R., 3, *10*, 84, 85, 88, 90, 96, 97, 98, 99, 101, *102*, *103*
Van Baalen, C., 57, 58, 59, 65, *66*
Van der Veen, R., 398, *411*
Van Niel, C. B., 130, *143*, 211, *214*, 387, 388, 391, 392, *393*
Varma, T. N. R., 210, *214*
Vennesland, B., 382, *385*
Vernon, L. P., 325, 329, 333, *334*, 335
Verworn, M., 109, *126*
Vevers, H. G., 15, 16, *29*, *30*, 228, *255*
Virgin, H. I., 200, *201*, 259, 260, *275*
Vischer, W., 109, *126*
Viscontini, M., 54, *66*
Vishniac, W., 325, *334*, 382, *385*, 391, *394*
Völker, O., 24, *28*, *30*
Voerkel, H., 114, *126*
Volkenshtein, M. V., 94, *102*
Von Buddenbrock, W., 279, *293*

Von Euler, H., 96, *102*
Von Uexküll, J., 279, *293*
Von Wettstein, D., 254, *256*
Vorobeva, L. M., 42, *51*

W

Wacker, A., 55, 61, *66*
Wald, G., 20, 22, 28, *30*, 158, 159, 160, 161, *166*
Waldmann, H., 62, *66*
Walker, A., 54, 55, *65*, 80, *81*
Wallace, R. H., 73, *81*
Wallenfels, K., 288, *293*
Walter, H., 169, 171, 172, 173, *178*
Warburg, O., 74, *81*, 268, *275*, 378, 379, 380, *385*
Wassink, E. C., 173, *179*, 187, *203*, 339, 340, 344, *359*, 387, 388, *394*, 397, *411*
Wasteneys, H., 109, *126*
Watson, C. J., 199, *203*
Watson, E., 406, 407, *411*
Weaver, K. S., 122, *126*
Weber, G., 395, 396, 398, 401, 403, 404, 405, 408, *411*
Weedon, B. C. L., 1, *10*
Weigl, J. W., 382, *385*
Weller, A., 175, *179*
Wells, J. W., 14, *29*
Welsh, J. H., 284, *293*
Wenderlein, H., 173, *179*
Wenderoth, H., 177, *178*
Went, F. W., *319*
Wessels, J. S. C., 398, *411*
Westall, R. G., 232, *256*
Whatley, F. R., 327, *334*
White, J. W., 408, *411*
Whitelock, O. V., ed., 145, *166*
Whittaker, J. R., 101, *103*
Wickman, F. E., 83, *103*
Wiese, C. E., 15, *30*
Willimot, S. G., 17, *30*

Willstätter, R., 172, 177, *179*
Winfield, M. E., 330, 332, *335*
Wisbar, G., 84, *103*
Wittenberg, J., 230, 231, 235, 248, *256*
Wolf, D. E., 56, *66*
Wolf, F. T., 56, 57, 58, *66*
Wolff, J. B., 259, 265, *275*
Wolken, J. J., 5, 8, *10*, 21, *30*, 109, 111, *126*, 145, 146, 149, 150, 152, 153, 154, 155, 156, 157, 158, 159, 160, 161, 162, 163, 165, *165*, *166*
Wolstenholme, G. E. W., 54, *66*
Wong, P. S., 208, 209, *213*, *214*
Work, E., 130, *144*
Wright, G. T., 408, *411*
Wu, C. C., 88, 96, *103*
Wurster, C. F., 266, *275*

Y

Yamagutchi, S., 324, *335*
Yaoi, H., 324, *335*
Yokoyama, H., 208, 210, *213*, *214*
Yocum, C., 343, 346, 352, *359*, 388, *393*
Yoshida, M., 13, *29*, 280, 282, 283, 284, 287, 288, 289, 290, 291, 292, *293*
Young, E. C., 189, *202*
Young, V. M. K., 42, *50*, 200, *201*, 259, 264, *275*, 340, 347, 348, 352, *358* (*see also* Koski, V. M.)

Z

Zalokar, M., 4, 6, *10*
Zechmeister, L., 16, 18, 25, *29*, *30*, 96, *103*
Ziegler-Günder, I., 54, *66*
Zscheile, F. P., 63, *65*
Züllig, H., 96, 98, 99, 100, *103*
Zurzycka, A., 114, 116, *126*
Zurzycki, J., 116, *126*
Zussman, H., 160, 161, *166*

Subject Index

A

Absorption spectra,
 of bleached chloroplast suspensions, 42
 of *Chlorobium* chlorophyll, 173
 of chlorophyll *in vivo*, 40
 of chlorophyll *a* in monolayers, 270
  in *Ochromonas danica*, 38, 42, 49-50
 of chlorophyll *d*, 170-171
 of chlorophylls, 34
 of colored retinal globules, 161
 of 2-desvinyl-2-formyl chlorophyll *a*, 171
 of 2-desvinyl-2-formyl pheophytin *a*, 171
 of ethyl chlorophyllide *a* in ether, 174
 of ethyl pheophorbide *a* in ether, 175
 of *Euglena* chloroplasts, 152
 of *Euglena gracilis*, 44-48
 of extracted *Hemiselmis virescens* chlorophylls, 37
 of green bacteria, 71
 of "inactive" chlorophyll, 42
 infrared, of *Chlorobium* chlorophyll, 177
 of *Micrasterias* pigments, 121
 of new leaf pigment, 74, 75, 78
 of phycocyanins, 182
 of phycoerythrins, 182-185
 of products of permanganate oxidation of chlorophyll *a*, 172
 of protochlorophylls, 263
Accessory pigments, 373
 in photosynthesis, 339-360
Action spectra,
 anthocyanin formation, 316
 chromatic transients, 369-372
 enhancement, 355-357
 photoperiodic flowering control, 305
 photoperiodism, 304-306
 seed germination, photoperiodic, 305
 shadow reaction of *Diadema antillarum*, 287
 vision of *Dendrocoelem lacteum*, 157
 vision of eye-color mutants of *Drosophila melanogaster*, 159

Action spectra of phototaxis,
 blue-green algae, 110
 Chilomonas, 108
 Chlamydomonas, 108
 chloroplasts, 114-115, 117
 Dunaliella salina, 108
 Eudorina elegans, 108
 Euglena, 108-109
 Euglena gracilis, 155
 Gonium, 108
 Micrasterias, 120-124
 Oscillatoria formosa, 110
 Pandorina, 108
 purple bacteria, 111
 Spondylomorum, 108
 Ulva, 108
 Volvocales, 108
 Volvox, 108
ALA-dehydrase, 233-234
Algal mutants,
 ζ-carotene in, 5-6, 8
 carotenoid biosynthesis in, 5-6, 8-9
 of *Chlamydomonas reinhardi*, 6
 of *Chlorella pyrenoidosa*, 5-6
 of *Chlorella vulgaris*, 5-6
 phytoene in, 5-6
 phytofluene in, 5-6, 8
 xanthophylls of, 5-6, 8-9
Algal pigments,
 separation of, 216-222
Allophycocyanin, 182
δ-Aminolevulinic acid, 233
 synthesis, 235-236
Anabaena, 349
Anabaena variabilis,
 phototaxis in, 110
Anacystis nidulans, 42-43
 biliprotein content, 182
 pteridines of, 58-60
 spectrum of red chlorophyll peak, 43
Antheroxanthin, 224
 in marine sediments, 98
Anthocyanin,
 action spectrum for formation, 316
 formation in apple peeling, 315

Anthraquinones,
  in red crinoids, 14
  in scale-insects, 14
Artemia salina,
  absence of vitamin A, 21
Ascophyllum nodosum,
  β-carotene in gametes of, 4
Aspergillus oryzae,
  pteridines in, 56
Astaxanthin, 2
  in ascidians, 19
  in *Beryx decadactylus,* 19
  in blue whale, 18
  from β-carotene, 23
  in colored retinal globules, 161
  conjugation with proteins, 27
  in crustacean carapace, 18
  in Crustaceans, 22
  in flagellates, 4-5
  in *Haematococcus pluvialis,* 4
  in locusts, 20
  in optic cushion of *Marthasterias,* 16
  reduction to vitamin A, 23
  in red variants of *Euglena,* 155
  stimulating effects on trout sperm, 15
  in *Velella lata,* 26
  in vision, 27
  in vision of invertebrates, 22
Azotobacter,
  pteridines in, 55

B

Bacillariophyceae, see Diatoms
Bacteria,
  α,ε-diaminopimelic acid in, 130
Bacteriochlorophyll, 69, 254
Bacteriovirdin, see Chlorobium chlorophyll
Bacterium photometricum, 107
Barnacles,
  absence of vitamin A, 21
Beryx decadactylus,
  astaxanthin in, 19
Biliproteins, 136, 140, 181-203
  amino acid analyses, 187
  in cryptomonads, 181
  fluorescence, 200-201
  hydrolysis, 198
  isoelectric points, 185
  as phylogenetic markers, 136-137
  protein-chromophore attachment, 200-201
  terminal groups, 190
Biliviolin, 192
Biochemical evolution, 84-85
Bleaching,
  of colored retinal globules, 161
Blue-green algae, 2, 3, 140
  action spectrum of phototaxis, 110
  carotenoids of, 2-3
  chlorophyll *a* in, 35
  chlorophylls in, 34
  α,ε-diaminopimelic acid in, 130
  effect of light intensity on biliprotein composition, 181
  photosynthetic action spectra, 345, 347-349
  phototactic responses in, 110
Bombyx mori,
  assimilation of carotenoids, 19
Brown algae, 1-2
  carotenoids of, 1-2
  chlorophylls in, 34
  chlorophyll *c* in, 35
  major xanthophyll of, 1
  photosynthetic action spectra, 344
  phototaxis, 110

C

Canary-xanthophyll, 24
Carminic acid, 14
Carotenes,
  assimilation of by *Octopus bimaculatus,* 19
α-carotene, 5, 8-9
  in *Coccinella septempunctata,* 18
  in *Cryptomonas ovata,* 136
  in Cryptophyceae, 2, 4
  in *Dunaliella salina,* 12
  in red algae, 136
  in sediments, 98
  in Siphonales, 2, 5
  vitamin A from, 23
  as xanthophyll precursor, 9
β-carotene, 1-9, 133, 140, 207, 341
  in chloroplasts, 148
  in chlorotic substrains of *Euglena gracilis* v. *bacillaris,* 6-7
  in *Coccinella septempunctata,* 18
  in cryptomonads, 140

in *Cyanidium caldarium*, 4
in *Dunaliella salina*, 4, 12
in ear-wax of cattle, 19
in eyespot of *Euglena gracilis*, 155
in gametes of *Ascophyllum nodosum*, 4
in gametes of *Fucus*, 4
in *Hematococcus pluvialis*, 12
in honey, 18
in locusts, 20
metabolism by canaries, 24
in *Ochromonas danica*, 3
in optic cushion of *Marthasterias*, 16
in *Prymnesium parvum*, 3
in sediments, 98
in seston, 96
stimulating effects on sperm of trout, 15
in *Trentepohlia aurea*, 4
vitamin A from, 23
as xanthophyll precursor, 9
γ-carotene, 2, 4,
in gametes of *Ulva*, 4
in green sulfur bacteria, 133
vitamin A from, 23
ζ-carotene, 7-9, 207, 211
in algal mutants, 5-6, 8
Carotene selectors, 16, 17
Carotenemia, 18
Carotenoid biochromes,
in grapefruit, 11
in *Neurospora* mutants, 11
in poppies, 11
in ripening fruits, 12
in tomatoes, 11
Carotenoid excluders, 17
Carotenoid storage, 16-18
nonselective, 16-17
selective, 16-17
Carotenoid synthesis,
in *Chlorella vulgaris*, 7
effect of light on, 7-9
in *Euglena gracilis*, 7-8
Carotenoids, 1-10
in animals, 14-28, 205
assimilation and storage, 16-18, 19, 20
biochemical modification of dietary, 20
biosynthesis, 205-214, 215-225
in chlorophyll formation, 153-155
in chloroplasts, 148

in *Chromatium* chromatophores, 384
in Chrysophyceae, 2, 3
in cocoons, 19
in *Colias philodice*, 18
conversion into other carotenoids, 16
in *Cryptomonas ovata*, 4
of Cryptophyceae, 2-4
in *Ctenosaura hemilopha*, 19
in *Cyanidium caldarium*, 4
destruction, 16, 20, 25
discharge of, 16
disposition of, 16-18
distribution in blue-green algae, 3
in eggs, 18, 25
extraplastidic, 4-5
in flamingo feathers, 24-25
formation of in algae, 7-9
in fossil organic matter, 96, 98-101
in freshwater and marine sediments, 96, 98-100
in mammals, 17-18, 19, 20
in marine sediments, 96
mutations involving, 207
oxidation of, 16, 17, 20, 23, 24
in photosynthesis, 342
in photosynthetic organisms, 133
photosynthetic oxygen transport, 215
as phylogenetic markers, 133-136
in plants, 205
protective effect on photo-oxidation of chlorophyll, 212
in red algae, 3
red carotenoids, 24
secretion, 18-20
stability, 98
structure, 206-207
Carotenoprotein,
in brown algae and diatoms, 1, 3
Cellulose,
in fossil deposits, 84
Chaetopterus variopedatus,
coproporphyrin in, 228
Charge transfer reactions, 331-332
Chemicals,
effect on phototaxis, 112-113
Chilomonas,
phototactic action spectrum, 108
Chlamydomonas,
action spectrum of phototaxis, 108
chloroplast organization of, 152

mutants of, 38-39
phototaxis in eyespotless mutant of, 111
Chlamydomonas reinhardi mutants, 6
Chlorella,
  action spectrum of chromatic transients, 371
  mutants of, 231
  in *Paramecium bursaria*, 107
  photosynthetic action spectra for, 342-343, 362-363
Chlorella ellipsoidea,
  α,ε-diaminopimelic acid in, 130
Chlorella pyrenoidosa mutants, 5, 6
  absorption spectrum of chlorophyll *a* in, 43-44
  lack of chlorophyll *b*, 39, 43-44, 49
  photosynthesis, 49
Chlorella vulgaris, 5-6, 7
  carotenoid synthesis in, 7
  mutants of, 5-6
Chlorin-e_6-trimethyl ester, 169
Chlorobium chlorophyll, 69-72, 173-178, 251
  absorption spectra, 173
  conjugated carbonyl group, 176-177
  ester group, 177
  infrared absorption spectrum of, 177
  magnesium, 176
  methanolysis, 175
  oxidation, 176
  partial formula, 178
  phase test, 71, 174
  products of chromic acid oxidation, 177-178
  reduction, 176
  relation to chlorophyll *a*, 71
  zinc salt, 176
Chlorobium chlorophyll-650, 70-72
Chlorobium chlorophyll-660, 70-72
Chlorobium limicola,
  cytochrome from, 325
Chlorobium thiosulfatophilum, 69-72
  chlorophyll from, 169, 173
Chloromonadophyceae,
  chlorophylls in, 34
Chlorophyll, 33-52, 169-178
  chemical participation in photosynthesis, 377-386

of *Chlorobium thiosulfatophilum*, 169
of chloroplasts, 148
of *Cyanidium caldarium*, 39
direct carboxylation, 378
of *Dunaliella salina*, 12
of fossil organic matter, 85, 96, 97
of green bacteria, 69-72
possible oxidation-reduction reactions, 382
as phylogenetic markers, 132-133
of seston, 86-87
spectral shift of newly formed, 265-266
synthesis of, 152-153
Chlorophyll *a*, 33-35, 37-50, 140, 219, 229, 341, 344, 345, 350
  absorption spectrum, 34
  biosynthesis, 215-225
  in early Paleozoic plants, 95
  in vivo spectrum, 48
  in vivo variations in absorption spectrum, 41-48
  in marine sediments, 88
  monolayer absorption spectra, 270
  oxidation, 169
  in sediments, 89
  in seston, 85-86
Chlorophyll *a* "forms,"
  as functional entities in photosynthetic organisms, 48-49
Chlorophyll-*a* holochrome,
  absorption spectrum, 269
  fluorescence polarization, 271-273
Chlorophyll *b*, 33-35, 39-40, 43, 49, 219, 341
  absorption spectrum, 34
  biosynthesis, 215-225
  in euglenoids and green algae, 133
  in lake sediments, 89
  in marine sediments, 88
  origin, 261-262
  in sediments, 89
Chlorophyll *c*, 33-40, 344, 350
  absorption spectrum, 34
  in brown algae, 133, 141
  in cryptomonads, 140-141
  in *Cryptomonas ovata*, 36
  in diatoms, 141
  in dinoflagellates, 141

in *Hemiselmis virescens*, 36
porphin structure, 37
in *Prymnesium parvum*, 36, 39
in *Rhodomonas lens*, 36
Chlorophyll *d*, 33-35, 171, 345
  absorption spectrum, 34, 170-171
  infrared spectrum, 171
  structure of, 169
Chlorophyll degradation products, 88
  quantitative determination, 90
  in sediment cores, 90
  in surface sediments, 90
Chlorophyll formation,
  action spectrum, 153
Chlorophyll photo-oxidation,
  carotenoid protection, 212
Chlorophyll spectral shift,
  in corn mutants, 266
Chlorophyllide *a*,
  in lake sediments, 89
  phytylation, 265
Chlorophyta, *see* Green algae
Chloroplast phototaxis,
  action spectra, 114-115
  effect of light intensity on, 115-116
  in *Mougeotia*, 113, 116
Chloroplast structure, 148-152
Chloroplastin,
  bleaching, 162-163
  from chloroplasts, 161-162
  molecular weights of, 163
  physiological activity, 162
Chloroplasts, 1, 3, 5-8, 147-148
  of blue-green algae, 3
  chemical composition, 148
  lamellar structure, 149-152
  molecular model, 150-151
  phototaxis, 113-117
Chromatic transients,
  cause, 373-375
  in *Porphyra thuretti*, 369
  in *Ulva lobata*, 367
Chromatium, 333
  carotenoids in chromatophores of, 384
Chromatophores, 147
Chromoplasts, 181
Chroococcus, 349
Chrysomonads,
  chlorophyll *c* in, 34-37

Chrysophyceae,
  carotenoids of, 2, 3
  chlorophylls in, 34
Coccinella septempunctata,
  α-carotene in, 18
  β-carotene in, 18
  lycopene in, 18
Coilodesme,
  photosynthesis in, 345
Colias philodice,
  carotenoids in, 18
  tetrapyrrole in, 18
Colored retinal globules,
  in birds, 159-161
Comatula,
  rhodocomatulin in, 14
Coproporphyrin,
  in annelids, 228
Corn mutants,
  chlorophyll shift in, 266
Corynebacterium diptheriae,
  pteridines in, 54
  xanthopterin in, 54
Crithidia factor, 54, 55, 56, 57, 58, 61
Crithidia fasciculata,
  pteridine bio-assay with, 54
Cryptomonads,
  accessory pigments, 35
  chlorophyll *c* in, 34-36
  as primitive metaprotists, 140
Cryptomonas ovata,
  absorption spectrum, 36
  α-carotene in, 136
  carotenoids in, 4
  chlorophyll *c* in, 36, 351
Cryptonemia,
  enhancement in, 356
Cryptophyceae,
  carotenoids of, 2-4
  chlorophylls of, 34
  photosynthetic action spectra, 350-352
Cryptopleura crispa,
  enhancement in, 356
Cryptoxanthin,
  vitamin A from, 23
Ctenosaura hemilopha,
  carotenoids of, 19
Cyanidium caldarium, 4, 136
  β-carotene in, 4
  carotenoids of, 4

chlorophyll of, 39
photosynthetic action spectrum, 349, 351
c-phycocyanin in, 39
unidentified xanthophyll in, 4
zeaxanthin in, 4, 39
Cyanophyceae, *see* Blue-green algae
Cyanophyta, *see* Blue-green algae
Cylindrospermum lichiniforme,
phototaxis in, 110
Cytochromes, 148
amino acid sequences, 328
from *Chlorobium limicola,* 325
from obligate anaerobes, 324
from *Rhodospirillum rubrum,* 325
Cytochrome *f,* 324
from *Euglena,* 325
from red, brown, and blue-green algae, 325

D

Daphnia,
absence of vitamin A, 21
Dendrocoelem lacteum,
action spectrum for vision of, 157
Derivative spectrophotometer, 40
Desoxophylloerythrin, 93
in fossil materials, 92
Desoxophylloerythroetioporphyrin, 93
in fossil materials, 92
from pheophytin *a,* 94
Desulfovibrio desulfuricans,
cytochromes in, 324
Deuteroetioporphyrin,
in fossil materials, 92
Diadema,
pigmentation, 284, 286
Diadema antillarum,
photoreceptive elements, 283
photosensitivity of, 279
Diatoms, 1-2
carotenoids of, 1-2
chlorophylls in, 34
chlorophyll *c* in, 35
major xanthophyll of, 1
photosynthetic action spectra, 344
phototaxis, 109
5,6-5′6′-diepioxy-β-carotene,
vitamin A from, 23

Dinoflagellates,
carotenoids of, 2
chlorophyll *c* in, 35
photosynthetic action spectra, 344-345
phototactic action spectra, 108
Dinophyceae, *see* Dinoflagellates
Dinoxanthin, 136
as precursor to astaxanthin, 24
Diphenylamine,
effect on carotenoid synthesis, 210
Draparnaldia glomerata,
phototaxis in, 107
Drosophila melanogaster,
action spectrum for vision of eye-color mutants of, 159
lack of vitamin A requirement, 21
Dunaliella,
carotenoids in, 343
Dunaliella salina,
α-carotene in, 12
β-carotene in, 4, 12
chlorophyll in, 12
lutein in, 12
phototactic action spectrum, 108

E

Echinenone, 136, 140
in blue-green algae, 2, 3
in chlorotic substrains of *Euglena gracilis* v. *bacillaris,* 7
in marine invertebrates, 3
in sediments, 98
in seston, 96
vitamin A from, 23
Echinochromes, 286
in crinoids, 13
in the sea otter, 13
Eisenia,
protoporphyrin in, 228
Electron microscopy, 145
Elongation, 317
Emerson effect, 340, 352-357, 361-364, 374
Endarachne Binghamiae, 355
Enhancement,
in *Cryptonemia,* 356
in *Cryptopleura crispa,* 356
in *Porphyra perforata,* 356

Enterococcus,
  pteridines in, 55
Enteromorpha,
  action spectrum of chromatic transients, 371
Eremothecium ashbyii,
  pteridines from, 57
  riboflavin from, 57
Erythropterin,
  from *Mycobacterium tuberculosis,* 54
Escherichia coli,
  pteridines in, 55
Ethyl chlorophyllide *a,* 174
Ethyl pheophorbide *a,* 175
Eudorina elegans,
  phototactic action spectrum, 108
Euglena,
  absorption spectra of chloroplasts of, 152
  absorption spectrum, resolution into components, 44
  astaxanthin in red variants of, 155
  chloroplast structure, 148-152
  cytochrome *f* from, 325
  eyespots in, 155
  phototactic action spectra, 108-109
  phototactic photoreceptor in, 108
Euglena gracilis,
  absorption spectrum of, 44-48
  action spectrum for phototaxis of, 155
  β-carotene in chlorotic substrains of, 6-7
  β-carotene in eyespot of, 155
  carotenoid synthesis in, 7-8
  chloroplasts of, 149
  conjugated pteridine in, 60
  echinenone in chlorotic substrains of, 7
  lutein in chlorotic substrains of, 7
  phototaxis, 111
  phytofluene in chlorotic substrains of, 6-7
  xanthophylls of chlorotic substrains of, 7
  zeaxanthin in chlorotic substrains of, 7
Euglena granulata,
  chloroplasts of, 149
Euglenineae,
  carotenoids, 1-2, 4

Euglenophyceae, 341
  carotenoids in, 1-2, 4
  chlorophylls in, 34
Euglenophyta, *see* Euglenophyceae
Euphausids,
  vitamin A in, 15, 21, 22, 23
Evolution,
  protistan, 130
Excitation energy transfer, 407-408
Eyespots,
  in *Euglena,* 155

F

Ferryl ion, 331
Flamingos,
  carotenoid metabolism, 24-25
Flavacin,
  in seston, 96
Flavins, 12-13
Flavoproteins,
  in fishes, 13
Flavorhodin,
  in marine sediments, 98
Flowering, 317-318
Fluorescence, 394-411
  excitation spectrum, 398-399
  lifetime of the excited state, 399-403
  polarization, 403-407
  polarization in chlorophyll, 271-273
  quantum yield, 395-398
Folic acid,
  in leaves, 61
Fossil pigments, 83-105
Fossil porphyrins,
  from chlorophylls, 92
  from iron-porphyrin compounds, 92
Fucoxanthin, 344
  conversion into zeaxanthin, 1
  in marine sediments, 98
  in *Ochromonas danica,* 3
  in photosynthesis, 339
  photosynthetic activity of, 344-345
  in *Prymnesium parvum,* 3
Fucus,
  β-carotene in gametes of, 4
Funaria hygrometrica,
  phototactic chloroplast movements in, 114, 116
Fundulus parvipinnis,
  xanthophylls from β-carotene in, 24

G

Gaffkya homari,
  pteridine produced by, 54-55
Galloxanthin,
  in colored retinal globules, 161
Gambusia holbrooki,
  vitamin A deficiency in, 22-23
Germination, 317
Ghost gene, 207
Gillichthys mirabilis,
  xanthophylls from β-carotene in, 24
Glycymerin,
  in marine sediments, 98
Gonium,
  phototactic action spectrum, 108
Gonyaulax catenella,
  phototactic action spectrum, 108
Gonyaulax polyedra,
  photosynthesis in, 345-346
Green algae, 341
  carotenoids of, 1-2, 5
  chlorophylls, 34-35
Green pigments,
  in freshwater and marine sediments, 88-91
Green sulfur bacteria,
  γ-carotene in, 133
  rubixanthin in, 133

H

Haematococcus,
  carotenoids, 343
Haematococcus pluvialis,
  astaxanthin in, 4
  β-carotene in, 12
Hallachrome, 77, 79-80
Helianthus annus,
  green, water-soluble pigment from, 73-81
  white mutant, 73
Hematin compounds,
  in chemosynthetic bacteria, 327
  in chloroplasts and chromatophores, 323
  distribution patterns, 326-328
  immunochemistry, 330
  photochemical mechanisms, 330
  in photosynthesis, 323-337
  physicochemical characteristics, 328-329
  potentials, 332

Hematococcus, see *Haematococcus*
Hematoporphyrin, 250
Heme,
  biosynthesis, 249-251
Hemiselmis,
  photosynthetic action spectra, 351
Hemiselmis virescens,
  absorption spectra of extracted chlorophylls, 37
  absorption spectrum, 36
  chlorophyll c in, 36, 351
  photosynthetic action spectrum, 353
  phycocyanin of, 184
Heterokontae,
  carotenoids, 1-2
Higher plants,
  pigments in, 341
Hill reaction, 323
Hydrogen donors,
  aromatic compounds as, 388-390
β-hydroxy-β-methylglutaric acid (HMG), 208

I

Indigoids, 12
Infrared,
  effect on chloroplast movement, 113, 115
Ionone,
  effect on carotenoid synthesis, 209
Isopheophytin d, 170, 171

K

Kermesic acid, 14

L

Laccaic acid, 14
Lactuca sativa,
  seed germination in, 305-306
Lemna trisulca,
  phototactic chloroplast movements in, 114, 116
Lepidium virginicum,
  seed germination in, 305-306
Leprotene,
  in marine sediments, 98
Leucine,
  effect on β-carotene production, 208
Leucoplasts, 8
Leucovorin,
  in leaves, 61

Lumbricus,
  protoporphyrin in, 228
Luminescence,
  of algae and chloroplasts, 408
  relation to fluorescence, 408
Lutein, 1-2, 5-8, 215, 223
  in chloroplasts, 148
  in chlorotic substrains of *Euglena gracilis* v. *bacillaris*, 7
  in colored retinal globules, 161
  in *Dunaliella salina*, 12
  metabolism by canaries, 24
  in sediments, 98
  in seston, 96
  in silk, 19
Lycopene, 18, 207, 211
  in *Coccinella septempunctata*, 18
  in lake sediments, 98
Lyngbya, 349

M

Magnesium protoporphyrin IX, 257
Magnesium vinyl pheoporphyrin a_5, 257-258
Marthasterias, 15-16
  astaxanthin in optic cushion of, 16
  β-carotene in optic cushion of, 16
  vitamin A absence from, 16
Melanins, 12, 13, 14, 26
Mesobilipurpurin, 192-193
Mesobilirubinogen, 192
Mesobiliverdin, 192
Mesobiliviolin, 192
Mesochlorophyll *a*, 170
Mesoetioporphyrin,
  in fossil materials, 92
Mesoporphyrin,
  in fossil materials, 92
Metalloporphyrins,
  in petroleum, 91-96
  in sedimentary rocks, 91
  in shales and crude oils, 94
Methylheptenone,
  effect on carotenoid synthesis, 209
Methylpheophorbide *a*, 172
  in lake sediments, 89
Methyl pyropheophorbide *a*, 169
Metridium senile,
  photosensitive pigment, 295
  reaction time, 296-300
  sensitivity to light, 295-301

Mevalonate, 208, 210
  incorporation into β-carotene, 7
Miscrasterias,
  absorption spectra of pigments of, 121
  negative phototaxis in, 123
  phototactic action spectra, 120-124
  phototactic orientation in, 124
  phototactic thresholds, 108
Micrasterias rotata,
  phototaxis, 117-125
Microspectrophotometer, 152, 160
Microsporum canis,
  pteridines in, 57-58
Microsporum gypseum,
  pteridines in, 57-58
Mnium,
  phototactic chloroplast movements in, 114
Monostroma,
  action spectrum of chromatic transients, 371
Mougeotia,
  chloroplast phototaxis in, 113, 116
  chloroplasts of, 149
Mycobacterium phlei,
  myxoxanthophyll in, 3
Mycobacterium tuberculosis,
  erythropterin from, 54
  xanthopterin from, 54
Myriogramme,
  photosynthetic action spectra, 347
Myxicola infundibulum,
  coproporphyrin in, 228
Myxoxanthophyll, 136, 140
  in *Mycobacterium phlei*, 3
  in sediments, 98
  in sediments of Swiss lakes, 99
  in seston, 96

N

Naphthoquinones,
  in animals, 13-14
  as photoeffectors, 14
  in plants, 13
  as respiratory stimulants, 13
  as sperm activators, 13
Navicula, 354
  phototaxis in, 109
Navicula minima,
  photosynthetic quantum yield, 344
Neoxanthin, 1-2, 5-8

Nereis diversicolor,
   coproporphyrin in, 228
Neurospora,
   carotenoid biochromes in mutants of, 11
Neurosporene, 211
New leaf pigment, 73-82
   chemical nature, 79
   extraction and purification, 75-79
   role in plant metabolism, 79-80
Nickel-porphyrins,
   in petroleum, 92
Nostocaceae,
   phototaxis in, 110

O

Ochromonas danica,
   absorption spectrum, 40-42, 46-47
   absorption spectrum, resolution into components, 47-48
   β-carotene in, 3
   changes in chlorophyll absorption, 45
   chlorophyll absorption spectra, 38, 42, 49-50
   fucoxanthin in, 3
   lack of chlorophyll *c*, 36
   photosynthesis, 49-50
   pteridines in, 60
Ochromonas malhamensis,
   absorption spectrum of chlorophyll *a* in, 50
   lack of chlorophyll *c*, 36
   photoreduction in, 391
   pteridines, 60
Octopus,
   eyes of, 159
Octopus bimaculatus,
   assimilation of carotenes in, 19
   as carotene selector, 17
   xanthophylls in, 19
Ommatidia, 157-159
Open-chain tetrapyrroles, *see* Phycobilins
Organic matter,
   chemical nature of fossil, 83-84
   total mass produced by green plants, 83
Oscillatoria, 349
   anaerobic metabolism, 391
Oscillatoria formosa,
   phototactic action spectrum, 110

Oscillatoria mougeotii,
   phototaxis in, 110
Oscillatoria rubescens, 99
Oscillatoriae,
   phototaxis in, 110
4-oxo-β-carotene, 3
"Oxo"-urobilin, 192-193

P

Pandorina,
   phototactic action spectrum, 108
Panulirus interruptus,
   astaxanthin precursors in, 20
Paramecium bursaria,
   Chlorella in, 107
Peridinin, 136, 344
   in photosynthesis, 345
   in seston, 96
Peridinium trochoideum,
   phototactic action spectra, 108
Petaloxanthin,
   in marine sediments, 98
Phaeophyceae, *see* Brown algae
Pheophorbide *a,*
   in lake sediments, 89
   in marine sediments, 88
Pheophorbide *b,*
   in lake sediments, 89
   in marine sediments, 88
Pheophytin *a,*
   in lake sediments, 89
   in marine sediments, 88
Pheophytin *b,*
   in lake sediments, 89
   in marine sediments, 88
Pheophytin *d,* 170, 172
   infrared spectrum, 172
   in marine sediments, 88
Phobotaxis, 110
Phormidium ectocarpi,
   action spectrum for photosynthesis, 348
Phormidium uncinatum,
   phototaxis in, 110
Photoinactivation,
   of tyrosinase, 13
Photokinesis,
   carotenoids in, 15
   flavoproteins in, 13
   riboflavin in, 13

Photomorphogenic pigment,
  assay and separation, 314
  concentration, 313
  detection, 314
Photo-oxidation,
  of hematin compounds, 333
Photoperiodic effects,
  reversibility, 307-308
Photoperiodism, 303-321
  action spectra, 304-306
  dark reaction, 308-309
  photoreaction, 309-310
Photophosphorylation, 327
Photoreactive pigments,
  as phylogenetic markers, 129-144
Photoreception,
  flavoproteins in, 13
  riboflavin in, 13
Photoreceptors, 145-167
  invertebrates, 155-159
  plant, 147-148
Photoreduction,
  in *Ochromonas malhamensis*, 391
Photosynthesis,
  chromatic transients, 367-376
  effect of phototactic chloroplast movements on, 116
  in *Gonyaulax polyedra*, 345-346
  hydrogen donor, 390-392
  and phototaxis, 116
  in tobacco leaves, 116
  wavelength dependence, 340
Photosynthetic action spectra,
  automatic recording, 361-365
  Bacillariophyceae, 344
  brown algae, 344
  Chlorella, 342-343
  Chlorophyta, 341
  Cryptophyceae, 350-352
  Cyanidium caldarium, 349, 351
  Cyanophyta, 345, 347-349
  Dinophyceae, 344-345
  Hemiselmis, 351
  Hemiselmis virescens, 353
  Myriogramme, 347
  Phormidium ectocarpi, 348
  Porphyra nereocystis, 346
  Porphyridium aerugineum, 350
  Porphyridium cruentum, 348
  Rhodomonas, 351

Rhodomonas lens, 352
Rhodophyta, 345-348
Schizymenia, 347
Ulva, 342
Photosynthetic bacteria,
  chromatophores of, 1
Photosynthetic bacteria oxidation,
  utilization of aromatic compounds, 387-394
Photosynthetic quantum yield,
  in *Navicula minima*, 344
Phototactic action spectra,
  Dinophyceae, 108
  Gonyaulax catenella, 108
  Peridinium trochoideum, 108
  Prorocentrum micans, 108
Phototactic orientation,
  Micrasterias, 124
Phototaxis, 107-126
  in achlorophyllous *Euglena gracilis*, 111
  in *Draparnaldia glomerata*, 107
  effect of chemicals on, 112-113
  effects of environmental factors, 112-113
  effect of temperature on, 112
  eyespotless mutant of *Chlamydomonas*, 111
  in *Micrasterias rotata*, 117-125
  negative in *Micrasterias*, 123
  in *Platymonas subcordiformis*, 112
  in *Polytoma*, 111
  in *Polytomella*, 111
  relation to metabolism, 116-117
  role of the stigma, 111
  in *Ulothrix subtilis*, 107
  in *Ulothrix zonata*, 113
Phycobilins, 3, 190-200, 345, 350
Phycocyanins, 181-196, 200-201
  in chloroplasts, 148
  chromophores, 193
  in photosynthesis, 339
  protein structure, 187-190
C-phycocyanin, 182, 184
  in *Cyanidium caldarium*, 39
  isoelectric point, 189-190
  photosynthetic effectiveness, 349
  sulfur content, 189
R-phycocyanin, 182, 184
  amino acid analyses, 189

chromophore, 195-196
  molecular weight, 186
Phycocyanobilin, 193-196
  absorption spectrum, 194
  from allophycocyanin, 195
  from C-phycocyanin, 194-195
Phycoerythrins, 181-192, 196-201
  in chloroplasts, 148
  chromatography, 187
  chromophores, 193
  dissociation, 190
  in photosynthesis, 339
  protein structure, 187-190
  in *Rhodymenia palmata*, 198
B-phycoerythrin, 183
  photosynthetic effectiveness, 347
C-phycoerythrin,
  photosynthetic activity, 347-348
R-phycoerythrin,
  amino acid analyses, 187
  isoelectric point, 189-190
  molecular weight, 186
  sulfur content, 189
  urobilinoid phycobilin in, 199
Phycoerythrobilin, 193, 196-200
  absorption spectra, 196-197
  conversion to urobilinoid pigment, 198
  from phycoerythrins, 196
  from R-phycocyanin, 195-196
Phycomyces, 208-209
  β-carotene production, 208
Physarum polycephalum,
  pteridines of, 58
Phytoene, 207
  in algal mutants, 5-6
Phytofluene, 211
  in algal mutants, 5-6, 8
  in chlorotic substrains of *Euglena gracilis* v. *bacillaris*, 6-7
Phytylation,
  of chlorophyllide *a*, 265
Pigment granules,
  in *Planaria*, 156
Pigment, photomorphogenic, 308
Pinnularia,
  metabolism, 391
Planaria,
  pigment granules in, 156
  sensory cells in, 156-157

Plankton,
  green pigments present in, 85-86, 88
Plant carotenoids,
  in animals, 27
Plastid pigments, 1-4
Platymonas subcordiformis,
  phototaxis, 112
Polarographic technique, 340-341
Polytoma,
  phototaxis, 111
Porphobilinogen,
  formation, 232-233
  utilization, 232-233
Porphyra naiadum var. *australis*,
  phycocyanin of, 183
Porphyra naiadum var. *naiadum*,
  B-phycoerythrin from, 183
Porphyra nereocystis,
  photosynthetic action spectrum for, 346
Porphyra perforata,
  action spectra for photosynthesis, 364
  action spectrum of chromatic transients, 370
  enhancement action spectrum, 357
  enhancement in, 356
Porphyra thuretii,
  chromatic transients, 369
Porphyria, 230, 232, 238
Porphyridium aerugineum, 349
  photosynthetic action spectrum, 350
Porphyridium cruentum,
  action spectrum for photosynthesis, 348
  phototaxis, 109
  phycoerythrin of, 182
Porphyrins,
  extraction from fossil materials, 92
Porphyrins,
  free in petroleum, 91
  free in sedimentary rocks, 91
  in pre-Cambrian deposits, 84
Porphyrinogens, 250
Production of oxidized carotenoids, 17
Prolycopene, 207
  isomerization in light, 9
Prorocentrum micans,
  phototactic action spectra, 108
Protetrahydrolycopene,
  isomerization in light, 9

Protista,
  monophylogeny of, 137
Protochlorophyll, 154
  biosynthesis, 227-256, 257-259
  continuous precursor to chlorophyll, 259-260
  formation, 251-254
  phototransformation, 259-263
  from *Rhodopseudomonas spheroides*, 262-263
Protochlorophyll-chlorophyll transformation, quantum yield, 273
Protochlorophyll holochrome,
  absorption spectrum, 268
  carotenoids, 269
  isolation, 267-268
  particle weight, 269
Protochlorophyll "holochrome", 165
Protochlorophyll transformation,
  effect of light intensity, 263-264
  effect of temperature, 264
  effect of wavelength, 264
Protoporphyrin,
  in *Eisenia*, 228
  in *Lumbricus*, 228
Protoporphyrin IX,
  biosynthesis, 229-231, 249-251
Prymnesium parvum,
  absorption spectrum of, 39
  β-carotene in, 3
  chlorophyll *c* in, 36, 39
  fucoxanthin in, 3
Pteridines, 12, 53-65
  absorption spectrum, 53-54
  of algae, 58-60
  of *Anacystis nidulans*, 58-60
  of animals, 54
  of *Aspergillus oryzae*, 56
  of *Azotobacter*, 55
  of bacteria, 54-55
  bio-assay with *Crithidia fasciculata*, 54
  biosynthesis, 57
  of chloroplasts, 61, 63-64
  conjugated, 53
  of *Corynebacterium diptheriae*, 54
  of *Enterococcus*, 55
  of *Eremothecium ashbyii*, 57
  of *Escherichia coli*, 55
  of fungi, 56-58
  of *Gaffka homari*, 54-55

  of higher plants, 60-65
  isolation from higher green plants, 61-64
  of *Microsporum canis*, 57-58
  of *Microsporum gypseum*, 57-58
  of *Ochromonas danica*, 60
  of *Ochromonas malhamensis*, 60
  peas, formed in darkness by, 63
  peas, formed in light by, 63
  as photoreceptors, 58
  of *Physarum polycephalum*, 58
  physiochemical properties, 53
  possible participation in photosynthesis, 60
  of *Sarcina lutea*, 55
  substituents, 53
  of *Tinea capitis*, 57
  unconjugated, 53
Pterins, see Pteridines
Pteroylglutamic acid, 53
Punctaria occidentalis,
  action spectrum of chromatic transients, 372
Purple bacteria,
  acyclic carotenoids in, 133
  phototactic action spectra, 111
Pyocyanine, 79-80
Pyrrophyta,
  chlorophylls in, 34

Q

Quinones, 79-81

R

Red algae,
  carotenoids of, 1-3
  chlorophylls in, 34
  chlorophyll *d* in, 35
  photosynthetic action spectra, 345-348
  phototaxis, 109-110
Red carotenoid,
  of wild bishop birds, 24
Retina, 159
Retinal rods,
  molecular model, 159
  rhodopsin in, 159
Retinene$_1$,
  in honeybees, 21
  in honeybee and housefly eyes, 158
  in lobster eyes, 158

Rhabdomeres, 157
  of cephalopods, 159
Rhizopterin,
  in leaves, 61
  in *Rhizopus nigricans*, 56
Rhizopus nigricans,
  rhizopterin from, 56
Rhodocomatulin,
  in *Comatula*, 14
Rhodomonas,
  photosynthetic action spectra, 351
Rhodomonas lens,
  action spectrum for photosynthesis, 352
  chlorophyll c in, 36, 351
Rhodophyceae, see Red algae
Rhodopseudomonas, 388
  carotenoid transformation in, 211
Rhodopseudomonas spheroides, 233
  protochlorophyll from, 262-263
Rhodopsin,
  molecular weights of, 163
  in retinal rods, 159, 161
Rhodopurpurin,
  in marine sediments, 98
Rhodospirillum,
  carotenoid transformation in, 211
Rhodospirillum rubrum,
  cytochrome from, 325
Rhodoviolascin,
  in sediments, 98
Rhodymenia palmata,
  phycoerythrin from, 198
  R-phycoerythrin from, 183
Riboflavin, 12-13
  in *Eremothecium ashbyii*, 57
Rubixanthin,
  in green sulfur bacteria, 133

S

Sarcina lutea,
  pteridines in, 55
Scenedesmus,
  hydrogen donors, 391
Schizymenia,
  photosynthetic action spectra, 347
Sea urchins,
  photosensitivity, 279-293
Semiconductivity, 408-410

Sensory cells,
  in *Planaria*, 156-157
Sepia,
  eyes of, 159
Seston,
  green pigments present in, 85-88
Siphonaxanthin, 136
  in Siphonales, 2, 5
Siphonein, 136
Smithora, see *Porphyra*
Spectrophotometry, 146
Spirilloxanthin, 211
Spirogyra,
  phototactic chloroplast movements, 116
Spondylomorum,
  phototactic action spectrum, 108
Stercobilin, 199
Sulcatoxanthin,
  in marine sediments, 98
Synechococcus, 349

T

Taraxanthin,
  as precursor to astaxanthin, 23-24
  in silk, 19
Temperature,
  effect on phototaxis, 112
Tetrapyrrole,
  in *Colias philodice*, 18
  formation, 236-248
Thiamine,
  effect on carotenoid pigmentation, 208
Tinea capitis,
  pteridines in, 57
Topotaxis, 110
Torulene,
  in marine sediments, 98
Trentepohlia,
  carotenoids, 343
  lack of sterols, 131
Trentepohlia aurea,
  β-carotene in, 4
Tribonema, 356
2,2,6-trimethylcyclohexane-1-carboxylic acid,
  from β-carotene, 99
Tyrosinase,
  photoinactivation of, 13

U

Ulothrix subtilis,
  phototaxis in, 107
Ulothrix zonata,
  phototaxis, 113
Ulva,
  γ-carotene in gametes of, 4
  photosynthetic action spectra for, 342
  phototactic action spectra, 108
Ulva lactuca,
  phototactic action spectrum, 108
Ulva lobata,
  action spectrum of chromatic transients, 371
  chromatic transients, 367
Urobilin, 192
Uroporphyrin,
  in pearl oyster, 227
  in planarians, 227
Uroporphyrinogen,
  biosynthetic hypotheses, 243
  decarboxylation, 249-251
Uroporphyrinogen I, 236-242
Uroporphyrinogen III, 242-248

V

Vanadium-porphyrins,
  in lake sediment, 90-91
  in marine sediment core, 91
  in petroleum, 92
Velella,
  zooxanthellae of, 26-27
Velella lata,
  astaxanthin in, 26
Violaxanthin, 1-2, 5-8, 215, 223
  metabolism by canaries, 24
  in silk, 19
Vision,
  flavoproteins in, 13
  riboflavin in, 13
Vitamin A, 14-15, 16, 17, 18, 19, 20-24, 27, 28
  absence from *Marthasterias*, 16
  from carotene, 23
  conversion of carotene into, 15, 23
  deficiency in *Gambusia holbrooki*, 22-23
  in euphausiids, 21, 22
  in houseflies, 21
  lack in many invertebrates, 21
  storage in eye-stalks of euphausiids, 15

Vitamin A_1,
  from housefly, 158
Volvocales,
  phototactic action spectra, 108
Volvox,
  phototactic action spectrum, 108

X

Xanthophyceae,
  carotenoids of, 1-2
  chlorophyll *a* in, 35
  chlorophylls in, 34
Xanthophylls,
  of algal mutants, 5-6, 8-9
  from β-carotene in *Fundulus parvipinnis*, 24
  from β-carotene in *Gillichthys mirabilis*, 24
  in chloroplasts, 148
  in chlorotic substrains of *Euglena gracilis* v. *bacillaris*, 7
  in fishes, 17, 24, 27
  function in fishes, 27
  function in invertebrates, 27
  need for oxygen in formation of, 8-9
  in *Octopus bimaculatus*, 19
  phylogenetic origin, 136
  synthesis of, 211
  unidentified in *Cyanidium caldarium*, 4
Xanthophyll selectors,
  birds, 16
Xanthopterin,
  in butterfly wings, 53
  in *Corynebacterium diptheriae*, 54
  in *Mycobacterium tuberculosis*, 54

Z

Zeaxanthin, 224
  in chloroplasts, 148
  in chlorotic substrains of *Euglena gracilis* v. *bacillaris*, 7
  in colored retinal globules, 161
  in cryptomonads, 140
  in Cryptophyceae, 2, 4
  in *Cyanidium caldarium*, 4, 39
  in marine sediments, 98
  metabolism by canaries, 24
Zooxanthellae,
  of *Velella*, 26-27